Data Engineering
for Data Science

Gilles Dejaegere · Alberto Abelló · Kristian Torp ·
Alkis Simitsis

Editors

Data Engineering for Data Science

 Springer

Editors
Gilles Dejaegere 
Université Libre de Bruxelles
Bruxelles, Belgium

Kristian Torp
Aalborg Universitet
Aalborg, Denmark

Alberto Abelló
Department of Service and Information
System Engineering (ESSI)
Universitat Politècnica de Catalunya
Barcelona, Spain

Alkis Simitsis
Athena Research Center
Information Management Systems Institute
Athens, Greece

Preface

In today's interconnected world, data has become a vital economic and strategic asset, driving innovations in areas ranging from personalized healthcare to smart cities. As organizations increasingly depend on data for informed decision-making and innovation, much of its potential remains underutilized, often due to challenges in collecting, integrating, managing, and analyzing raw data. To address these challenges, two complementary fields have emerged. *Data Science* focuses on extracting insights, patterns, and knowledge from data using methods from statistics, machine learning, and data visualization. *Data Engineering* focuses on designing, building, and maintaining the infrastructure and systems that support large-scale data collection, storage, processing, and access. Together, these fields form the backbone of modern data-driven systems.

This collective book, titled *Data Engineering for Data Science*, aims to synthesize and integrate the research challenges in this interdisciplinary area. It offers a comprehensive survey of the entire data management stack, from scalable and explainable data analytics to traceable data workflows. By providing a consistent framework, this book facilitates a thorough understanding of the data science lifecycle, from basic definitions to state-of-the-art concepts and techniques. It is designed to be a valuable resource for both researchers and professionals seeking to leverage data engineering for data science applications.

Target Audience

This book is designed for a diverse audience of researchers, professionals, and practitioners in the fields of Data Science and Data Engineering. It aims to serve as a resource for Ph.D. students and academics engaged in cutting-edge research, offering a comprehensive framework to navigate the complexities of the data management lifecycle. Industry professionals, including data analysts, engineers, and managers, will find practical insights and state-of-the-art techniques that can be applied to real-world challenges, making it an essential guide for leveraging data engineering to drive innovative solutions. Whether you are a seasoned expert or a budding professional, this book provides the tools and knowledge needed to stay at the forefront of data-driven advancements.

Purpose of the Book

The purpose of this book is to present the state of the art and highlight open research challenges across data science, data engineering, and data management more broadly. It addresses key issues involved in transforming raw data into meaningful insight and actionable knowledge.

Aimed at bridging theory and practice, the book offers both conceptual frameworks and practical perspectives. It explores current methodologies, emerging techniques, and unresolved problems, to help researchers, professionals, and students grasp the breadth of the field and identify promising directions for future work.

How to Read This Book

The book is divided into four parts, each focusing on a different aspect of the data management lifecycle: (1) **Governance**, (2) **Storage and Processing**, (3) **Preparation**, and (4) **Analysis**.

Each part of the book is organized to provide a coherent conceptual framework and is divided into multiple chapters, each focusing on a specific topic, but together offering a comprehensive overview of the state of the art and the key challenges in each area. While the parts and chapters follow a logical sequence, each chapter is designed to be self-contained and can be read independently.

Chapters include references for further reading and deeper exploration. Most also provide concrete examples or use cases to make the material more accessible. In addition, many chapters introduce a taxonomy to break down complex research areas into manageable components, highlighting the core directions and developments within each domain.

Looking Ahead

As the fields of Data Engineering and Data Science continue to evolve at a rapid pace, we find ourselves at the intersection of exciting opportunities and complex challenges. By bringing together diverse perspectives, methodologies, and real-world applications, we hope this book serves as both a reference and a source of inspiration. Whether you are deepening your expertise or entering the field for the first time, we believe the content will provide valuable insights into the technical foundations, open problems, and future potential of data engineering for data science. Happy reading!

Bruxelles, Belgium   Gilles Dejaegere
Barcelona, Spain   Alberto Abelló
Aalborg, Denmark   Kristian Torp
Athens, Greece   Alkis Simitsis
September 2025

Acknowledgements

The editors extend their sincere gratitude to all chapter authors and reviewers for their contributions and collaboration, which were essential to the completion of this volume.

The work presented in this collective book has been supported by the H2020 MSCA-ITN-EJD DEDS project, funded by the European Union's Horizon 2020 research and innovation program under the Marie Sklodowska-Curie grant agreement No 861198.

Introduction and Book Summary

The Era of Data-Driven Insights

In today's interconnected world, data has become an indispensable asset. The 2030 Agenda for Sustainable Development underscores the necessity of "quality, accessible, timely, and reliable disaggregated data" to drive actions and monitor achievements.[1] The value of data lies in its ultimate use, with significant benefits going to those who can aggregate data uniquely and provide valuable analytics.[2] The European Political Strategy Centre (EPSC) emphasizes that *"Data is rapidly becoming the lifeblood of the global economy. It represents a key new type of economic asset. Those that know how to use it have a decisive competitive advantage in this interconnected world, through raising performance, offering more user-centric products and services, fostering innovation–often leaving decades-old competitors behind. ... Data analytics will soon be indispensable to any economic activity and decision-making process, both public and private."*[3]

Despite the recognized importance of data, the full economic value it can offer remains largely untapped. McKinsey reports that *"Most companies are capturing only a fraction of the potential value from data and analytics. ... manufacturing, the public sector, and health care have captured less than 30 percent of the potential value we highlighted five years ago."* One of the primary challenges is the difficulty in unlocking value from raw data. Before data can be useful, it must undergo extensive preprocessing, integration, cleaning, storage, and preparation for analysis. Modern data usage extends beyond simple querying or data mining to include exploration, recommendation, and explanation, often requiring special treatment of temporal and spatial characteristics. Unfortunately, the initial quality of incoming data is low, requiring significant management and processing efforts to enhance its value.[2]

---

[1] https://sustainabledevelopment.un.org/content/documents/21252030%20Agenda%20for%20Sustainable%20Development%20web.pdf.

[2] https://www.mckinsey.com/business-functions/mckinsey-analytics/our-insights/the-age-of-analytics-competing-in-a-data-driven-world.

[3] https://ec.europa.eu/epsc/sites/epsc/files/strategic_note_issue_21.pdf.

The data value creation chain involves multiple disciplines and roles,[4] with Data Science and Data Engineering being pivotal. Data Science is an *"interdisciplinary field that uses scientific methods, processes, algorithms and systems to extract knowledge and insights from data in various forms, both structured and unstructured."*[5] Data scientists focus on deriving meaningful insights through analytics, supported by data engineers who *"design and build the data ecosystem that is essential to analytics. Data engineers are responsible for the databases, data pipelines, and data services that are prerequisites to data analysis and data science."*[6] They face challenges posed by the extreme characteristics of Big Data, defined as data of a very large size that presents significant logistical challenges in manipulation and management. Addressing these challenges requires innovative technological solutions for data management, encompassing *"architectures, policies, practices and procedures that properly manage the full data lifecycle needs of an organisation"*).[7]

Content and Structure of This Book

To provide a coherent journey through the data engineering landscape, the book is structured into four interconnected parts. Each part builds on the previous one: understanding governance and integration lays the foundation for efficient storage and processing; well-structured and accessible data then enables effective preparation; finally, prepared data supports advanced analysis and actionable insights. This sequence reflects the natural lifecycle of data, guiding readers from raw data to decision-making applications.

Part I: Governance and Integration. The first part on data governance and integration contains three chapters. Chapter 1 covers the integration of text data. There is a very large quantity of such unstructured data that does not follow a predefined schema. The chapter looks at how to integrate text data with both structured and unstructured data to provide unified data access to end users. The chapter looks at techniques from the Semantic Web (SW), Machine Learning (ML), Natural Language Processing (NLP), and Knowledge Representation (KR).

The next chapter (Chap. 2) looks at data fusion. This topic is relevant because many organizations struggle with massive, overlapping data from multiple potentially conflicting sources. This leads to inconsistencies and reliability issues. Data fusion addresses these issues by integrating different data sources into a single, consistent, and comprehensive source. The chapter looks at managing and evaluating both conflicts, and accuracy. It takes a practical approach, outlining the

---

[4] https://www.forrester.com/report/Data+Engineers+Have+Become+More+Important+Than+Data+Scientists/-/E-RES140511.

[5] V. Dhar. *Data science and prediction.* Commun. ACM 56 (12), 2013.

[6] https://www.eckerson.com/articles/data-engineering-coming-of-age.

[7] http://www.intellectussolutions.com.au/data-management.

key steps of the data fusion process and highlighting effective techniques for each stage.

The final chapter (Chap. 3) in this part looks into privacy because sensitive data is heavily restricted under GDPR. This can significantly limit access to data. The idea of data synthesis offers a promising solution to limited (or no) data access. This is done by generating anonymized datasets that preserve statistical properties while protecting privacy. However, scaling the synthesis to complex relational data in the medical domain introduces challenges in guaranteeing privacy. This chapter explores methods for scalable, privacy-aware data synthesis, aiming to enable a GDPR compliant sharing of medical datasets.

Part II: Storage and Processing. The second part of the book covers data storage and processing. The first chapter in this part (Chap. 4) covers feature selection (FS). This is an important early step in a machine learning pipeline as it can improve model accuracy, interpretability, and efficiency. The core idea is that FS identifies the most relevant part of the data. The chapter presents a comprehensive overview of FS, introducing a taxonomy of techniques and practical examples. The taxonomy categorizes existing FS methods by data from six perspectives, such as label dependency, search strategies, and feature interactions. The chapter discusses important aspects such as scalability, stability, and cost-effectiveness.

Chapter 5 looks at time-series data. Within the renewable energy data domain, such data can be high-frequency sensor data from wind turbines and solar panels. Such data creates the foundation for optimizing renewable energy production. However, the data size can be very large to the extent of overwhelming traditional RDBMSs. For this reason, Time Series Management Systems (TSMSs) have been developed. A TSMS can efficiently handle the scale, velocity, and analytics needs of time-series data across the entire pipeline. This chapter surveys academic and industrial TSMSs using classification criteria such as architecture, deployment, maturity, scalability, and approximation methods.

The next chapter (Chap. 6) in this part addresses ML operations (MLOps). ML pipelines have become increasingly complex and resource-intensive to build and maintain. This has created a demand for systems that (nearly) automate the design, execution, and evaluation of an ML pipeline. MLOps systems address this demand by providing tools for pipeline automation, provenance tracking, and collaboration. A general problem is the tradeoff between automation and cost-efficiency, where MLOps systems always favor the former. This chapter reviews the existing MLOps landscape using a taxonomy that covers design, execution, and tracking.

Chapter 7 investigates techniques for heterogeneous CPU-GPU processing. To efficiently use heterogeneous resources, data processing systems depend on appropriate scheduling and workload placement strategies to assign compute to the right processors, a non-trivial task that deals with various complex and conflicting tradeoffs. The chapter reviews state-of-the-art strategies for workload placement and scheduling on heterogeneous CPU-GPU architectures, along with runtime prediction techniques and methods to support multi-device code. It concludes with a discussion on open issues and promising future research directions.

Part III: Preparation. The third part of the book covers data preparation. Chapter 8 deals with Federated Learning (FL), which enables collaborative decentralized model training, i.e., without copying sensitive data to a central server. The chapter looks at blockchain-based FL because this has emerged as a promising solution that enhances transparency, integrity, and trust among the distributed and decentralized data owners. This chapter surveys architectural approaches again, using a newly developed taxonomy to structure the chapter. The chapter highlights both theoretical foundations and practical implementations.

The next chapter (Chap. 9) looks more closely at eXplainable AI (XAI), which has emerged as an interesting research direction because classic ML/AI may raise concerns about trust and accountability. XAI addresses these concerns by providing insights into model behavior, supporting debugging, bias detection, and data refinement. This chapter presents a taxonomy of XAI methods and surveys model-based, feature-based, and especially example-based methods. For each of these, strengths and weaknesses are discussed.

The last chapter (Chap. 10) in this part looks at data lakes. Such repositories have created new opportunities for flexible storage and exploration. However, a new major challenge is finding relevant data efficiently. Using an elaborated running example, this chapter explores table understanding, table search, and indexing techniques for data lakes. Combined, these techniques support tasks such as data integration and data augmentation. At the same time, new major challenges must be addressed, such as various types of table-relatedness and improving semantic accuracy. The chapter surveys existing work using a taxonomy for table understanding, table search techniques, and table indexing techniques.

Part IV: Analysis. The last part of the book covers data analysis and contains four chapters. The first chapter, Chap. 11, looks at adversarial ML for fraud detection. Adversarial ML is used to reveal critical vulnerabilities in data-driven systems. However, most research focuses on image recognition, leaving other security-sensitive domains relatively underexplored. This chapter examines adversarial attacks in the context of credit card fraud detection. From an ML perspective, the challenges are feature sparsity, semantic constraints, and delayed feedback. A hierarchical representation of threats is presented along with a set of metrics that measure the success of attackers.

The next chapter (Chap. 12) looks at inference queries, where AI models evaluate predicates over unstructured data such as text, images, and video. The cost of these queries is often dominated by the high computational costs of invoking deep models or human annotators at scale. First two use cases are introduced, and then using a taxonomy of approximate methods for inference queries, the chapter surveys methods for making inference queries more efficient, including proxy models, data similarity indexing, and approximation techniques.

Chapter 13 investigates the analysis of free-range trajectory data. An example use case is provided with the Automatic Identification System (AIS) from ships.

The widespread deployment of the AIS has created several very large collections of ship trajectory data. Such collections offer new opportunities for analyzing mobility patterns and support new downstream applications of the data. This chapter examines two examples from the maritime domain: detecting domain-specific activities such as fishing and estimating CO_2 emissions from maritime transport. A key challenge of AIS and free-range trajectory data lies in the frequent spatial and temporal gaps, which require methods such as trajectory imputation and integration.

The final chapter (Chap. 14) looks at network-constrained trajectory data. An example use case is Global Navigation Satellite System (GNSS) data from cars. The increasing availability of such data has made it possible to analyze vehicle movements constrained by road networks both at a large scale and with high precision. This chapter explores how network-constrained trajectory data can support urban planning, with applications in road traffic management and sustainable mobility. We highlight methods for monitoring congestion, modeling traffic dynamics, eco-driving, and eco-routing.

Contents

Part II   Storage and Processing

5 Current Systems for Managing Massive High Frequency Time Series 111

Abduvoris Abduvakhobov, Søren Kejser Jensen,
Christian Thomsen, and Esteban Zimányi

6 MLOps Systems for Developing ML Pipelines 141

Antonios Kontaxakis, Dimitris Sacharidis, Alkis Simitsis,
Alberto Abelló, and Sergi Nadal

Part I
Governance and Integration

Chapter 1
Text Data Integration

Md. Ataur Rahman◉, Dimitris Sacharidis◉, Oscar Romero◉,
and Sergi Nadal◉

Abstract Data comes in many forms. From a shallow perspective, they can be viewed as being either in structured (e.g., as a relation, as key-value pairs) or unstructured (e.g., text, image) formats. So far, machines have been fairly good at processing and reasoning over structured data that follows a precise schema. However, the heterogeneity of data poses a significant challenge on how well diverse categories of data can be meaningfully stored and processed. *Data Integration*, a crucial part of the data engineering pipeline, addresses this by combining disparate data sources and providing unified data access to end-users. Until now, most data integration systems have leaned on only combining structured data sources. Nevertheless, unstructured data (a.k.a. free text) also contains a plethora of knowledge waiting to be utilized. Thus, in this chapter, we firstly make the case for the integration of textual data, to later present its challenges, state of the art and open problems.

Keywords Text data integration · Data discovery · Data augmentation · Data enrichment · Natural language processing

1.1 Introduction

In this chapter, we will motivate why there is a need to integrate textual data with structured sources. First, we discuss why integrating text remains challenging despite being the most widely available type of data. Next, we highlight the critical role that textual data can play in integration scenarios. In particular, we describe how textual data can mitigate data sparsity, enable data discovery, and enhance integration through data augmentation. Each section is accompanied by clear, motivating examples to concretely demonstrate the impact of integrating textual data.

Md. Ataur Rahman (✉) · O. Romero · S. Nadal
Universitat Politècnica de Catalunya, Barcelona, Spain
e-mail: md.ataur.rahman@upc.edu; md.ataur.rahman@ulb.be

Md. Ataur Rahman · D. Sacharidis
Université libre de Bruxelles, Brussels, Belgium

G. Dejaegere et al. (eds.), *Data Engineering for Data Science*,
https://doi.org/10.1007/978-3-032-18765-9_1

1.1.1 Text Everywhere, Yet Hard to Integrate

Humans generate roughly 2.5 quintillion bytes of digitized data every day [1]. Apart from being in "structured" sources; these data can be found in numerous other "unstructured" formats such as web pages, weather forecasts, file servers, product reviews, scientific articles, commute maps, patient records, disease surveys, etc. Answering questions from a single source of data is relatively easy and suitable for small databases. However, a complex query we want to answer, could be requiring access over several heterogeneous data sources. Data integration plays a vital role in combining these distinct sources and providing the user with a unified view and query interface in such scenarios.

The task of data integration mainly falls into the domain of data management and engineering. But, when it comes to integrating data that are in unstructured formats such as texts, it is essential to incorporate ideas from several other domains such as the Semantic Web (SW), Machine Learning (ML), Natural Language Processing (NLP), and Knowledge Representation (KR) techniques. Again, each of the domains mentioned above solves a particular task in a different way. Hence, there is a need to develop a common framework for data integration techniques to benefit from all the aforementioned domains and yield solutions for heterogeneous data sources. On top of this, rather than following any schema, most state-of-the-art integration platforms that include textual sources store text data as plain instances [2]. This leads to the same chicken-and-egg problem where some extra manual efforts are needed to extract structured information from those text instances before they can be utilized and integrated with other structured sources.

To aid this, a common way of representing disparate data sources is needed. As humans, we are good at inferring in such scenarios mainly because we perceive and represent information in our brains through conceptualization. *Conceptualization* can be thought of as an abstract perspective of representing the knowledge we retain regarding the world around us [3]. Every concept is expressed and linked in terms of articulated relations with some other concepts in this representation. Each concept, in turn, could be associated with its own real-world examples (e.g., attributes, causes and effects), and might also form hierarchical relations. Knowledge Graphs (KGs) are well known for their ability to store contextual information with data in a way that resembles the above scenarios. They can represent complex relationships between entities while maintaining semantic clarity through well-defined structures. This semantic richness, combined with their inherent capability to integrate heterogeneous data sources through common vocabularies, makes them particularly suitable for large-scale data integration tasks. Therefore, by creating a concrete and explicit manifestation of conceptualization for machines, KGs could lead to a better unification model for diversified sources of data.

1.1.2   The Critical Role of Text in Data Integration

Unstructured data preserved in the form of natural language text is undoubtedly an invaluable source of knowledge. The integration of structured and unstructured data has become a critical area of research in recent years [4]. Adaptation by the leading technological industries like Google, Microsoft, and Facebook emphasizes the significance of harvesting textual information with other structured sources in an enterprise. For example, Google integrates textual data from websites with structured databases (i.e., KGs) to enhance search results and provide richer answers in response to user queries. Similarly, Microsoft leverages textual data from emails and documents to improve their Office products and services, while Facebook combines textual content from posts with structured data to improve targeted advertisements and user recommendations. Integrating structured and unstructured data is important because structured data alone often captures only limited aspects of a topic, missing the contextual details that textual data naturally provides. Textual data can enrich structured datasets with domain knowledge, and implicit relationships that are typically absent in structured sources. Hence, integrating multi-modal data creates a more comprehensive view, enabling better understanding, accurate analysis, and sophisticated data-driven decision-making through improved inference over texts. Manually converting all the heterogeneous data types into a single homogeneous structure involves numerous manual efforts and is often impossible, so we need to address the limitations of human intervention, ensuring adaptability to large-scale, real-world scenarios. Thus, the challenge is developing a scalable and automated data integration framework that can efficiently handle textual information alongside other structured sources.

In the following, we justify the need of such a framework. Indeed, text could offer three key benefits when integrated with structured data: *(i) mitigating data sparsity*, *(ii) data discovery*, and *(iii) data augmentation*. We will discuss these scopes in the following sections, providing motivating examples to clarify their significance and impact.

1.1.2.1   Mitigating Data Sparsity

The data integration process often results in a large number of missing values (i.e., NULL), as each source provides only a partial view of the data. This issue, commonly referred to as the *data sparsity problem*, arises when schemas across sources mismatch or the available structured data lacks sufficient context to fill in the missing values. Traditional value imputation techniques fail to address the heteroscedasticity (i.e., variance inconsistency) and the non-independent and identically distributed (Non-IID) nature of data integrated from different sources [5]. Moreover, these methods are limited to structured data sources and cannot effectively utilize unstructured textual data, which often contains valuable information for completing sparse

Table 1.1 Structured data sources containing joinable information

D1: Complication dataset

Disease	Complication
Tuberculosis	Seizures
Acne	Skin cancer

D2: Anatomy dataset

Disease	Anatomy
Acne	Skin
Acoustic neuroma	Nervous system

Table 1.2 Integrated data, (a) before and (b) after imputing missing values via text

(a) Integrated data with missing values ($\perp$)

Disease	Anatomy	Complication
Tuberculosis	$\perp$	Seizures
Acne	Skin	Skin cancer
Acoustic Neuroma	Nervous System	$\perp$

(b) Enriched integrated data

Disease	Anatomy	Complication
Tuberculosis	Lungs	Seizures
Acne	Skin	Skin cancer
Acoustic Neuroma	Nervous System	Unsteadiness

datasets. This challenge can be mitigated by leveraging external textual sources to enrich integrated datasets through *text conceptualization.*

Motivating Example-1: Consider the two structured datasets containing health-related information of Table 1.1. The first dataset, **D1**, provides information about diseases and their complications. The second dataset, **D2**, records diseases and their anatomical associations. When these datasets are integrated (Table 1.2), missing values ($\perp$) appear due to differences in their schema (Table 1.2a). To address this sparsity, we turn to unstructured textual data, such as clinical texts. By integrating **conceptualized texts** with the structured data, we can enrich the integrated dataset with the missing values (Table 1.2b). For instance, the missing *Anatomy* value for *Tuberculosis* can be filled using textual data describing its effect on the *lungs*. Similarly, the missing *Complication* values for *Acoustic Neuroma*, such as *hearing loss* and *unsteadiness*, can also be identified from the textual descriptions.

> **Conceptualized Text:** "**Acoustic neuroma** (Disease) is a slow-growing non-cancerous **brain tumor** (Complication) effecting main (vestibular) **nerve** (Anatomy) leading from your **inner ear** (Anatomy) to your **brain** (Anatomy). It can cause **hearing loss** (Complication), and **unsteadiness** (Complication). In contrast, **Tuberculosis** (Disease) generally damages the **lungs** (Anatomy), leading to **empyema** (Complication)."

1.1.2.2   Data Discovery

A common challenge in data integration is combining independently generated datasets that lack explicit connections or shared identifiers between them. These datasets, created in isolation from one another, often have no predetermined join keys or obvious relationships that would facilitate their integration. Data discovery through external textual sources introduces a new possibility for integrating such sources. By utilizing text, we could find conceptual links to produce join-paths across datasets. These conceptual links can be derived from textual sources discussing the same data elements. We can then identify implicit relationships between seemingly disparate data that traditional schema matching methods might miss. Thus, extending the integration scope beyond directly joinable tables.

Motivating Example-2: Consider the scenario in Fig. 1.1, where we start with two disjoint structured datasets **D1** (blue) and **D2** (yellow):

- **Disease Dataset**: Contains diseases, diagnoses, and surgeries.
- **Complication Dataset**: Records complications and prescribed drugs.

These datasets lack explicit connections. However, textual data (bottom) from clinical books could help discover the inferred dataset **D*** (in red) along with new concepts and relationships (data model in the middle) such as:

- **Anatomy**: Terms like *heart* are extracted as anatomical entities.

- **Organ**: Specific organs such as *aortic valve*.

- **Join-paths**: The identified join-paths via the inferred concepts/relations:

$$\textit{Disease} \rightarrow \textit{Diagnosis} \rightarrow \textit{Organ} \rightarrow \textit{Anatomy} \rightarrow \textit{Complication}$$

Thus textual data could uncover join-paths via inferred concepts and relationships, enabling discovery and integration of disjoint datasets.

1.1.2.3   Data Augmentation

Data augmentation is essential for data integration as it helps address the limitations of individual structured sources, such as incomplete schemas, sparse attributes, or missing relationships. By introducing additional concepts and relationships, augmentation ensures more comprehensive integration and enhances the overall utility of the combined data. This augmentation is crucial when combining two or more sources that do not contain any common attributes. For example, consider two tables from the same domain that could be integrated with a new relationship, or by simply adding a new column to one of the tables, leading to better integration. Textual data serves as a valuable source for identifying new concepts and relationships that can effectively augment and integrate structured datasets. This augmentation might involve

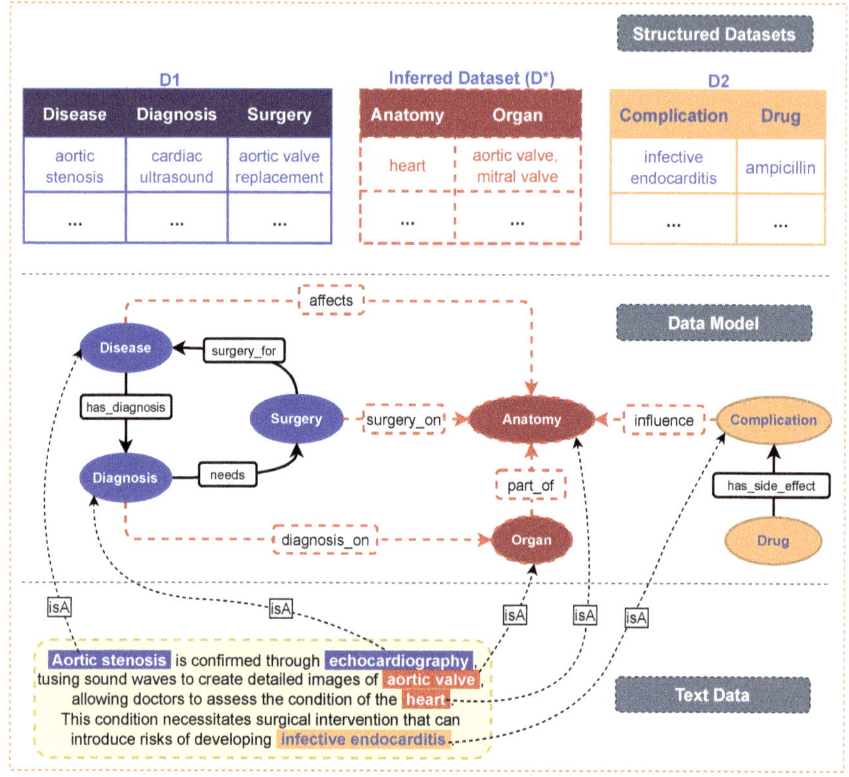

Fig. 1.1 Discovering join-paths accross disjoint datasets via text

Table 1.3 Source tables with disjoint structured data

Patients table

PatientID	Name	Age
P1	Alice Johnson	34
P2	Bob Smith	28

Medications table

MedicationID	Drug	Dosage
M1	Lisinopril	10 mg
M2	Metformin	500 mg

extending the schema and/or enriching the instances of the structured datasets. In addition, schema evolution methods could enable the integration system to adapt to previously unknown information.

Motivating Example-3: Consider the two tables from medical domain (in Table 1.3), where the *Patients Table* contains basic patient details such as ID, Name, and Age; whereas the *Medications Table* lists Drug and amount of Dosage. On their own, these two tables do not share any common attributes or connections. However, unstructured *medical notes* could provide valuable relationships linking patients to medications. For example:

Table 1.4 Augmented tables showing integrated relationships

Associative table: prescription

PatientID	MedicationID
P1	M1
P2	M2

Integrated data

Name	Age	DrugName	Dosage
Alice Johnson	34	Lisinopril	10 mg
Bob Smith	28	Metformin	500 mg

"Patient Alice Johnson *presents with* hypertension. **Prescribed** Lisinopril 10mg *once daily. Advised to monitor blood pressure regularly."*

" Bob Smith*,* **diagnosed with** type 2 diabetes mellitus. **Started on** Metformin 500mg *twice daily. Recommended lifestyle modifications."*

By extracting these relationships from text, we can bridge the gap between the two tables and create an associative *Prescription Table* linking patients to their medications. This augmentation will allow queries to produce a unified view, such as retrieving patient information alongside their prescriptions (shown in Table 1.4). This demonstrates how textual data can lead to a comprehensive integration with through data augmentation.

1.1.3 Challenges in Integrating Textual Data

Data sources vary in formats and applications. Structured and semi-structured data models such as relational data, JSON, or CSV may be useful for some use cases. Meanwhile, textual sources hold a wealth of information to combine with the former sources. Interpreting or querying over multiple such heterogeneous sources complicates the situation even more. Thus, integrating unstructured and structured data presents several challenges:

1. **Heterogeneity of Data**: Structured data adheres to specific schemas, whereas unstructured data lacks such formalization. Extracting meaningful concepts and aligning them with existing schemas requires advanced processing techniques.
2. **Semantic Ambiguity**: Textual data is inherently ambiguous, as words or phrases may carry different meanings depending on the context. Disambiguating these terms to identify their correct sense or underlying concept is a critical challenge.
3. **Scalability**: Conceptualizing and integrating large volumes of textual data requires computationally efficient algorithms and representations.
4. **Schema Evolution**: Static schema cannot accommodate new concepts and relationships extracted from texts. Evolving the schema dynamically to include new entities remains a critical open problem.

5. **Knowledge Representation**: Representing extracted textual knowledge in a machine-understandable form that supports efficient inferencing and querying is essential yet challenging.

To address these challenges, research in *Information Extraction (IE)*, and *KGs* has gained momentum, specially with the advent of *Large Language Models (LLMs)*. KGs and LLMs, in particular, could provide a way of integrating both structured and unstructured data by allowing entities, concepts, and relationships to be represented in an interconnected graph format. We will discuss these possibilities in detail throughout Sect. 1.2.

1.1.4 Objectives

This chapter presents a comprehensive overview of the field of **Text and Structured Data Integration**, focusing on the following key aspects:

1. **Understanding the Challenges**: Identifying the fundamental challenges in integrating structured and unstructured data, including heterogeneity, ambiguity, schema evolution, and knowledge representation.
2. **Key Techniques and Approaches**: Highlighting state-of-the-art methods in data integration, information extraction, and ontology learning that enable the conceptualization of textual data.
3. **Existing Solutions, Gaps and Open Problems**: Identifying prominent research contributions and the gaps that remain unresolved, particularly in text conceptualization and information extraction domains.

Rather than proposing a specific framework, the focus remains on presenting existing literature, highlighting its contributions, and identifying future research directions. Additionally, we aim to provide a thorough understanding of the challenges, approaches, and advancements related to structured and unstructured data integration.

1.2   State of the Art

In order to achieve data integration that considers natural language text with structured data, we have to combine methods and techniques from several disciplines such as KGs, LLMs, NLP, IR, DI, and Ontology Learning (OL). Text integration requires solutions from all these fields, since they are not enough on their own to solve the task. Thus, we will arrange this section in a way that reflects the relationship of textual integration with the most prominent literature related to these domains (shown in Fig. 1.2).

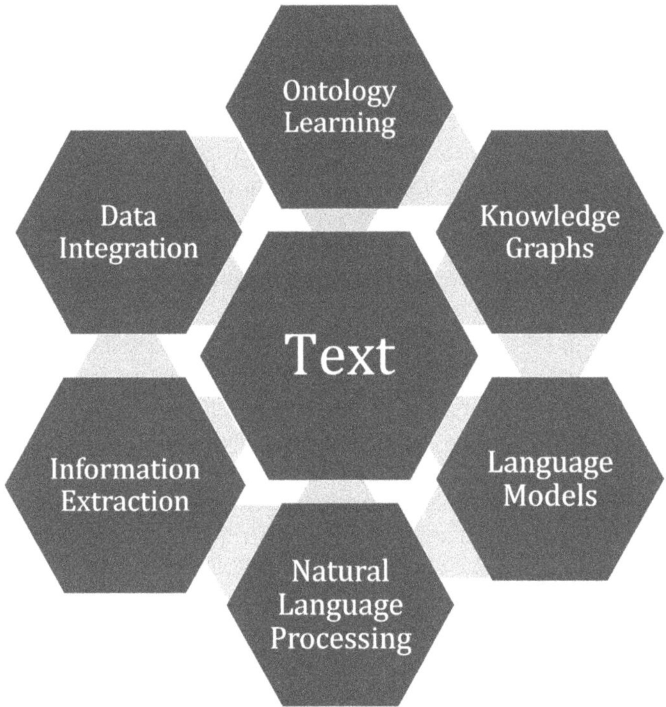

Fig. 1.2 Relevant fields for data integration with text

1.2.1 Data Integration

Traditional approaches to data integration were predominantly focused on ETL (Extract, Transform and Load) processes [6, 7] for ingesting and cleaning the data before loading them into a data lake or data warehouse [8]. However, one of the pitfalls of these classical approaches to data integration is—dealing with the data *variety* challenge at a large scale. In solving the variety issue, one must assume that, integration must be done over a multitude of data formats (i.e., text, xml, csv, relational) coming from structured, semi-structured and unstructured sources.

The integration of multiple data sources can be either based on a *physical* or a *virtual* approach. While the former consolidates the original sources, the latter however, keeps the sources intact through modeling an integrated view (a.k.a. mediated schema) [9]. Thus, individual local schemas of the associated data sources are mapped together to form a single virtual global schema. The modeling of different sources to the integrated view could be achieved via either Global-As-View (GAV) [10, 11] or Local-As-View (LAV) mappings [12]. While the constructs of the global schema are represented as views over the local schemas in the GAV-based approach, the opposite applies to the LAV mapping. However, a better choice is to combine both approaches, which is commonly referred to as Both-As-View (BAV) [13] or Global-and-Local-As-View (GLAV) [14] mapping. As the data sources adhere to different data models, part of the integration process involves transforming them into a common data model. This is typically done through a Hypergraph-based Data Model (HDM) [15]. One such implementation of HDM uses KGs by utilizing Resource Description Framework (RDF) triples [16, 17]. In this way, each source schema is mapped into a local RDF schema which sometimes is referred to as the *local graph*. Eventually, the local graphs are used to form the integrated *global graph* (which is also an RDF named graph) through the use of a (semi-) automatic global schema builder [18, 19].

Discrepancies in schemas and vocabularies used by different data sources while modeling the same domain of interest are termed as conceptual or **semantic hetero-geneity** [20]. Resolving this type of heterogeneity is considered as one of the major hurdles in data integration [21]. In dealing particularly with unstructured data such as texts, traditional fixed schema-based integration techniques are incapable of providing a satisfactory result while resolving the semantic heterogeneity, mainly due to the schemaless nature of that data [22, 23]. For this reason, a popular way is to first construct an intermediate form of structure or schema out of text sources [24]. This task is formally known as *Ontology Learning* (OL) [25] and nowadays often interchangeably referred to as *Knowledge Graph Construction* (KGC) from the text.

1.2.2 *Ontology Learning and Knowledge Graph Construction*

An *ontology* can be described as an 'explicit specification of conceptualization' [26] that encodes domain-specific implicit knowledge by utilizing some form of a semantic structure. In layman's terms, we may view an ontology as a conceptualization of a particular domain. The terms OL [27] and KGC are often used in the same context based on their underlying knowledge representation principles. While these two terms are seen to be denoting the same formalism, there is a vague but subtle difference between them. Ontologies are highly structured knowledge that facilitates queries and reasoning over complex relationships and can be used to infer new knowledge. They are fundamentally designed to capture very complex relationships between classes and individuals. On the other hand, KGs are often used to represent or instantiate ontologies in their simplest form without worrying too much about the integrity constraints and formal semantics. Thus, by combining the power of semantic web technologies with NLP and IE [28], ontologies in the form of KGs can be (semi-) automatically constructed from textual data [29–32] in order to extract and represent knowledge contained within the text [33, 34].

Figure 1.3 depicts an extended framework that includes *Concept and Relation Representation* (originally proposed by Buitelaar, 2005 [33]) for different layers of outputs that are commonly targeted (partially or fully based on the task and domain at hand) while constructing KGs from textual sources. In the following, we correlate

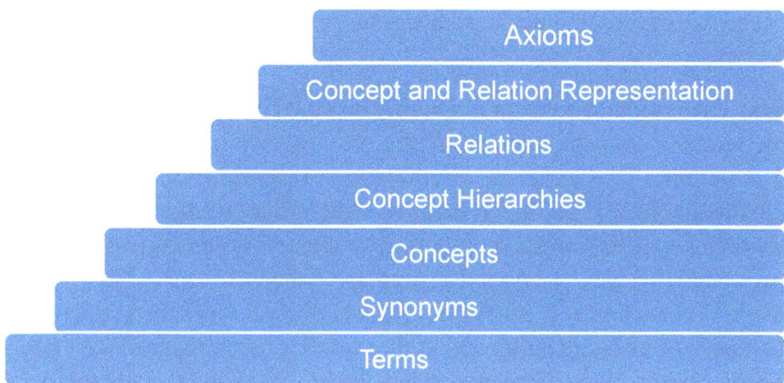

Fig. 1.3 Extended layer cake for OL, KGC and KR from text (extended from [33])

the tasks and approaches to achieve these layers of outputs with the help of the most prevalent practices within this domain.

Term: A term can be defined as a word or phrase from the natural language text, having a meaning associated with it. Term extraction is the initial requirement for KGC. Terms are the individual entities that act as a basis for identifying concepts and relations. Extracting terms from textual sources mainly uses NLP techniques. Approaches such as *Named Entity Recognition (NER)*, *sentence parsing*, *part-of-speech (POS) tagging* [35] and *morphological analysis* are commonly used in identifying the terms of interest that are associated with a domain [36]. Again, the above can be achieved via either purely rule-based or following statistical and probabilistic ways [37]. Recent advances [38] in NLP and ML techniques also pushed the boundaries of the aforementioned approaches.

Synonyms: Although linguistically synonyms are words or tokens that denote the same meaning, in OL, it is mostly a phase where terms that are associated with similar concepts are clustered together. While terms can be seen as purely syntactic properties of language, synonyms are associated with the semantics of those terms. Earlier work for this intent uses synsets such as WordNet [39], or EuroWordNet [40] which links words through semantic relations in order to represent synonyms. However, as there could be different meanings associated with a particular term depending on the context and domain at hand, it is important to determine the relevant sense. *Word-sense disambiguation* [41] tries to resolve this issue in general. However, for OL and KGC, it is important to extract and disambiguate domain-specific sense of a term [42–44]. Other techniques from information retrieval and text mining domain such as *heuristic* [45] and *conceptual clustering* [46] in parallel with *latent semantic indexing* (such as LSA, LSI) [47] and *topic modeling* (such as LDA) algorithms [48] are shown to be effective in domain-specific semantic disambiguation. A more recent approach within the NLP community, however, follows the use of *word embedding* techniques such as Word2Vec [49], FastText [50] or GloVe [51] in order to solve the *semantic textual similarity* (STS) task. STS can help determine the word, phrase as well as sentence-level semantic similarities more effectively, even for a very specific domain [52].

Concepts: The notion of concept is a highly disputed philosophical topic.[1] In purely linguistic terms, a concept is anything that we conceive in our mind when we hear or read something. It is a mental representation of entities that exist in our brain. Although the literature [33, 53, 54] in the context of our discussion, profiles concepts as purely being entities formed by grouping or clustering semantically similar terms (i.e., synonyms) [46, 55], we do not distinguish concepts from entities, or attributes at this stage. The idea is to identify and extract concepts from the text that will, later on, be categorized as being either one of the two. For the purpose of data integration, the concepts might come from the global graph of already integrated structured sources. This also narrows down the domain for the specification of what might be the viable concepts. The most basic approach to concept extraction in the form of entities involves the use of NER and *co-reference resolution* [56, 57]. Another way is to use *syntactic parsing* techniques to identify domain-specific noun phrases (NPs) as concepts [58, 59]. Entity and relations could also be jointly extracted from text using *universal dependency parsing* where subject-predicate-object triples are identified [56, 60]. However, a more recent practice to such concept extraction utilizes neural *Language Models* (LMs) based on the *transformer architecture* along with *attention mechanisms* (e.g., BERT, XLNet, RoBERTa, T5) in a supervised way [61, 62].

Concept Hierarchies: The process of obtaining a more general, higher-level concept from a set of lower-level concepts can be defined as the formation of concept hierarchy [63]. For instance, *Vegetables* and *Fruits* could be combined to form a higher-level concept, namely *Food*. It is often termed as *taxonomic relations*, which defines the hierarchical relationships among concepts. Lexico-syntactic patterns [64, 65] are one of the most popular ways of obtaining such hierarchies in the form of hypernym/hyponym relationships. It uses syntactic and grammatical rules within the same sentence, such as '<NP>, <NP> and/or other <NP>' (e.g., 'Oranges, Apples and other Fruits') to classify the taxonomic relationship. A different paradigm uses *distributional semantics* or *co-occurrence analysis* with the assumption that "a word is characterized by the company it keeps" [66–68]. Word embedding is originally inspired by this distributional hypothesis and can be utilized to build semantic hierarchies of concepts given a large corpus to learn from [69–71].

Relations: The goal here is to identify non-taxonomic relationships between the concepts discovered in earlier stages. Non-hierarchical textual relations are mostly denoted by attributes, thematic/semantic roles (i.e., agent, patient, location, source, or goal), meronymy (part-of relation), causality (e.g., smoking causes lung cancer), and possession (i.e., my, your, his, her, have) [72]. As mentioned earlier, the task of concept (entities) and relation extraction can be done separately or jointly [73] and is considered a research area itself. There are many ways to discover relations from the text [74]. Extracted potential relations might take the form of either *binary* ($n = 2$)

---

[1] https://plato.stanford.edu/entries/concepts/.

or *n-ary* ($n \geq 3$) types. Subject-verb-object *triple extraction* [60, 75] is an example of a binary relation extraction task. Traditional unsupervised approaches [56, 76] to such triple extraction often make use of syntactic and semantic parsing along with some feature engineering such as POS tagging. Some of the other ways are— *association rule mining* using seed *knowledge base* (KB) [27], *dependency structure analysis* [77], *thematic roles extraction* for relation phrase association, *semantic role labeling*, *discourse-based* relation extraction, *frame parsing*, domain-specific verb pattern matching [78], and utilizing *semantic templates* as seeds [79]. On the other hand, *distant supervision* techniques [80, 81] and LMs [61, 76, 82] are shown to be effective in this area as well. For conceptualizing text, the schema at hand (e.g., global graph) could either contain this concept or could be a completely new one. Based on these criteria, relationships might either have zero, one, or two new concepts associated with it. Depending on the type of extracted concepts, we can categorize relationship types differently (as shown in Fig. 1.4).

Concept and Relation Representation: As of now, semantic web, and data management technologies have made huge progress in data storage, representation, and retrieval. Yet, they failed to provide a system capable of learning and producing high-level representations, the sort of concepts humans exploit during language usage and comprehension. Humans are capable of using those higher-order concepts in order to generalize and reason in robust ways. It is something even an infant could do while the machine struggles with it. KGs are grounded on the principle of providing a layer of abstraction to data through graph-based configurations. A recent trend [83, 84] shows that KGs are being extensively deployed in scenarios that require integrating and extracting information from multiple, possibly diverse data sources at a large scale. Apart from this, it is easier to manage and *query* over KGs, making it a perfect candidate while representing text for data integration purposes [85]. However, the most notable works in this field try to embody texts in the form of a *data graph* (instances represented by nodes and edges) and do not follow any schema while doing so [2, 86–88]. This approach of keeping text *instances* as it is [89], fails to facilitate reasoning over them. Instead, a more reasonable approach should be to model data as a KG having dynamic *schemata*, the intention of which is to specify a high-level hierarchical structure with semantics [90]. Thus, an evolving semantic schema could aid in specifying the meaning of high-level terms (a.k.a. terminology or vocabulary) manifested by the KG, which will eventually facilitate reasoning over the data [91]. This might give KGs the supremacy we sought for modeling human-like high-level representation of abstract concepts and reasoning, the kind that we use in language understanding or while solving complex problems. For this purpose, representation languages such as RDF, RDFS, and OWL could be used.

Axioms: Axioms are rules and constraints that assert and govern the interaction between concepts and relations by adding expressivity into the ontology [92]. Yielding a set of rules and axioms that explicitly captures knowledge in a particular domain is the foremost aim of OL [54]. Frequently expressed by first-order logic or description logics, axioms are the top-level output in the OL process (see Fig. 1.3). The task can be divided into *discovering axioms* and *learning axioms*. Automatic derivation

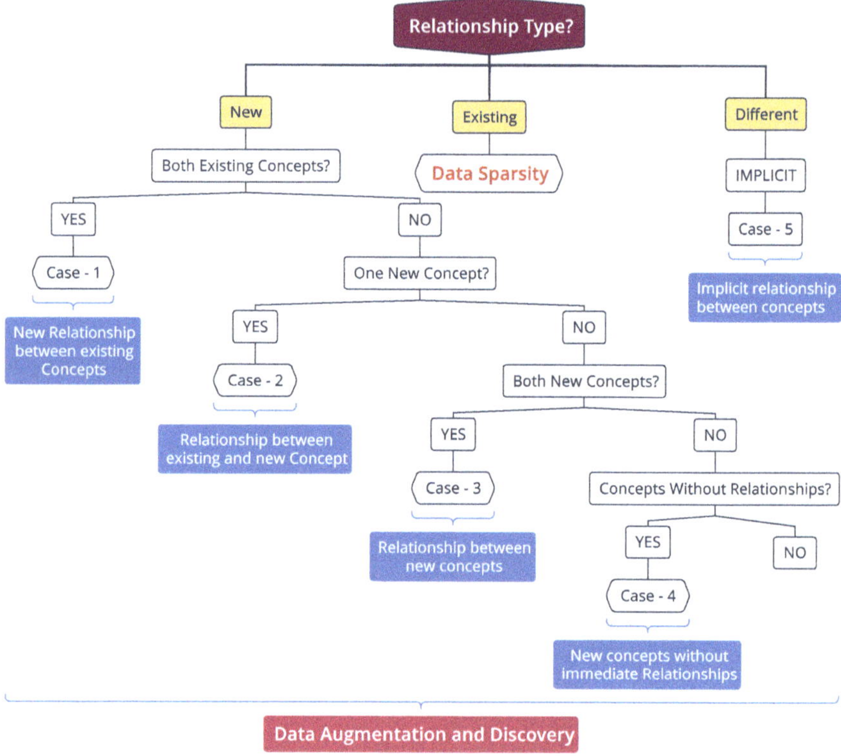

Fig. 1.4 Types of relationships depending on the underlying concepts

of such rules from text is still a challenging and less explored area [37, 92, 93]. Classic approaches to extract axioms include *inductive logic programming* [94, 95] or the usage of *axiom templates* in the form of linguistic patterns [96].

1.2.3 Information Extraction

One of the biggest challenges of integrating structured and unstructured data is extracting the information from the unstructured sources into a common structure. The problem can be approached from a textual information extraction perspective by conceptualizing unstructured text documents using the concepts present in the integrated schema. This could also aid the data sparsity problem for data generated from an integration system. Data integration focuses on providing a unified view of data over a set of disparate and heterogeneous sources [97], and typically combines the underlying datasets with operators that allow for partial matches, such as *outer join* [98] or *full disjunction* [99]. The consequence, however, is the generation of a large number of missing values (a.k.a. *labeled nulls*), which are reported to account for 15% of the values [100], since each data source captures a different set of instances (particularly, if these datasets have been independently generated). Further, each data source provides a partial view of the data of interest, thus even after combining them the resulting integrated data presents an incomplete view. This is commonly referred to in the literature as the *data sparsity problem* [101, 102].

Missing value imputation is a very common data cleaning technique in the data integration pipeline to enhance data quality [4]. Focusing on repairing structured data, such cleaning approaches can be categorized into constraint-based and learning-based ones. Systems in the former category (e.g., LLUNATIC [103], NADEEF [104], or HORIZON [105]), aim to model and enforce data dependencies or business rules modeled as denial constraints [106]. Alternatively, those in the latter (e.g., Holo-Clean [107], Baran [108], or Garf [109]), automatically generate data repairs leveraging pre-trained probabilistic models or models trained from the available structured data. Despite their effectiveness, the approaches above are limited to the available structured sources to learn and compute repairs. A significant portion of data organizations hold resides in texts [110], which is largely unused. The idea of enhancing data cleaning methods with external information has already been adopted in systems such as KATARA [111], using knowledge bases and crowdsourcing, or Cleenex [112], that leverages user feedback to define an iterative process. Yet, to the best of our knowledge no missing value imputation solution exists that exploits raw textual data (i.e., without annotations). On the one hand, the data management community has proposed several information extraction techniques [113], aiming to extract relational views from textual sources. However, the generation of structured views from textual data is limited to those extraction rules that can be defined over the text [114]. On the other hand, NLP community has proposed text conceptualization approaches [115] which aim to automatically label text span with a predefined label.

State of the art techniques for text conceptualization include (a) transformer-based LMs to perform entity recognition from text, and (b) zero-shot approaches with LLMs. However, the former require domain-specific large annotated data with rich context [116]. At the same time, they also need to be re-annotated and re-trained when the reference schema changes, while they also suffer from bias towards the most frequent entity, inconsistent performance, and demand vast resources to be trained. In such a data integration setting, LMs perform poorly when trained with structured data only [117]. The latter approaches, despite their massive popularity, often fail due to their uncertainty and tendency to hallucinate [118, 119]. LLMs struggle to maintain recall and precision when prompted with large text and complex concept categories [120]. Because of their attention mechanism, they often overlook fine-grained instances in the text, which are crucial for IE. Furthermore, resource-wise, it is not feasible to train or fine-tune an LLM like GPT-4 on a case-by-case basis, a situation that often pertains to organizations typically dealing with their own data for integration purposes [121].

Recent advances in data science have enabled novel data integration techniques that extract and integrate information from diverse data sources on a large scale [19]. Despite being a well-researched area in data engineering, the task of mitigating incomplete data during data integration is often overlooked due to its inherent challenges [122]. To tackle this problem, we can apply existing techniques to extract, organize and enrich [123] structured information from unstructured or semi-structured data. One such strategy is *Entity-Centric Slot-Filling* [124, 125] where incomplete information is extracted via conceptualization with regard to the concepts defined within the data integration system.

Statistical Data Imputation: Most conventional techniques [126] to impute data rely on statistical measures such as mean substitution, frequent category (mode) imputation, and maximum likelihood estimates; they are predominantly focused on either categorical, numeric, or quantitative data [127, 128]. Although most techniques perform quite well for numeric data [129, 130], their performance decreases for qualitative data [131]. From a statistical point of view, data imputation in data integration scenarios is more challenging than data imputation in single-database scenarios [122], since traditional statistical techniques do not deal with the *heteroscedasticity* (i.e., variance inconsistency) and *non-independent and identically distributed* (i.e., Non-IID) nature of data integrated from different sources [5]. As a consequence, a common practice is to simply ignore missing values [132]. Methods that try to mitigate qualitative missing values are either too limited in the sense that they only consider a fixed number of variables [133] or heavily dependent on domain-specific large datasets [131].

Entity Recognition: The problem of extracting information from textual data has been deeply studied in the IE and NLP communities. IE techniques encompass several tasks, being one of them *entity recognition* (ER) [56, 134], which intends to identify entities mentioned in natural language text into pre-defined categories. Yet, traditional ER techniques involve creating lexicons or dictionaries containing a list of entities with specific tagging rules [135]. Although these approaches are useful on small domains, they often tend to fall short in scenarios having complex entity types. Developing those rules also requires strong domain knowledge; hence, these systems are not generic. Alternatively, machine learning (ML) methods, such as Support Vector Machines (SVM), Decision Trees, Hidden Markov Models (HMM) [136], and Conditional Random Fields (CRFs) [137–139], have also been employed for this task. These are trained in a supervised fashion using annotated text samples with entity labels and ad-hoc feature engineering. However, since these models solely rely on training examples, they struggle to generalize to new, unseen data.

Most advanced state-of-the-art ER techniques use (L)LMs based on transformer architectures with attention mechanisms (e.g., BERT, RoBERTa, T5, XLNet, GPT, LlaMA, UniversalNER). LMs exploit contextual information (i.e., neighboring words) and distributional semantics (words with similar distributions have similar meanings) [140] via an unsupervised pre-training phase. Then, they apply a supervised strategy called *masked language modeling* (MLM) [141–143]. These techniques, in the presence of a large corpus of annotated text, outperform previous ER techniques [61, 62, 120, 144–146]. Regardless of the technique used, we find two main approaches when predicting the entity label in an ER technique. Either identify the label by only considering the text at hand or the text and a reference schema. The former is a common practice in the knowledge base population and enrichment [147–152], where they extract information from text in the form of *subject-verb-object* relations and use entity linking [153] techniques to enrich the KB with the extracted subjects and objects [2, 86–88, 154, 155]. These techniques rely on syntactic categories (e.g., noun phrases) and tend to generate long-tail entities that require further processing before being used [89]. For this reason, the current trend is to conceptualize entities in the textual data following a reference schema [90]. State-of-the-art techniques [28, 156, 157] can be used in this regard and convert the information embedded in the text into a structured representation via conceptualization. This is also the chosen approach to facilitate the enhancement [151, 158] and construction [29–31, 159, 160] of structured knowledge sources. *Slot-Filling* [161–163] is another well-known technique that utilizes ER in order to reduce data sparsity via conceptualized text [164]. Slot filling involves populating entity-specific templates (e.g., *[Acoustic Neuroma, causes, <slot>]*) with information extracted from text. It can be either *document-centric*, focusing on entities represented by a single document, or *entity-centric*, where information about a concept is spread across multiple documents in a corpus. By leveraging external text sources, missing information can be filled in, enhancing the quality and usefulness of the integrated data.

1.2.4 Advanced Language Modeling: Going Beyond

Previous attempts at automatically extracting concepts have proven challenging since they almost always relied on a fixed set of concept categories (i.e., static schema) [165], making it impossible to build KGs from raw text without knowing the associated concepts in advance [53]. With the cutting-edge advancement of (Large) Language Models, it is now possible to materialize this task. We examine these approaches in detail below:

Contextual Embeddings: Contextual embeddings such as ELMo and BERT [141] have changed how words are represented based on surrounding texts. These methods often perform better than static embeddings because they take into account polysemy and contextual details, which makes concept modeling more accurate across a wide range of domains [166]. Techniques like pooled contextualized embeddings [167] achieved state-of-the-art ER results, while domain-specific approaches [168] improved tasks like clinical concept extraction. Unifying contextualized span representations [169] with dynamic entity types [170] could thus be a promising direction towards flexible schema evolution.

Large Language Models: Pre-trained Language Models (PLMs) like BERT [141] and RoBERTa [142] are usually the go-to options for downstream tasks like concept extraction [171]. However, PLMs require task-specific fine-tuning [172]. They also struggle to adapt to new domains or entity types without additional labeled data

[173, 174]. In contrast, Large Language Models excel at these tasks because they can be instruction-tuned with simple prompts without fine-tuning [175]. Unlike PLMs, which encode static token-level representations, LLMs operate with a larger parameter space (ranging from a few billion to hundreds of billions) and utilize in-context learning [176]. This means that by simply providing a few examples or instructions in the input prompt, these models can adjust their behavior to recognize new entity types and handle domain-specific terminology [177]. In zero- or few-shot situations, even smaller LLMs (under 7B parameters) often outperform PLMs since they can connect loosely related concept hierarchies [178]. This makes information extraction pipelines more generic.

Retrieval-Augmented Generation (RAG): RAG-based techniques can improve LLM output for concept and entity recognition by directly integrating external knowledge sources into the retrieval, generation, and augmentation process [124, 179, 180]. RAG dynamically retrieves domain-specific information to address hallucinations and outdated knowledge [181]. This grounding in external data makes predictions more accurate and credible [182]. It also enables better handling of rare concepts and entities for flexible schema evolution through continuous updates [183]. More recently, Retrieval-Augmented LLMs have demonstrated even greater refinement capabilities, ensuring adaptability in dynamic and knowledge-intensive scenarios [184].

1.3 Open Problems and Future Directions

To sum up, in order to facilitate data integration that benefits from both structured and unstructured sources, we have to resort to a number of techniques from a multitude of domains. KGs facilitate a provision for virtually managing and representing data in the form of a concept map or semantic network that follows a particular schema [91]. On the other hand, unstructured data such as text does not follow any specific structure or schema. *Information Extraction* [28, 156, 157] techniques from *NLP* can be used in this regard to convert the information embedded in text into a structured representation. Thus facilitating the construction and enhancement [151, 158] of structured knowledge sources such as KG [29–31, 159]. IE typically consists of a sequence of tasks, comprising linguistic pre-processing, co-reference resolution [57], named entity recognition or concept extraction [56], and relation Extraction. More recent practice to such concept extraction utilizes neural *Language Models* based on transformer and attention mechanisms (i.e., BERT, XLNet, RoBERTa, T5, GPT, Llma, UniversalNER) [61, 62, 120, 145, 146]. In contrast, popular approaches to RE can be categorized as bootstrapping [185–187], distant supervision [80, 188–190], open-IE or unsupervised [154, 191], and supervised [61, 76, 161, 192, 193] methods. All these approaches can be useful to extract and integrate information from unstructured data sources like text and to populate KGs with structured data.

However, most advanced KGC techniques, essentially, follow the same principles as LM-based techniques for ER or RE and suffer from the same problems: (**i**) the need for a domain-specific large annotated data with rich context, (**ii**) the need to re-annotate and re-train when the reference schema changes, (**iii**) bias towards the most frequent entity type, (**iv**) inconsistent performance and hallucination and (**v**) resource-hungry training. Unfortunately, these assumptions do not hold in data integration scenario, as available data is predominantly structured with limited context (e.g., tabular data), making it difficult for LMs to work properly. Also, despite organizations having lots of textual data, those are not annotated. In the absence of these requirements, the performance of LMs quickly decreases [194]. Further, in a data integration system, the integrated schema evolves, which would require re-annotating the corpus and re-training the model. Indeed, adapting these methods in the presence of structured data, which by definition have limited context, is nowadays an open research area [146, 195, 196]. Alternatively, zero-shot LLMs such as GPT [197], despite their massive popularity, are unsuited here due to their unpredictable nature [118], inconsistent performance [119] and tendency to hallucinate [198]. LLMs also struggle to maintain recall and precision with large text corpora in the presence of complex concept categories [120]. Furthermore, it is not feasible in terms of resources (i.e., compute power and time) to train or fine-tune a large model like GPT on a case-by-case basis [121], a situation that often pertains to organizations typically dealing with their own data for integration purposes.

1.4 Conclusion

In recent years, there have been significant advances in the use of NLP, SW, and KG technologies for data integration and unstructured data processing. These advances have made this field of research timely and relevant, as organizations increasingly need to integrate and analyze structured and unstructured data to gain insights and make data-driven decisions. The use of natural language processing techniques such as named entity recognition and relation extraction has become crucial for extracting structured information from unstructured data sources like text. These techniques enable the processing and analysis of large volumes of text data, which is essential for integrating unstructured data into structured data models. Semantic web technologies provide a structured and formal vocabulary that facilitates the integration of data from different sources. The representation of data in a machine-understandable format allows for automated processing and reasoning over data. KGs are structured representations of knowledge that enable the representation of entities and relationships in a graph-like structure, which can be queried and analyzed using graph algorithms and other analytical techniques. With the help of recent NLP breakthroughs in language modeling via LLMs, KGs can be populated with data extracted from a variety of sources, including unstructured data such as text. This will allow a more complete and accurate integration of the multi-modal data, which can be queried and analyzed more effectively. Thus, benefiting a wide range of applications in sectors such as healthcare, business, and government.

References

1. Lemahieu W, vanden Broucke S, Baesens B (2018) Principles of database management: the practical guide to storing, managing and analyzing big and small data. Cambridge University Press
2. Anadiotis AC, Balalau O, Conceicao C, Galhardas H, Haddad MY, Manolescu I, Merabti T, You J (2022) Graph integration of structured, semistructured and unstructured data for data journalism. Inf Syst 104:101846
3. Guizzardi G (2006) On ontology, ontologies, conceptualizations, modeling languages, and (meta)models. In: Vasilecas O, Eder J, Caplinskas A (eds) Databases and information systems IV—selected papers from the seventh international baltic conference, DB&IS. IOS Press, Lithuania, pp 18–39
4. Ilyas IF, Chu X (2019) Data cleaning. ACM
5. Chen S, Yang S, Kim JK (2022) Nonparametric mass imputation for data integration. J Surv Stat Methodol 10(1):1–24
6. Nath RPD, Hose K, Pedersen TB, Romero O, Bhattacharjee A (2020) SETLBI: an integrated platform for semantic business intelligence. In: Seghrouchni AEF, Sukthankar G, Liu T, van Steen M (eds) Companion of The 2020 web conference. ACM / IW3C2, Taiwan, pp 167–171
7. Deb Nath RP, Romero O, Pedersen TB, Hose K (2022) High-level etl for semantic data warehouses. Semantic Web 13(1):85–132
8. Alqarni AA, Pardede E (2012) Integration of data warehouse and unstructured business documents. In: Barolli L, Taniar D, Enokido T, Rahayu JW, Takizawa M (eds) 15th international conference on network-based information systems. NBiS, IEEE Computer Society, Australia, pp 32–37
9. Williams D, Poulovassilis A (2003) Combining data integration with natural language technology for the semantic web. In: Tablan V, Bontcheva K, Maynard D (eds) Proceedings of the workshop on human language technology for the semantic web and web services at ISWC, CEUR, USA, pp 35–42
10. Chawathe S, Garcia-Molina H, Hammer J, Ireland K, Papakonstantinou Y, Ullman JD, Widom J (1994) The tsimmis project: integration of heterogeneous information sources. In: Proceedings of the conference of the information processing society of Japan (IPSJ), (ed) of Japan IPS. Information Processing Society of Japan, Tokyo, Japan, pp 7–18
11. Roth MT, Schwarz PM (1997) Don't scrap it, wrap it! A wrapper architecture for legacy data sources. In: Jarke M, Carey MJ, Dittrich KR, Lochovsky FH, Loucopoulos P, Jeusfeld MA (eds) Proceedings of 23rd international conference on very large data bases. VLDB, Morgan Kaufmann, Greece, pp 266–275
12. Levy AY, Rajaraman A, Ordille JJ (1996) Querying heterogeneous information sources using source descriptions. In: Vijayaraman TM, Buchmann AP, Mohan C, Sarda NL (eds) Proceedings of 22th international conference on very large data bases. VLDB, Morgan Kaufmann, India, pp 251–262
13. McBrien P, Poulovassilis A (2003) Data integration by bi-directional schema transformation rules. In: Dayal U, Ramamritham K, Vijayaraman TM (eds) Proceedings of the 19th international conference on data engineering. ICDE, IEEE, India, pp 227–238
14. Friedman M, Levy AY, Millstein TD et al (1999) Navigational plans for data integration AAAI/IAAI 1999:67–73
15. Poulovassilis A, McBrien P (1998) A general formal framework for schema transformation. Data Knowl Eng 28(1):47–71
16. Manola F, Miller E, McBride B et al. (2004) Rdf primer. W3C Recommendation 10(1–107):6
17. Augenstein I, Padó S, Rudolph S (2012) Lodifier: generating linked data from unstructured text. In: Simperl E, Cimiano P, Polleres A, Corcho Ó, Presutti V (eds) The semantic web: research and applications—9th extended semantic web conference. ESWC, Springer, Greece, pp 210–224

18. Touma R, Romero O, Jovanovic P (2015) Supporting data integration tasks with semi-automatic ontology construction. In: Song I, Garcia-Alvarado C, Ordonez C (eds) Proceedings of the eighteenth international workshop on data warehousing and OLAP, DOLAP, ACM, Australia, pp 89–98
19. Jovanovic P, Nadal S, Romero O, Abelló A, Bilalli B (2021) Quarry: a user-centered big data integration platform. Inf Syst Front 23(1):9–33
20. Wang L (2017) Heterogeneous data and big data analytics. Autom Control Inf Sci 3(1):8–15
21. Golshan B, Halevy AY, Mihaila GA, Tan W (2017) Data integration: after the teenage years. In: Sallinger E, den Bussche JV, Geerts F (eds) Proceedings of the 36th ACM symposium on principles of database systems. PODS, ACM, USA, pp 101–106
22. Doan A, Naughton JF, Ramakrishnan R, Baid A, Chai X, Chen F, Chen T, Chu E, DeRose P, Gao B et al (2009) Information extraction challenges in managing unstructured data. ACM SIGMOD Rec 37(4):14–20
23. Adnan K, Akbar R, Khor SW, Ali ABA (2020) Role and challenges of unstructured big data in healthcare. In: Sharma N, Kumar A, Garg H, Balas VE (eds) Data management. Analytics and innovation. Springer Singapore, pp 301–323
24. Buitelaar P, Cimiano P, Frank A, Hartung M, Racioppa S (2008) Ontology-based information extraction and integration from heterogeneous data sources. Int J Hum Comput Stud 66(11):759–788
25. Gómez-Pérez A, Manzano-Macho D (2004) An overview of methods and tools for ontology learning from texts. Knowl Eng Rev 19(3):187–212
26. Gruber TR (1993) A translation approach to portable ontology specifications. Knowl Acquis 5(2):199–220
27. Maedche A, Staab S (2000) Mining ontologies from text. In: Dieng R, Corby O (eds) 12th international conference on knowledge acquisition. Modeling and management, EKAW. Springer, France, pp 189–202
28. Martinez-Rodriguez JL, Hogan A, Lopez-Arevalo I (2020) Information extraction meets the semantic web: a survey. Semantic Web 11(2):255–335
29. Kertkeidkachorn N, Ichise R (2018) An automatic knowledge graph creation framework from natural language text. IEICE Trans Inf Syst 101(1):90–98
30. Pujara J, Singh S (2018) Mining knowledge graphs from text. In: Chang Y, Zhai C, Liu Y, Maarek Y (eds) Proceedings of the eleventh international conference on web search and data mining. WSDM, ACM, USA, pp 789–790
31. Martinez-Rodriguez JL, Lopez-Arevalo I, Rios-Alvarado AB (2018) Openie-based approach for knowledge graph construction from text. Expert Syst Appl 113:339–355
32. Hofer M, Obraczka D, Saeedi A, Köpcke H, Rahm E (2024) Construction of knowledge graphs: current state and challenges. Information 15(8):509
33. Buitelaar P, Cimiano P, Magnini B (2005) Ontology learning from text: methods, evaluation and applications. IOS press
34. Buitelaar P, Cimiano P (2008) Ontology learning and population: bridging the gap between text and knowledge. Ios Press
35. Abney S (1997) Part-of-speech tagging and partial parsing. In: Bloothooft G, Young S (eds) Corpus-based methods in language and speech processing. Springer, pp 118–136
36. Shamsfard M, Barforoush AA (2004) Learning ontologies from natural language texts. Int J Hum Comput Stud 60(1):17–63
37. Khadir AC, Aliane H, Guessoum A (2021) Ontology learning: grand tour and challenges. Comput Sci. Rev. 39:100339
38. Ruder S (2020) NLP-progress: tracking progress in natural language processing. https://nlpprogress.com
39. Miller GA (1995) Wordnet: a lexical database for english. Commun ACM 38(11):39–41
40. Vossen P (1998) A multilingual database with lexical semantic networks. Kluwer Academic Publishers, Dordrecht. https://doi.org/10.978-94
41. Wang Y, Wang M, Fujita H (2020) Word sense disambiguation: a comprehensive knowledge exploitation framework. Knowl-Based Syst 190:105030

42. Magnini B, Strapparava C (2000) Experiments in word domain disambiguation for parallel texts. In: Ide N, Véronis J (eds) ACL workshop on word senses and multi-linguality. ACL, China, pp 27–33
43. Buitelaar P, Sacaleanu B (2002) Extending synsets with medical terms. In: Vossen P, Fellbaum C (eds) Proceedings of the first international wordnet conference (GWC 2002). Central Institute of Indian Languages, Mysore, India, pp 107–117
44. Turcato D, Popowich F, Toole J, Fass D, Nicholson D, Tisher G (2000) Adapting a synonym database to specific domains. In: Ide N, Véronis J (eds) ACL workshop on recent advances in natural language processing and information retrieval. ACL, China, pp 1–11
45. Lin D, Pantel P (2001) Induction of semantic classes from natural language text. In: Lee D, Schkolnick M, Provost FJ, Srikant R (eds) Proceedings of the seventh ACM international conference on Knowledge discovery and data mining. SIGKDD, ACL, USA, pp 317–322
46. Lin D, Pantel P (2002) Concept discovery from text. In: Tseng SC, Chen TE, Liu YF (eds) 19th international conference on computational linguistics. COLING, Morgan Kaufmann, Taiwan, pp 577–583
47. Landauer TK, Dumais ST (1997) A solution to plato's problem: the latent semantic analysis theory of acquisition, induction, and representation of knowledge. Psychol Rev 104(2):211
48. Blei DM, Ng AY, Jordan MI (2003) Latent dirichlet allocation. J Mach Learn Res 3:993–1022
49. Mikolov T, Sutskever I, Chen K, Corrado GS, Dean J (2013) Distributed representations of words and phrases and their compositionality. In: Burges CJC, Bottou L, Ghahramani Z, Weinberger KQ (eds) 27th annual conference on neural information processing systems. Curran Associates Inc, USA, pp 3111–3119
50. Bojanowski P, Grave E, Joulin A, Mikolov T (2017) Enriching word vectors with subword information. Trans ACL 5:135–146
51. Pennington J, Socher R, Manning CD (2014) Glove: Global vectors for word representation. In: Moschitti A, Pang B, Daelemans W (eds) Proceedings of the conference on empirical methods in natural language processing. EMNLP, ACL, Qatar, pp 1532–1543
52. Harispe S, Ranwez S, Janaqi S, Montmain J (2015) Semantic similarity from natural language and ontology analysis. Morgan & Claypool Publishers
53. Al-Aswadi FN, Chan HY, Gan KH (2020) Automatic ontology construction from text: a review from shallow to deep learning trend. Artif Intell Rev 53(6):3901–3928
54. Wong W, Liu W, Bennamoun M (2012) Ontology learning from text: a look back and into the future. ACM Comput Surv (CSUR) 44(4):1–36
55. Gómez-Suta M, Echeverry-Correa JD, Mejía JAS (2020) Semi-automatic extraction and validation of concepts in ontology learning from texts in spanish. In: Chbeir R, Manolopoulos Y, Akerkar R, Mizera-Pietraszko J (eds) The 10th international conference on web intelligence. Mining and semantics, WIMS Biarritz, ACM, France, pp 7–16
56. Exner P, Nugues P (2012) Entity extraction: from unstructured text to dbpedia RDF triples. In: Rizzo G, Mendes PN, Charton E, Hellmann S, Kalyanpur A (eds) Proceedings of the web of linked entities workshop. CEUR-WS.org, USA, pp 58–69
57. El-Kilany A, Tazi NE, Ezzat E (2017) Building relation extraction templates via unsupervised learning. In: Desai BC, Hong J, McClatchey R (eds) Proceedings of the 21st international database engineering and applications symposium, IDEAS. ACM, United Kingdom, pp 228–234
58. Gamallo P, Gonzalez M, Agustini A, Lopes G, de Lima VS (2002) Mapping syntactic dependencies onto semantic relations. In: Aussenac-Gilles N, Maedche A (eds) Proceedings of the ECAI, workshop on machine learning and natural language processing for ontology engineering. INRIA, France, pp 15–22
59. Rajagopal D, Cambria E, Olsher D, Kwok K (2013) A graph-based approach to commonsense concept extraction and semantic similarity detection. In: Carr L, Laender AHF, Lóscio BF, King I, Fontoura M, Vrandecic D, Aroyo L, de Oliveira JPM, Lima F, Wilde E (eds) 22nd international world wide web conference, WWW. ACM, Brazil, pp 565–570
60. Liu Q, Xu K, Zhang L, Wang H, Yu Y, Pan Y (2008) Catriple: extracting triples from wikipedia categories. In: Domingue J, Anutariya C (eds) The semantic web, 3rd Asian semantic web conference ASWC. Springer, Thailand, pp 330–344

61. Hu W, Liu L, Sun Y, Wu Y, Liu Z, Zhang R, Peng T (2022) Nlire: a natural language inference method for relation extraction. J Web Seman 72:100686
62. Milosevic N, Thielemann W (2022) Relationship extraction for knowledge graph creation from biomedical literature. CoRR 1–12. arxiv:abs/2201.01647
63. Han J, Kamber M, Pei J (2011) Data warehousing and online analytical processing. In: Gray J (ed) Data mining: concepts and techniques. Morgan Kaufmann, pp 123–157
64. Hearst MA (1992) Automatic acquisition of hyponyms from large text corpora. In: Boitet C (ed) 14th international conference on computational linguistics. COLING, ACL, France, pp 539–545
65. Buitelaar P, Olejnik D, Sintek M (2004) A protégé plug-in for ontology extraction from text based on linguistic analysis. In: Bussler C, Davies J, Fensel D, Studer R (eds) The semantic web: research and applications. First European semantic web symposium, ESWS. Springer, Greece, pp 31–44
66. Harris ZS (1954) Distributional structure. Word 10(2–3):146–162
67. Firth JR (1957) A synopsis of linguistic theory, 1930–1955. In: Firth JR (ed) Studies in linguistic analysis. Basil Blackwell, pp 1–32
68. Sahlgren M (2008) The distributional hypothesis. Ital J Disabil Stud 20:33–53
69. Fu R, Guo J, Qin B, Che W, Wang H, Liu T (2014) Learning semantic hierarchies via word embeddings. In: Toutanova K, Wu H (eds) Proceedings of the 52nd annual meeting of the ACL. ACL, USA, pp 1199–1209
70. Nguyen KA, Köper M, im Walde SS, Vu NT (2017) Hierarchical embeddings for hypernymy detection and directionality. In: Palmer M, Hwa R, Riedel S (eds) Proceedings of the conference on empirical methods in natural language processing. EMNLP, ACL, Denmark, pp 233–243
71. Alsuhaibani M, Maehara T, Bollegala D (2019) Joint learning of hierarchical word embeddings from a corpus and a taxonomy. In: McCallum A, Augenstein I, Singh S (eds) 1st conference on automated knowledge base construction. AKBC, OpenReview, USA, pp 1–9
72. Peters S, Westerståhl D (2013) The semantics of possessives. Language 89(4):713–759
73. Al-Aswadi FN, Chan HY, Gan KH (2020b) Extracting semantic concepts and relations from scientific publications by using deep learning. In: Saeed F, Mohammed F, Al-Nahari A (eds) Innovative systems for intelligent health informatics—data science. Health informatics, intelligent systems, smart computing, IRICT. Springer, Malaysia, pp 374–383
74. Alobaidi M, Malik KM, Sabra S (2018) Linked open data-based framework for automatic biomedical ontology generation. BMC Bioinform 19(1):1–13
75. Akter YA, Rahman MA (2019) Extracting rdf triples from raw text. In: Reza AW, Shahnaz C (eds) 2019 1st international conference on advances in science. Engineering and robotics technology (ICASERT). IEEE, Bangladesh, pp 1–4
76. Han X, Gao T, Yao Y, Ye D, Liu Z, Sun M (2019) Opennre: an open and extensible toolkit for neural relation extraction. In: Padó S, Huang R (eds) Proceedings of the conference on empirical methods in natural language processing and the 9th international joint conference on natural language processing, EMNLP-IJCNLP. ACL, China, pp 169–174
77. Qiu J, Chai Y, Liu Y, Gu Z, Li S, Tian Z (2018) Automatic non-taxonomic relation extraction from big data in smart city. IEEE Access 6:74854–74864
78. Kaushik N, Chatterjee N (2018) Automatic relationship extraction from agricultural text for ontology construction. Inf Process Agric 5(1):60–73
79. Yahya M, Whang S, Gupta R, Halevy AY (2014) Renoun: Fact extraction for nominal attributes. In: Moschitti A, Pang B, Daelemans W (eds) Proceedings of the conference on empirical methods in natural language processing, EMNLP. ACL, Qatar, pp 325–335
80. Mintz M, Bills S, Snow R, Jurafsky D (2009) Distant supervision for relation extraction without labeled data. In: Su K, Su J, Wiebe J (eds) Proceedings of the 47th annual meeting of the ACL and the 4th international joint conference on natural language processing of the AFNLP. ACL, Singapore, pp 1003–1011
81. Madaan A, Mittal AR, Mausam, Ramakrishnan G, Sarawagi S (2016) Numerical relation extraction with minimal supervision. In: Schuurmans D, Wellman MP (eds) Proceedings of the thirtieth conference on artificial intelligence AAAI. AAAI Press, USA, pp 2764–2771

82. Nakashole N, Theobald M, Weikum G (2011) Scalable knowledge harvesting with high precision and high recall. In: King I, Nejdl W, Li H (eds) Proceedings of the forth international conference on web search and web data mining, WSDM. ACM, China, pp 227–236

83. Ji S, Pan S, Cambria E, Marttinen P, Yu PS (2022) A survey on knowledge graphs: representation, acquisition, and applications. IEEE Trans Neural Netw Learn Syst 33(2):494–514

84. Tiwari S, Al-Aswadi FN, Gaurav D (2021) Recent trends in knowledge graphs: theory and practice. Soft Comput 25(13):8337–8355

85. Liang S, Stockinger K, de Farias TM, Anisimova M, Gil M (2021) Querying knowledge graphs in natural language. J Big Data 8(1):1–23

86. Clancy R, Ilyas IF, Lin J, Cheriton D (2019) Knowledge graph construction from unstructured text with applications to fact verification and beyond. In: Thorne J, Vlachos A, Cocarascu O, Christodoulopoulos C, Mittal A (eds) Proceedings of the second workshop on fact extraction and verification (FEVER). ACL, China, pp 3–7

87. Mao J, Yao Y, Heinrich S, Hinz T, Weber C, Wermter S, Liu Z, Sun M (2019) Bootstrapping knowledge graphs from images and text. Front Neurorobot 13:93

88. Smith E, Papadopoulos D, Braschler M, Stockinger K (2022) Lillie: information extraction and database integration using linguistics and learning-based algorithms. Inf Syst 105:101938

89. Melnyk I, Dognin PL, Das P (2022) Knowledge graph generation from text. In: Goldberg Y, Kozareva Z, Zhang Y (eds) Findings of the ACL: EMNLP. ACL, UAE, pp 1610–1622

90. Rossiello G, Chowdhury MFM, Mihindukulasooriya N, Cornec O, Gliozzo AM (2023) Knowgl: Knowledge generation and linking from text. In: Williams B, Chen Y, Neville J (eds) Thirty-seventh AAAI conference on artificial intelligence. AAAI Press, USA, pp 16476–16478

91. Hogan A, Blomqvist E, Cochez M, d'Amato C, de Melo G, Gutiérrez C, Kirrane S, Labra Gayo JE, Navigli R, Neumaier S, Ngonga Ngomo AC, Polleres A, Rashid SM, Rula A, Schmelzeisen L, Sequeda JF, Staab S, Zimmermann A (2021) Knowledge graphs. Morgan & Claypool

92. Browarnik A, Maimon O (2015) Ontology learning from text: why the ontology learning layer cake is not viable. Int J Signs Semiotic Syst (IJSSS) 4(2):1–14

93. Gillani Andleeb S (2015) From text mining to knowledge mining: an integrated framework of concept extraction and categorization for domain ontology. PhD thesis, Budapesti Corvinus Egyetem

94. Fleischhacker D, Völker J (2011) Inductive learning of disjointness axioms. In: Meersman R, Dillon TS, Herrero P, Kumar A, Reichert M, Qing L, Ooi BC, Damiani E, Schmidt DC, White J, Hauswirth M, Hitzler P, Mohania MK (eds) On the move to meaningful internet systems: OTM—confederated international conferences: CoopIS, DOA-SVI, and ODBASE, Proceedings, Part II. Springer, Greece, pp 680–697

95. Klarman S, Britz K (2015) Towards unsupervised ontology learning from data. In: Booth R, Casini G, Klarman S, Richard G, Varzinczak I (eds) Proceedings of the international workshop on defeasible and ampliative reasoning, DARe. CEUR, Argentina, pp 33–42

96. Sánchez D, Moreno A, Del Vasto-Terrientes L (2012) Learning relation axioms from text: an automatic web-based approach. Expert Syst Appl 39(5):5792–5805

97. Lenzerini M (2002) Data integration: a theoretical perspective. In: Popa L, Abiteboul S, Kolaitis PG (eds) Proceedings of the twenty-first ACM symposium on principles of database systems, PODS. ACM, USA, pp 233–246

98. Lacroix M, Pirotte A (1976) Generalized joins. SIGMOD Rec 8(3):14–15

99. Rajaraman A, Ullman JD (1996) Integrating information by outerjoins and full disjunctions. In: Hull R (ed) Proceedings of the fifteenth symposium on principles of database systems, SIGACT-SIGMOD-SIGART. ACM, Canada, pp 238–248

100. Stonebraker M, Ilyas IF (2018) Data integration: the current status and the way forward. IEEE Data Eng Bull 41(2):3–9

101. Allison B, Guthrie D, Guthrie L (2006) Another look at the data sparsity problem. In: Sojka P, Kopecek I, Pala K (eds) Text, speech and dialogue, 9th international conference. Czech Republic, TSD, Springer, pp 327–334

102. Xue AY, Qi J, Xie X, Zhang R, Huang J, Li Y (2015) Solving the data sparsity problem in destination prediction. VLDB J 24(2):219–243
103. Geerts F, Mecca G, Papotti P, Santoro D (2020) Cleaning data with llunatic VLDB 29(4):867–892
104. Dallachiesa M, Ebaid A, Eldawy A, Elmagarmid AK, Ilyas IF, Ouzzani M, Tang N (2013) NADEEF: a commodity data cleaning system. In: Ross KA, Srivastava D, Papadias D (eds) Proceedings of the ACM international conference on management of data, SIGMOD. ACM, USA, pp 541–552
105. Rezig EK, Ouzzani M, Aref WG, Elmagarmid AK, Mahmood AR, Stonebraker M (2021) Horizon: scalable dependency-driven data cleaning. Proc VLDB Endow 14(11):2546–2554
106. Fan W (2015) Data quality: from theory to practice. SIGMOD Rec 44(3):7–18
107. Rekatsinas T, Chu X, Ilyas IF, Ré C (2017) Holoclean: holistic data repairs with probabilistic inference. Proc VLDB Endow 10(11):1190–1201
108. Mahdavi M, Abedjan Z (2020) Baran: effective error correction via a unified context representation and transfer learning. Proc VLDB Endow 13(11):1948–1961
109. Peng J, Shen D, Tang N, Liu T, Kou Y, Nie T, Cui H, Yu G (2022) Self-supervised and interpretable data cleaning with sequence generative adversarial networks. Proc VLDB Endow 16(3):433–446
110. Chu X, Ilyas IF, Krishnan S, Wang J (2016) Data cleaning: overview and emerging challenges. In: Özcan F, Koutrika G, Madden S (eds) Proceedings of the 2016 international conference on management of data, SIGMOD. ACM, USA, pp 2201–2206
111. Chu X, Morcos J, Ilyas IF, Ouzzani M, Papotti P, Tang N, Ye Y (2015) KATARA: A data cleaning system powered by knowledge bases and crowdsourcing. In: Sellis TK, Davidson SB, Ives ZG (eds) Proceedings of the 2015 ACM international conference on management of data, SIGMOD. ACM, Australia, pp 1247–1261
112. Pereira JLM, Fonseca MJ, Lopes A, Galhardas H (2024) Cleenex: support for user involvement during an iterative data cleaning process. ACM J Data Inf Qual 16(1):6:1–6:26
113. Fagin R, Kimelfeld B, Reiss F, Vansummeren S (2016) A relational framework for information extraction. ACM SIGMOD Rec 44(4):5–16
114. Fagin R, Kimelfeld B, Reiss F, Vansummeren S (2015) Document spanners: a formal approach to information extraction. J ACM 62(2):12:1–12:51
115. Sarawagi S (2008) Information extraction. Found Trends Databases 1(3):261–377
116. Schneider ETR, Zavala RMR, Martínez P, Moro C, Paraiso EC (2022) UC3M-PUCPR at semeval-2022 task 11: an ensemble method of transformer-based models for complex named entity recognition. In: Emerson G, Schluter N, Stanovsky G, Kumar R, Palmer A, Schneider N, Singh S, Ratan S (eds) Proceedings of the 16th international workshop on semantic evaluation, SemEval NAACL. ACL, USA, pp 1448–1456
117. Keymanesh M, Benton A, Dredze M (2022) What makes data-to-text generation hard for pretrained language models? CoRR 1–16. arxiv:abs/2205.11505
118. Chen L, Zaharia M, Zou J (2023) How is chatgpt's behavior changing over time? CoRR 1–12. arxiv:abs/2307.09009
119. Arora D, Singh HG, Mausam (2023) Have llms advanced enough? a challenging problem solving benchmark for large language models. In: Bouamor H, Pino J, Bali K (eds) Proceedings of the 2023 conference on empirical methods in natural language processing, EMNLP. ACL, Singapore, pp 7527–7543
120. Zhou W, Zhang S, Gu Y, Chen M, Poon H (2024) Universalner: targeted distillation from large language models for open named entity recognition. In: Kim B, Yue Y, Chaudhuri S, Fragkiadaki K, Khan ME, Sun Y (eds) The twelfth international conference on learning representations. ICLR, OpenReview, Austria, pp 1–12
121. Yuan B, He Y, Davis J, Zhang T, Dao T, Chen B, Liang PS, Re C, Zhang C (2022) Decentralized training of foundation models in heterogeneous environments. Adv Neural Inf Process Syst 35:25464–25477
122. Yang S, Kim JK (2020) Statistical data integration in survey sampling: a review. Jpn J Stat Data Sci 3:625–650

123. Allen M, Cervo D (2015) Data quality management. In: Master M-D (ed) Allen M, Cervo D. Data management, advanced mdm and data governance in practice. Morgan Kaufmann, pp 131–160
124. Petroni F, Piktus A, Fan A, Lewis PSH, Yazdani M, Cao ND, Thorne J, Jernite Y, Karpukhin V, Maillard J, Plachouras V, Rocktäschel T, Riedel S (2021) KILT: a benchmark for knowledge intensive language tasks. In: Toutanova K, Rumshisky A, Zettlemoyer L, Hakkani-Tür D, Beltagy I, Bethard S, Cotterell R, Chakraborty T, Zhou Y (eds) Proceedings of the conference of the North American chapter of the ACL: human language technologies, NAACL-HLT. ACL, Online, pp 2523–2544
125. Glass MR, Rossiello G, Gliozzo A (2021) Zero-shot slot filling with DPR and RAG. CoRR 1–12. arxiv:abs/2104.08610
126. Pigott TD (2001) A review of methods for missing data. Educ Res Eval 7(4):353–383
127. Khan SI, Hoque ASML (2020) Sice: an improved missing data imputation technique. J Big Data 7(1):1–21
128. Jäger S, Allhorn A, Bießmann F (2021) A benchmark for data imputation methods. Front Big Data 4:693674
129. Jadhav A, Pramod D, Ramanathan K (2019) Comparison of performance of data imputation methods for numeric dataset. Appl Artif Intell 33(10):913–933
130. Bruni R, Daraio C, Aureli D (2021) Imputation techniques for the reconstruction of missing interconnected data from higher educational institutions. Knowl-Based Syst 212:106512
131. Bießmann F, Salinas D, Schelter S, Schmidt P, Lange D (2018) "deep" learning for missing value imputationin tables with non-numerical data. In: Cuzzocrea A, Allan J, Paton NW, Srivastava D, Agrawal R, Broder AZ, Zaki MJ, Candan KS, Labrinidis A, Schuster A, Wang H (eds) Proceedings of the 27th ACM international conference on information and knowledge management, CIKM. ACM, Italy, pp 2017–2025
132. Ren L, Wang T, Seklouli AS, Zhang H, Bouras A (2023) A review on missing values for main challenges and methods. Inf Syst 119:102268
133. Romero V, Salmerón A (2004) Multivariate imputation of qualitative missing data using bayesian networks. In: López-Díaz M, Gil MA, Grzegorzewski P, Hryniewicz O, Lawry J (eds) Soft methodology and random information systems. Springer, Berlin, pp 605–612
134. Fries JA, Wu S, Ratner A, Ré C (2017) Swellshark: A generative model for biomedical named entity recognition without labeled data. CoRR 1–12. arxiv:abs/1704.06360
135. Quimbaya AP, Múnera AS, Rivera RAG, Rodríguez JCD, Velandia OMM, Peña AAG, Labbé C (2016) Named entity recognition over electronic health records through a combined dictionary-based approach. Procedia Comput Sci 100:55–61
136. Zhao S (2004) Named entity recognition in biomedical texts using an HMM model. In: Collier N, Ruch P, Nazarenko A (eds) Proceedings of the international joint workshop on natural language processing in biomedicine and its applications, NLPBA/BioNLP. COLING, Switzerland, pp 87–90
137. Li D, Savova G, Schuler KK (2008) Conditional random fields and support vector machines for disorder named entity recognition in clinical texts. In: Demner-Fushman D, Ananiadou S, Cohen KB, Pestian J, Tsujii J, Webber BL (eds) Proceedings of the workshop on current trends in biomedical natural language processing, BioNLP. ACL, USA, pp 94–95
138. Rocktäschel T, Weidlich M, Leser U (2012) Chemspot: a hybrid system for chemical named entity recognition. Bioinformatics 28(12):1633–1640
139. Leaman R, Wei CH, Lu Z (2015) tmchem: a high performance approach for chemical named entity recognition and normalization. J Cheminform 7(1):1–10
140. Lenci A, Sahlgren M (2023) Distributional semantics. Cambridge University Press
141. Devlin J, Chang M, Lee K, Toutanova K (2019) BERT: pre-training of deep bidirectional transformers for language understanding. In: Burstein J, Doran C, Solorio T (eds) Proceedings of the 2019 conference of the North American chapter of the ACL: human language technologies, NAACL-HLT. ACL, USA, pp 4171–4186
142. Liu Y, Ott M, Goyal N, Du J, Joshi M, Chen D, Levy O, Lewis M, Zettlemoyer L, Stoyanov V (2019) Roberta: a robustly optimized BERT pretraining approach. CoRR 1–12. arxiv:abs/1907.11692

143. Alsentzer E, Murphy JR, Boag W, Weng W, Jin D, Naumann T, McDermott MBA (2019) Publicly available clinical BERT embeddings. CoRR 1–7. arxiv:abs/1904.03323
144. Bhowmick SS, Dragut EC, Meng W (2023) Globally aware contextual embeddings for named entity recognition in social media streams. In: Li C, Chen L, Manegold S (eds) 39th IEEE international conference on data engineering. ICDE, IEEE, USA, pp 1544–1557
145. Wang X, Jiang Y, Bach N, Wang T, Huang Z, Huang F, Tu K (2021) Automated concatenation of embeddings for structured prediction. In: Zong C, Xia F, Li W, Navigli R (eds) Proceedings of the 59th Annual Meeting of the ACL and the 11th international joint conference on natural language processing, ACL/IJCNLP, ACL, Virtual, pp 2643–2660
146. Wang X, Shen Y, Cai J, Wang T, Wang X, Xie P, Huang F, Lu W, Zhuang Y, Tu K, Lu W, Jiang Y (2022) DAMO-NLP at semeval-2022 task 11: a knowledge-based system for multilingual named entity recognition. In: Emerson G, Schluter N, Stanovsky G, Kumar R, Palmer A, Schneider N, Singh S, Ratan S (eds) Proceedings of the 16th international workshop on semantic evaluation, SemEval@NAACL. ACL, USA, pp 1457–1468
147. Lin X, Li H, Xin H, Li Z, Chen L (2020) Kbpearl: a knowledge base population system supported by joint entity and relation linking. Proc VLDB Endow 13(7):1035–1049
148. Ji H, Grishman R (2011) Knowledge base population: Successful approaches and challenges. In: Lin D, Matsumoto Y, Mihalcea R (eds) The 49th annual meeting of the ACL: human language technologies. ACL, USA, pp 1148–1158
149. Vannur LS, Ganesan B, Nagalapatti L, Patel H, Tippeswamy MN (2021) Data augmentation for fairness in personal knowledge base population. In: Gupta M, Ramakrishnan G (eds) Trends and applications in knowledge discovery and data mining. Springer, India, pp 143–152
150. Sui D, Wang C, Chen Y, Liu K, Zhao J, Bi W (2021) Set generation networks for end-to-end knowledge base population. In: Moens M, Huang X, Specia L, Yih SW (eds) Proceedings of the conference on empirical methods in natural language processing, EMNLP. ACL, Dominican Republic, pp 9650–9660
151. Xu D, Zhou J, Xu T, Xia Y, Liu J, Chen E, Dou D (2023) Multimodal biological knowledge graph completion via triple co-attention mechanism. In: Li C, Chen L, Manegold S (eds) 39th IEEE international conference on data engineering. ICDE, IEEE, USA, pp 3928–3941
152. Trajanoska M, Stojanov R, Trajanov D (2023) Enhancing knowledge graph construction using large language models. CoRR 1–12. arxiv:abs/2305.04676
153. Rao D, McNamee P, Dredze M (2013) Entity linking: finding extracted entities in a knowledge base. In: Poibeau T, Saggion H, Piskorski J, Yangarber R (eds) Multi-source. Multilingual information extraction and summarization. Springer, pp 93–115
154. Angeli G, Premkumar MJJ, Manning CD (2015) Leveraging linguistic structure for open domain information extraction. In: Zong C, Strube M (eds) Proceedings of the 53rd annual meeting of the ACL and the 7th international joint conference on natural language processing of the asian federation of natural language processing, ACL, Volume 1: Long Papers. The Association for Computer Linguistics, China, pp 344–354
155. Amer-Yahia S, Koutrika G, Braschler M, Calvanese D, Lanti D, Lücke-Tieke H, Mosca A, Mendes de Farias T, Papadopoulos D, Patil Y et al (2022) Inode: building an end-to-end data exploration system in practice. ACM SIGMOD Rec 50(4):23–29
156. Wilks Y (1997) Information extraction as a core language technology. In: Pazienza MT (ed) Information extraction: a multidisciplinary approach to an emerging information technology, international summer school, SCIE-97. Springer, Italy, pp 1–9
157. Dagdelen J, Dunn A, Lee S, Walker N, Rosen AS, Ceder G, Persson KA, Jain A (2024) Structured information extraction from scientific text with large language models. Nat Commun 15(1):1418
158. Fernàndez-Cañellas D, Rimmek JM, Espadaler J, Garolera B, Barja A, Codina M, Sastre M, Giró-i-Nieto X, Riveiro JC, Bou-Balust E (2020) Enhancing online knowledge graph population with semantic knowledge. In: Pan JZ, Tamma V, d'Amato C, Janowicz K, Fu B, Polleres A, Seneviratne O, Kagal L (eds) 19th international semantic web conference. ISWC, Springer, Greece, pp 183–200

159. Jurafsky D, Martin JH (2025) Speech and language processing: an introduction to natural language processing, computational linguistics, and speech recognition. https://web.stanford.edu/~jurafsky/slp3/
160. Fernandez RC, Elmore AJ, Franklin MJ, Krishnan S, Tan C (2023) How large language models will disrupt data management. Proc VLDB Endowment 16(11):3302–3309
161. Zhang Y, Zhong V, Chen D, Angeli G, Manning CD (2017) Position-aware attention and supervised data improve slot filling. In: Palmer M, Hwa R, Riedel S (eds) Proceedings of the conference on empirical methods in natural language processing, EMNLP. ACL, Denmark, pp 35–45
162. Glass MR, Rossiello G, Chowdhury MFM, Gliozzo A (2021) Robust retrieval augmented generation for zero-shot slot filling. In: Moens M, Huang X, Specia L, Yih SW (eds) Proceedings of the 2021 Conference on Empirical Methods in Natural Language Processing, EMNLP, ACL, Dominican Republic, pp 1939–1949
163. Zhou W, Chen M (2021) Learning from noisy labels for entity-centric information extraction. In: Moens M, Huang X, Specia L, Yih SW (eds) Proceedings of the conference on empirical methods in natural language processing, EMNLP. ACL, Dominican Republic, pp 5381–5392
164. Rahman MA, Nadal S, Romero O, Sacharidis D (2024) Mitigating data sparsity in integrated data through text conceptualization. In: Das G, Sellis T, He B (eds) 40th IEEE international conference on data engineering ICDE. IEEE, The Netherlands, pp 3490–3504
165. Fel T, Boutin V, Béthune L, Cadène R, Moayeri M, Andéol L, Chalvidal M, Serre T (2023) A holistic approach to unifying automatic concept extraction and concept importance estimation. In: Oh A, Naumann T, Globerson A, Saenko K, Hardt M, Levine S (eds) Advances in neural information processing systems 36: annual conference on neural information processing systems. NeurIPS, Curran Associates Inc., USA, pp 1–12
166. Taillé B, Guigue V, Gallinari P (2020) Contextualized embeddings in named-entity recognition: An empirical study on generalization. In: Jose JM, Yilmaz E, Magalhães J, Castells P, Ferro N, Silva MJ, Martins F (eds) Advances in information retrieval—42nd European conference on IR research, ECIR. Springer, Portugal, pp 383–391
167. Akbik A, Bergmann T, Vollgraf R (2019) Pooled contextualized embeddings for named entity recognition. In: Burstein J, Doran C, Solorio T (eds) Proceedings of the 2019 conference of the North American chapter of the ACL: human language technologies, NAACL-HLT 2019, Volume 1 (Long and Short Papers). ACL, USA, pp 724–728
168. Si Y, Wang J, Xu H, Roberts K (2019) Enhancing clinical concept extraction with contextual embeddings. J Am Med Inform Assoc 26(11):1297–1304
169. Wadden D, Wennberg U, Luan Y, Hajishirzi H (2019) Entity, relation, and event extraction with contextualized span representations. In: Inui K, Jiang J, Ng V, Wan X (eds) Proceedings of the 2019 conference on empirical methods in natural language processing and the 9th international joint conference on natural language processing, EMNLP-IJCNLP. ACL, China, pp 5783–5788
170. Martinelli G, Molfese F, Tedeschi S, Fernández Castro A, Navigli R (2024) CNER: concept and named entity recognition. In: Duh K, Gomez H, Bethard S (eds) Proceedings of the 2024 conference of the North American chapter of the ACL: human language technologies, NAACL. ACL, Mexico, pp 8336–8351
171. Fang Y, Zhang Y (2022) Data-efficient concept extraction from pre-trained language models for commonsense explanation generation. In: Goldberg Y, Kozareva Z, Zhang Y (eds) Findings of the ACL: EMNLP. ACL, United Arab Emirates, pp 5883–5893
172. Zhang Z, Liu B, Shao J (2023) Fine-tuning happens in tiny subspaces: Exploring intrinsic task-specific subspaces of pre-trained language models. In: Rogers A, Boyd-Graber JL, Okazaki N (eds) Proceedings of the 61st annual meeting of the association for computational linguistics. ACL, Canada, pp 1701–1713
173. Huang J, Li C, Subudhi K, Jose D, Balakrishnan S, Chen W, Peng B, Gao J, Han J (2021) Few-shot named entity recognition: An empirical baseline study. In: Moens M, Huang X, Specia L, Yih SW (eds) Proceedings of the conference on empirical methods in natural language processing, EMNLP. ACL, Dominican Republic, pp 10408–10423

174. Guo X, Yu H (2022) On the domain adaptation and generalization of pretrained language models: a survey. CoRR 1–12. arxiv:abs/2211.03154

175. Radford A, Wu J, Child R, Luan D, Amodei D, Sutskever I et al (2019) Language models are unsupervised multitask learners. OpenAI Blog 1(8):9

176. Kojima T, Gu SS, Reid M, Matsuo Y, Iwasawa Y (2022) Large language models are zero-shot reasoners. Adv Neural Inf Process Syst 35:22199–22213

177. Brown TB, Mann B, Ryder N, Subbiah M, Kaplan J, Dhariwal P, Neelakantan A, Shyam P, Sastry G, Askell A, Agarwal S, Herbert-Voss A, Krueger G, Henighan T, Child R, Ramesh A, Ziegler DM, Wu J, Winter C, Hesse C, Chen M, Sigler E, Litwin M, Gray S, Chess B, Clark J, Berner C, McCandlish S, Radford A, Sutskever I, Amodei D (2020) Language models are few-shot learners. In: Larochelle H, Ranzato M, Hadsell R, Balcan M, Lin H (eds) Advances in neural information processing systems 33: annual conference on neural information processing systems, NeurIPS. Curran Associates Inc, Virtual, pp 1877–1901

178. Christiansen JG, Gammelgaard M, Søgaard A (2023) Large language models partially converge toward human-like concept organization. In: Sanborn S, Shewmake C, Azeglio S, Miolane N (eds) Proceedings of the 2nd NeurIPS workshop on symmetry and geometry in neural representations. PMLR, USA, pp 346–365

179. Gao Y, Xiong Y, Gao X, Jia K, Pan J, Bi Y, Dai Y, Sun J, Guo Q, Wang M, Wang H (2023) Retrieval-augmented generation for large language models: a survey. CoRR 1–12. arxiv:abs/2312.10997

180. Lewis P, Perez E, Piktus A, Petroni F, Karpukhin V, Goyal N, Küttler H, Lewis M, Wt Y, Rocktäschel T et al (2020) Retrieval-augmented generation for knowledge-intensive nlp tasks. Adv Neural Inf Process Syst 33:9459–9474

181. Zhao P, Zhang H, Yu Q, Wang Z, Geng Y, Fu F, Yang L, Zhang W, Cui B (2024) Retrieval-augmented generation for ai-generated content: a survey. CoRR 1–15. arxiv:abs/2402.19473

182. Posedaru BS, Pantelimon FV, Dulgheru MN, Georgescu TM (2024) Artificial intelligence text processing using retrieval-augmented generation: applications in business and education fields. Proc Int Conf Bus Excellence 18:209–222

183. Guu K, Lee K, Tung Z, Pasupat P, Chang M (2020) Retrieval augmented language model pre-training. In: III HD, Singh A (eds) Proceedings of the 37th international conference on machine learning, ICML. PMLR, Virtual, pp 3929–3938

184. Fan W, Ding Y, Ning L, Wang S, Li H, Yin D, Chua T, Li Q (2024) A survey on RAG meeting llms: Towards retrieval-augmented large language models. In: Baeza-Yates R, Bonchi F (eds) Proceedings of the 30th ACM conference on knowledge discovery and data mining, KDD. ACM, Spain, pp 6491–6501

185. Agichtein E, Gravano L (2000) Snowball: extracting relations from large plain-text collections. In: Nürnberg PJ, Hicks DL, Furuta RK (eds) Proceedings of the fifth ACM conference on digital libraries. ACM, USA, pp 85–94

186. Etzioni O, Cafarella MJ, Downey D, Kok S, Popescu A, Shaked T, Soderland S, Weld DS, Yates A (2004) Web-scale information extraction in knowitall: (preliminary results). In: Feldman SI, Uretsky M, Najork M, Wills CE (eds) Proceedings of the 13th international conference on World Wide Web, WWW. ACM, USA, pp 100–110

187. Carlson A, Betteridge J, Kisiel B, Settles B, Jr ERH, Mitchell TM (2010) Toward an architecture for never-ending language learning. In: Fox M, Poole D (eds) Proceedings of the twenty-fourth conference on artificial intelligence. AAAI Press, USA, pp 1306–1313

188. Augenstein I, Vlachos A, Maynard D (2015) Extracting relations between non-standard entities using distant supervision and imitation learning. In: Màrquez L, Callison-Burch C, Su J, Pighin D, Marton Y (eds) Proceedings of the 2015 conference on empirical methods in natural language processing, EMNLP. The ACL, Portugal, pp 747–757

189. Yang Y, Chen W, Li Z, He Z, Zhang M (2018) Distantly supervised NER with partial annotation learning and reinforcement learning. In: Bender EM, Derczynski L, Isabelle P (eds) Proceedings of the 27th international conference on computational linguistics, COLING. ACL, USA, pp 2159–2169

190. Shang YM, Huang H, Sun X, Wei W, Mao XL (2022) A pattern-aware self-attention network for distant supervised relation extraction. Inf Sci 584:269–279
191. Fader A, Soderland S, Etzioni O (2011) Identifying relations for open information extraction. In: Barzilay R, Johnson M (eds) Proceedings of the 2011 conference on empirical methods in natural language processing, EMNLP. ACL, UK, pp 1535–1545
192. Soares LB, FitzGerald N, Ling J, Kwiatkowski T (2019) Matching the blanks: Distributional similarity for relation learning. In: Korhonen A, Traum DR, Màrquez L (eds) Proceedings of the 57th conference of the ACL. ACL, Italy, pp 2895–2905
193. Tian Y, Chen G, Song Y, Wan X (2021) Dependency-driven relation extraction with attentive graph convolutional networks. In: Zong C, Xia F, Li W, Navigli R (eds) Proceedings of the 59th annual meeting of the ACL and the 11th international joint conference on natural language processing, ACL/IJCNLP. ACL, Virtual, pp 4458–4471
194. Bender EM, Gebru T, McMillan-Major A, Shmitchell S (2021) On the dangers of stochastic parrots: Can language models be too big? In: Elish MC, Isaac W, Zemel RS (eds) FAccT '21: 2021 ACM conference on fairness, accountability, and transparency, virtual event. ACM, Canada, pp 610–623
195. Yin P, Neubig G, Yih W, Riedel S (2020) Tabert: pretraining for joint understanding of textual and tabular data. In: Jurafsky D, Chai J, Schluter N, Tetreault JR (eds) Proceedings of the 58th annual meeting of the association for computational linguistics. ACL, Online, pp 8413–8426
196. Sedlakova J, Daniore P, Horn Wintsch A, Wolf M, Stanikic M, Haag C, Sieber C, Schneider G, Staub K, Alois Ettlin D et al (2023) Challenges and best practices for digital unstructured data enrichment in health research: a systematic narrative review. PLOS Digital Health 2(10):e0000347
197. OpenAI (2023) GPT-4 technical report. CoRR 1–98. arxiv:abs/2303.08774
198. Chen Y, Fu Q, Yuan Y, Wen Z, Fan G, Liu D, Zhang D, Li Z, Xiao Y (2023) Hallucination detection: robustly discerning reliable answers in large language models. In: Frommholz I, Hopfgartner F, Lee M, Oakes M, Lalmas M, Zhang M, Santos RLT (eds) Proceedings of the 32nd acm international conference on information and knowledge management, CIKM. ACM, United Kingdom, pp 245–255

Chapter 2
Exploring the Landscape of Data Fusion

Yeasmin Ara Akter⬤, Alberto Abelló⬤, Petar Jovanovic⬤, Tomer Sagi⬤,
and Katja Hose⬤

Abstract In the digital world, organizations across different sectors such as education, retail, finance, and healthcare are struggling to manage massive amounts of data. A critical challenge is handling overlapping data from multiple sources, which often leads to inconsistencies and reliability issues. While combining data from multiple sources can strengthen the credibility when the information matches, it creates significant problems when different sources provide conflicting information about the same attributes or events. Data fusion has emerged as a solution to these challenges. This approach combines data from different sources to create consistent and comprehensive information. The key aspects of data fusion include managing the database relationship, resolving conflicts, determining accurate information from conflicting sources, and tracking data origins (provenance). Although many papers review data fusion in specific fields, this paper takes a practical approach. Instead of presenting another literature survey, we focus on explaining the key steps in the data fusion process and examining the most effective techniques for implementing each step.

Keywords Data quality · Truth discovery · Conflict resolution

2.1 Overview

Data are the most valuable assets nowadays to make any decision. From smart watches to human-like robots, various decision-making systems rely on the data. Without data, it would be impossible to live in the highly technological era we are in today. However, imprecise, irregular (e.g., missing data, anomalies), inconsistent,

Y. A. Akter (✉) · A. Abelló · P. Jovanovic
Universitat Politècnica de Catalunya, Barcelona, Spain
e-mail: yeasmin.ara.akter@upc.edu; yeasminaa@cs.aau.dk

Y. A. Akter · T. Sagi
Aalborg University, Aalborg, Denmark

K. Hose
TU Wien: Technische Universität Wien, Vienna, Austria

© The Author(s) 2026

G. Dejaegere et al. (eds.), *Data Engineering for Data Science*,
https://doi.org/10.1007/978-3-032-18765-9_2

<table>
<tr><td>Description</td><td>Problems</td><td>Impacts</td></tr>
<tr><td>1. Farmers depend on weather forecasting projections to make essential choices about planting, irrigation, and harvesting.</td><td>1. Humidity for Madrid on January 10, 2024 is absent.</td><td rowspan="4">Farmers who depend on these projections suffer from the following hardships which can result in crop and financial loss:

1. Planting crops at the incorrect time.
2. Irrigating their fields excessively or insufficiently.
3. Failing to harvest before adverse weather conditions.</td></tr>
<tr><td rowspan="3">2. Weather forecasting service companies include temperature, humidity, wind speed, and precipitation data such as Table 2.1 in weather forecasting models.</td><td>2. The prediction of precipitation for Valencia on January 10, 2024 is missing.</td></tr>
<tr><td>3. Assume that on January 13, 2024, one source predicts the temperature as 30°C for Barcelona. However, if another source predicts 21°C, conflicts arise.</td></tr>
<tr><td>Without these local data, the system might struggle to predict sudden weather events such as frost or heat waves. This could result in forecasts that are outdated or extremely vague.</td></tr>
</table>

Fig. 2.1 Weather prediction scenario

Table 2.1 Weather data (source 1)

Date	Location	Temperature (°C)	Humidity (%)	Wind speed (km/h)	Precipitation (mm)
2024–01–10	Barcelona	20	60	15	0
2024–01–10	Madrid	22	–	20	5
2024–01–10	Valencia	18	65	10	–
2024–01–11	Barcelona	–	58	15	0
2024–01–11	Madrid	23	62	–	10
2024–01–11	Valencia	19	70	12	0
2024–01–12	Barcelona	21	60	14	0
2024–01–12	Madrid	25	65	22	0
2024–01–12	Valencia	20	67	10	0

and incomplete data can harm the findings of decision-making tasks and businesses. Fig. 2.1 depicts a scenario of weather prediction and how farming decisions become dependent on the quality of the weather data shown in Table 2.1. The illustration demonstrates how weather prediction systems without reliable historical data can hamper farming decisions. Not only in farming but also in healthcare, education, retail, finance, and other sectors, decisions are impeded because of poor-quality data. Therefore, data must be of high quality before being ingested into decision-making systems to facilitate accurate predictions.

Due to the exponential growth of data, many sources have become available over the internet or in company intranets to collect different aspects of the same real-world entity. The opportunity to access different sources in a single interface has compelled the research community to work on data integration. However, integrating data from multiple sources faces two main challenges [1]:

1. **Heterogeneous data**: Heterogeneity occurs both at the schema and instance level. Schema-level heterogeneity happens when different data sources describe the same domain using different schemas. An example of weather data for the month of January 2024 in Spanish cities is shown in Table 2.1. The unit of temperature in Table 2.1 is *Celsius*, and the date is stored as *YYYY/MM/DD*. However, in another source, the unit of temperature could be *Kelvin*, and the date could be stored in multiple columns such as *Year, Month, and Day*, which creates schema heterogeneity. Instance-level heterogeneity arises when identical information is stored in different ways in different sources. For instance, in Table 2.1, the column *Location* can be named as *Area* in another source.
2. **Conflicting data**: Data contradiction occurs if different values are provided for the same feature of the same real-world entity. As shown in Fig. 2.1, the problems arising from the weather data presented in Table 2.1 are illustrated. Specifically, Problem 3 provides an example of conflicting data. Notice that this is independent of the schemas or formats used in the multiple representations of the entity.

Accordingly, data integration requires three main steps:

1. **Schema matching**: Identifies semantic correspondences between attributes and entity classes from different schemas. This step produces structural homogeneity among multiple sources.
2. **Entity resolution**: Identifies different records that refer to the same real-world entity. This step generates correspondences between records of the same instance.
3. **Data fusion**: Aims to determine the true value from conflicting data elements in records corresponding to the same instance by resolving conflicting attribute values to produce uniform, accurate, and valuable information. This process produces high-quality data by merging correct values from multiple sources rather than relying on a single data source [2].

Data conflicts emerge because of erroneous, vague, and outdated data. As shown in Fig. 2.1, data conflicts can lead to various problems. For example, one may visit the wrong address of a public library or call an out-of-date phone number. Data fusion was first introduced by [3] to deal with such conflicts in data. Since then, researchers have worked to develop solutions for these conflicts. Data fusion is now being applied in diverse fields with a common purpose: *to provide a unified view with reliable data*. Examples of such applications include sensor data fusion [4], monitoring and tracking [5], IoT [6, 7], linked data [8], information extraction [9], and ETL processes [10]. As a result, there are many review papers on data fusion based on different applications. Several of the more recent reviews are discussed in the following.

Lau et al. [11] review fusion methods applied to smart city applications. They classify the fusion work of smart city applications based on fusion objectives, input/output data types, fusion techniques, data source types, and fusion scales and discuss the open challenges in this domain. Meng et al. [12] review machine learning techniques applied as data fusion methods. They discuss supervised and unsupervised learning approaches that are used as fusion techniques. Gao et al. [13] review the deep learning architectures of multi-modal data fusion. The architectures include DBN, SAE, CNN, and RNN-based fusion methods. Gutiérrez et al. [2] survey the information fusion techniques used to improve information quality. They have addressed common IF techniques, IQ dimensions, metrics, system types, adaptability of the systems, application domains, data types, and validation processes. Shaik et al. [14] classify the fusion methods as rule-based, probability-based, optimization-based, and machine learning-based. They also discuss approaches for truth discovery and methods that account for data quality. Canalle et al. [15] classify the fusion methods into rule-based, probability-based, optimization-based, and machine learning-based. Lahat et al. [16] provide a guideline about how to approach data fusion based on multi-modality. They discuss the data and model-level challenges of fusion methods in detail. Li et al. [17] review the information fusion techniques. They discuss the strengths and limitations of probabilistic models, belief functions, Fuzzy set theory, and Rough set theory in terms of computational demand, training data requirements, explainability, and ability to handle uncertainty.

A common fusion challenge addressed by all of these papers is conflicting data, which is divided into two kinds: *uncertainty of missing values* and *contradiction*. Uncertainty of missing values is divided into two categories: *uncertainty quantification* and *uncertain data handling (i.e., uncertain data encoding)*. An example of uncertainty quantification is estimating the salary of a person to be between $45,000 and $55,000, quantifying the uncertainty in the actual salary value. On the other hand, an example of uncertainty encoding is representing the salary as 50% chance of $48,000, 30% chance of $50,000, and 20% chance of $52,000, encoding the uncertainty about the possible salary values into a probabilistic distribution. In the existing review papers, uncertainty quantification is discussed but not the uncertainty encoding techniques, as they are highly related to the research domain of incomplete databases. Consequently, a comprehensive guideline about how uncertain data encoding can be incorporated into fusion methods is absent.

Nevertheless, it is crucial to establish connections between uncertain data encoding and data fusion to provide a framework for developing an integration system that yields more complete and reliable output. We aim to gather models, techniques, and systems of the vast yet unconnected field of data fusion and present them systematically. The focus of this paper is to systematically highlight the key steps involved in the data fusion framework and to discuss the most common techniques for each step.

Specifically, this work provides the following contributions:

(i) A taxonomy to classify the related work according to the data conflicts they consider.
(ii) An analysis and comparison of the different characteristics of data fusion techniques.

2.2 The Landscape of Data Fusion

This section provides the necessary definitions to organize the works on data fusion. Fig. 2.2 depicts the different branches of research in this field. *Level of data fusion* defines different levels where data fusion can take place based on raw (e.g., low level) or processed (e.g., feature or decision level) data. *Data quality* determines which methods can be applied to identify the data relationship resolving discovered data conflicts. *Data type* gives an idea about what kind of data is used. *Evaluation* illustrates which metrics are used to show the fusion system performance.

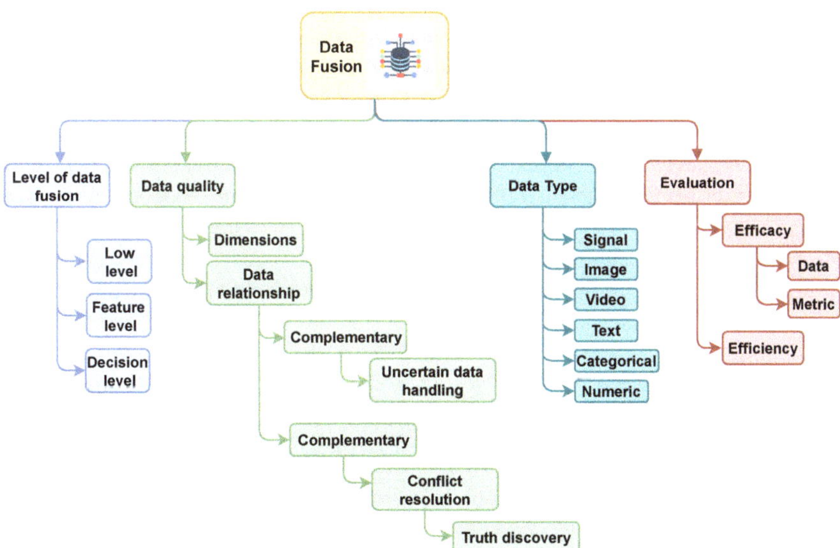

Fig. 2.2 Data fusion research branches

2.2.1 Level of Data Fusion

Fusion processes are categorized into three levels (Fig. 2.3) based on the stage at which fusion occurs: low level, feature level, and decision level.

1. **Low level**: Work under this category fuses raw data from multiple sources without any transformation. The goal is to produce data that is more useful and comprehensive than the original inputs.
   Example 1: Multiple sensors are placed in different cities of a country to accumulate their temperature. The data from different cities is further aggregated at the country level. Fusing the temperature of the cities from different sources without any modification of the raw data falls under low-level data fusion. Considering Table 2.1, if another source provides a temperature for Barcelona as 30 °C on 2024–01–12, fusing these two sources results in 26 °C if averaging is applied.

2. **Feature level**: Work under this category combines normalized feature subsets to produce a unified representation of different data types. The data may go through several transformations to extract the necessary features. This is useful in multi-modal data fusion, such as combining images and texts. *Example 2: Continuing the previous example, we can calculate the "Feels Like" feature for each city using data from multiple sensors and then fuse them. Table 2.2 shows the "Feels Like" feature values derived from the data in Table 2.1. If, for example, the "Feels Like" feature for Barcelona on 2024–01–10 is 15.3 °C in another source, fusion using averaging will result in 16.3 °C.*

3. **Decision level**: Work under this category integrates the results (or decisions) from multiple classifiers to reach a final decision about an event or activity. This approach is used when data from different sources (like sensors) cannot be easily combined into a single feature set, so each sensor independently uses its own classifier to analyze the data. The outputs from these classifiers are then merged to form the final decision.
   Example 3: Consider Table 2.3, which shows the weather prediction from two different sources (i.e., sensors). Based on the data from Table 2.1, assume sensor 1 measures temperature, humidity, and wind speed, while sensor 2 measures only precipitation. From sensor 1 data, the "Feels Like" feature can be derived using temperature and humidity. Meanwhile, from sensor 2 data, the "Rain Indicator"

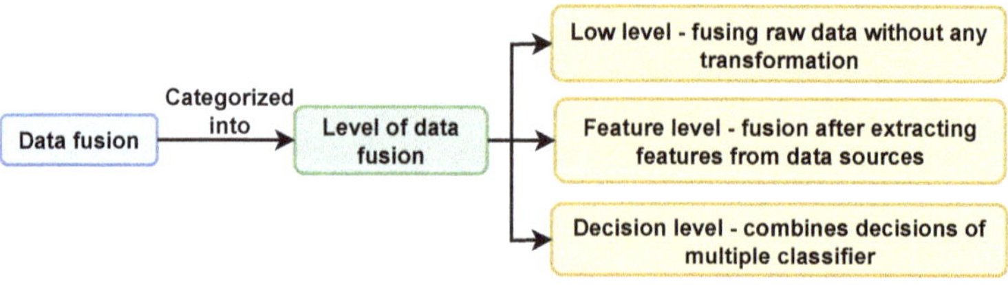

Fig. 2.3 Data fusion taxonomy: level of data fusion

Table 2.2 Weather data with a derived feature (source 1)

Date	Location	Feels like (°C)
2024–01–10	Barcelona	17.7
2024–01–10	Madrid	–
2024–01–10	Valencia	16.5
2024–01–11	Barcelona	–
2024–01–11	Madrid	22.9
2024–01–11	Valencia	17.8
2024–01–12	Barcelona	19.3
2024–01–12	Madrid	23.6
2024–01–12	Valencia	19.3

Table 2.3 Weather data prediction comparison

Date	Location	Feels like (°C) (Sensor 1)	Rain indicator (Sensor 2)	Weather prediction (Sensor 1)	Weather prediction (Sensor 2)
2024–01–10	Barcelona	17.7	0	Cloudy	Partly cloudy
2024–01–10	Madrid	–	1	Rainy	Rainy
2024–01–10	Valencia	16.5	1	Rainy	Rainy
2024-01–11	Barcelona	–	0	Sunny	Sunny
2024–01–11	Madrid	22.9	1	Rainy	Extremely rainy
2024–01–11	Valencia	17.8	0	Sunny	Sunny
2024–01–12	Barcelona	19.3	0	Sunny	Sunny
2024–01–12	Madrid	23.6	0	Cloudy	Partly cloudy
2024–01–12	Valencia	19.3	0	Partly cloudy	Sunny

feature can be derived from precipitation. Based on these feature vectors, two different classifiers are used in each sensor, resulting in weather predictions as shown in Table 2.3. If a similarity function is applied to compare the predictions, the fusion will result in "Rainy" on 2024–01–11 in Madrid.

2.2.2 Data Quality

One of the goals of data fusion is to enhance the quality of the data. Data quality, defined as "fit for use" [18], is primarily assessed using four dimensions: *accuracy, completeness, timeliness,* and *consistency* [19]. However, other dimensions, such as *accessibility, correctness, availability,* and *distortion,* are also considered in data

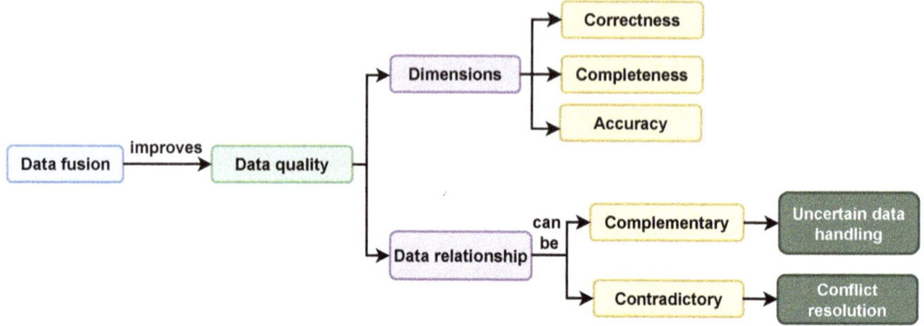

Fig. 2.4 Data fusion taxonomy: data quality

fusion systems, depending on the data type and application domain [2]. In this paper, we use the following dimensions to categorize the related works (Fig. 2.4):

Accuracy: This is the degree to which the data correctly represent the real-life objects that they are intended to model [2]. The intended metric for this dimension is calculating the difference between the real and recorded values.

Correctness: Refers to the "clean" data that is free of structural and semantic errors. This is a qualitative dimension.

Example 4: *Accuracy shows how accurate the temperature measurement is, while correctness tells if the temperature is measured in a correct way or not.*

Completeness Refers to the absence of missing data in a database. Therefore, the most common metric for this dimension is the ratio of non-null values to the total number of expected values.

In fusion, two types of data relationships are considered to improve the data quality: *complementary* and *contradictory* data.

1. **Complementary data**: Refers to uncertainty in missing values (i.e., if one source provides a non-null value and another source provides nulls for the same data element, uncertainty arises).
   Example 5: *Consider the example data given in Tables 2.4 and 2.5. Two sources are available to collect sea surface temperature according to time and spatial dimensions. Both of these sources have missing values (i.e., nulls in row 3 of source 1 and row 2 of source 3). Although they alternatively provide data for these two rows, fusing these rows will resolve this uncertainty.*
2. **Contradictory data:** Data becomes contradictory when two sources provide different values for the same attribute of the same instance.
   Example 6: *Consider Tables 2.4 and 2.5. Sea surface temperature is different for row 1 in these two sources, which creates contradictory data. Data fusion aims to determine which data should be stored in the master record.*

The difficulties in fusing these two types of data relationships are addressed differently in fusion techniques, namely *uncertain data handling* and *conflict resolution*.

Table 2.4 Climate data source

Rowid	Date	Latitude	Longitude	Sea surface temperature (K)
1001	2023–01–01	30.125	−18.26	293.14
1002	2023–01–02	30.125	−18.26	292.56
1003	2023–01–03	30.125	−18.26	null
1004	2023–01–04	30.125	−18.26	291.45

Table 2.5 Copernicus data source

Rowid	Date	Latitude	Longitude	Sea surface temperature (K)
1001	2023–01–01	30.125	−18.26	293.90
1002	2023–01–02	30.125	−18.26	null
1003	2023–01–03	30.125	−18.26	291.67
1004	2023–01–04	30.125	−18.26	291.45

2.2.2.1 Uncertain Data Handling

Uncertain data makes data sources imperfect and of poor quality. Hence, characterization and quantification of uncertain data are necessary to improve the quality and usefulness of the datasets. Approaches dealing with uncertainty typically consider several components (Fig. 2.5).

(a) **Symbolic representation**: Uncertain data is encoded using symbols or possible worlds in the database to facilitate query processing over uncertain data. Papers employ representation techniques such as C-tables, s-tables, attribute bounds, and others for encoding.

  (i) **Encoding with a C-table**: Imielinski and Lipski [20] first introduced a conditional table or C-table to simplify the projection, positive selection, renaming, and union operations in the queries. This table includes an additional conditional/Boolean expression to determine the best guess of a missing value in query processing.

  Example 7: *An example of a C-table is given in Table 2.6, where the column "condition" shows the Boolean expressions representing the value of a missing instance. The Boolean expression $x = 292.56 \lor x \neq 292.56$ indicates whether the missing value can be 292.56 or not and this is determined by a classifier. A C-table specifically represents missing or null values.*

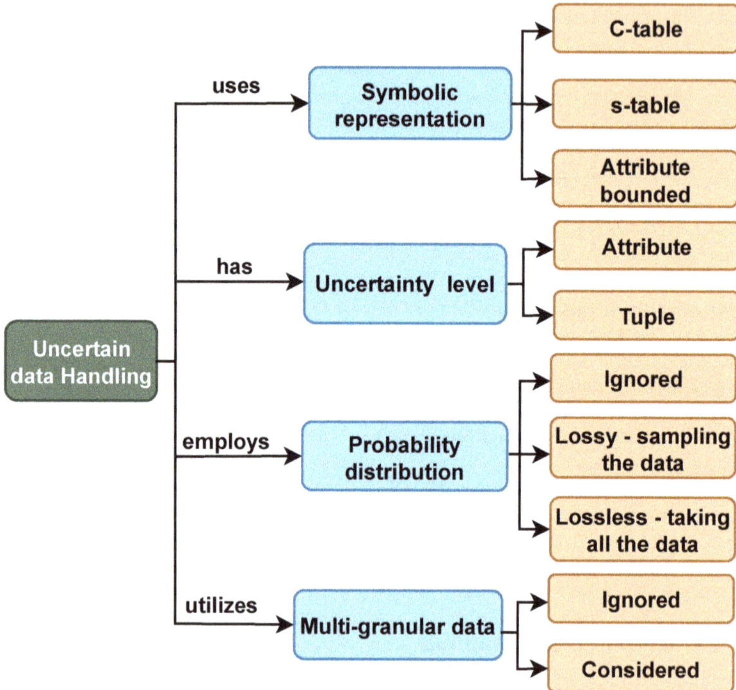

Fig. 2.5 Data fusion taxonomy: uncertain data handling

Table 2.6 Copernicus data source encoding with a C-table

Rowid	Date	Latitude	Longitude	Sea surface temperature (K)	Conditions
1001	2023–01–01	30.125	−18.26	293.90	
1002	2023–01–02	30.125	−18.26	Null	x = 292.56 ∨ x ≠ 292.56
1003	2023–01–03	30.125	−18.26	291.67	
1004	2023–01–04	30.125	−18.26	291.45	

(ii) **Encoding with an s-table**: Missing values or errors in the collected values of a source are encoded in s-table using symbolic expressions (e.g., linear expressions) [21]. For example, an expression can be $v \cdot (1 + x)$ where v represents the value and x represents the error in the recorded values or any missing value. ***Example 8***: *An example of s-table encoding is shown in Table 2.7, which provides a symbolic representation of Table 2.5.*

(iii) **Encoding with attribute bounds**: Uncertain data are stored in the database with bounds on the attribute [22]. In this case, missing values are predicted by

Table 2.7 Copernicus data source encoding with an s-table

Rowid	Date	Latitude	Longitude	Sea surface temperature (K)
1001	2023–01–01	30.125	–18.26	$293.90\,(1 + x_1)$
1002	2023–01–02	30.125	–18.26	x_2
1003	2023–01–03	30.125	–18.26	$291.67\,(1 + x_3)$
1004	2023–01–04	30.125	–18.26	$291.45\,(1 + x_4)$

Table 2.8 Copernicus data source encoding with attribute bounds

Rowid	Date	Latitude	Longitude	Sea surface temperature (K)
1001	2023–01–01	30.125	–18.26	[291.45, 293.90, 293.90]
1002	2023–01–02	30.125	–18.26	[291.45, 293.90, 293.90]
1003	2023–01–03	30.125	–18.26	[291.45, 293.67, 293.90]
1004	2023–01–04	30.125	–18.26	[291.45, 293.45, 293.90]

the probabilistic methods and possible worlds are created. The bound for each attribute is in the form of a triple [lower bound, selected guess, upper bound], where the lower bound is the lowest value for a data element among all the possible worlds, the selected guess comes from a selected guess world, and the upper bound is the highest value among all possible worlds.

Example 9: An example of attribute bounds encoding is shown in Table 2.8. In the Copernicus data source, the lowest value is 291.45 and the highest value is 293.90. For Rowid 1002, the null value could be any number between this range, creating possible worlds (PWs). Therefore, the bounds are created as [lower bound, selected guess, upper bound]. For example, PW1 may contain 293.85, whereas PW2 may contain 293.60, and so on. Applying any selected guess method, if PW2 is chosen, the encoded data with attribute bounds is shown in Table 2.8.

(b) **Uncertainty level**: Uncertainty may occur at different levels of data collection, (i.e., attribute and tuple levels). As discussed earlier, attribute-level uncertainty is whether a data value is null or non-null, while tuple-level uncertainty is whether a tuple will be in the query result or not.

Example 10: Attribute level uncertainty is shown in Tables 2.4 and 2.5, where one of the values in both data sources is missing. However, if a query is applied to these sources such that "Select sea surface temperature where sea surface temperature is less than 292", it will be difficult to determine whether record 3

Table 2.9 Lossy probability distribution of Copernicus data source

Rowid	Date	Latitude	Longitude	Sea surface temperature (K)	PR
1001	2023–01–01	30.125	−18.26	293.90	1.0
1002	2023–01–02	30.125	−18.26	291.90	0.6

in Table 2.4 and record 2 in Table 2.5 will be in the final result or not which creates uncertainty of tuples.

(c) **Probability distribution**: Encoding of uncertain data collection employing probabilistic distribution is categorized into two:

    (i) **Lossy**: Represents a finite subset of the possible worlds but results in a reduction of information. In other words, the probability is distributed over a sample of the possible worlds that cause meaningful information loss (Table 2.9).
Example 11: *Table 2.9 shows an example of a lossy distribution, which is a sample of Table 2.5. Two rows are selected in the sample that truncates the domain of the whole data source. Consequently, the imputation of the missing value remains under the truncated data domain, but the actual value can fall outside this domain.*

    (ii) **Lossless**: Refers to including independent properties to factorize the set of possible worlds, enabling data utilization without any information loss. In general, lossless techniques use the whole dataset instead of sampling.

(d) **Multi-granular data**: This term refers to operating on data at different levels. Some works use multi-granular data [21, 23], while others ignore them to avoid more complex operations.
Example 12: *An Example of multi-granular data is shown in Tables 2.10 and 2.11. Table 2.10 provides sea surface temperature (SST) at a resolution of 0.20° latitude and longitude, while Table 2.11 provides SST at a resolution of 0.05° latitude and longitude.*

2.2.2.2 Conflict Resolution

Reconciliation of conflicts enhances the reliability of the data and evaluates the trustworthiness of the sources as well. The reconciliation process employs truth discovery methods. Improving reliability likewise improves data quality. Truth discovery

Table 2.10 Copernicus data source 2 (resolution 0.20°)

Rowid	Date	Latitude	Longitude	Sea surface temperature (K)
1001	2023–01–01	30.20	−18.20	292.56
1002	2023–01–01	30.40	−18.40	291.67
1003	2023–01–01	30.60	−18.60	293.90
1004	2023–01–01	30.80	−18.80	291.70

Table 2.11 Copernicus data source 3 (resolution 0.05°)

Rowid	Date	Latitude	Longitude	Sea surface temperature (K)
1001	2023–01–01	30.05	−18.05	292.12
1002	2023–01–01	30.10	−18.10	291.45
1003	2023–01–01	30.15	−18.15	293.23
1004	2023–01–01	30.20	−18.20	291.67

Table 2.12 Climate data source 2

Rowid	Date	Latitude	Longitude	Sea surface temperature (K)	Precipitation (mm)
1001	2023–01–01	30.125	−18.26	293.14	null
1002	2023–01–02	30.125	−18.26	292.56	null
1003	2023–01–03	30.125	−18.26	null	null
1004	2023–01–04	30.125	−18.26	291.45	5

methods determine the actual true value for a data item from the conflicting ones, considering the domain knowledge and source dependency, computing the truth, and utilizing an iterative process using source weights. Fig. 2.6 depicts what are the techniques used in different truth discovery methods.

(a) **Domain knowledge**: Refers to the understanding of different categories of sources. Domain knowledge impacts on overall source trustworthiness because one source may not have the same data quality in all the categories or attributes.

Example 13: *Table 2.12 represents an example where two categories/attributes are recorded in Climate data sources. From the table, it is clear that the completeness of "Precipitation" is much lower than the completeness of "Sea Surface Temperature". Hence, the trustworthiness of this source is difficult to determine without considering the domain knowledge.*

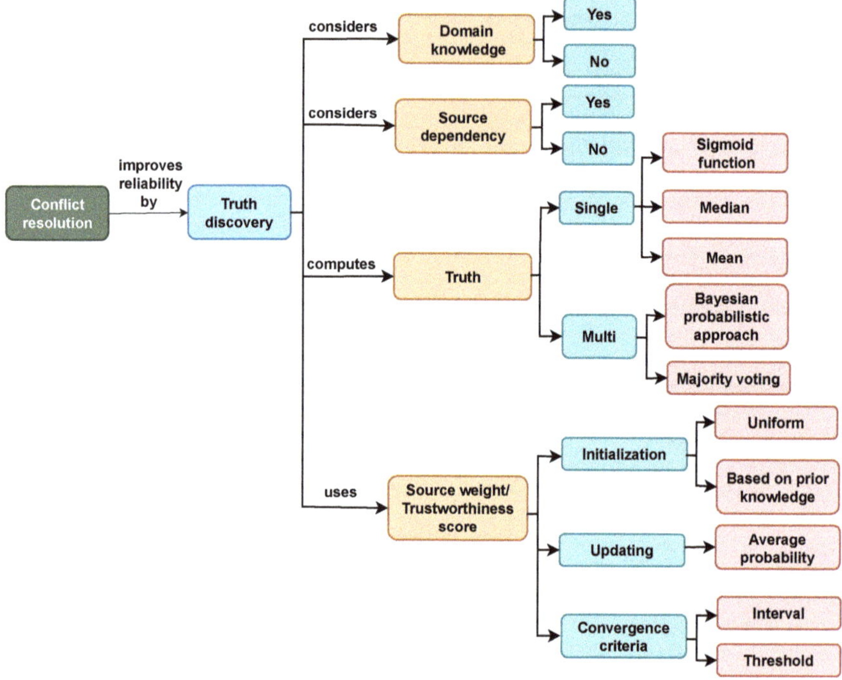

Fig. 2.6 Data fusion taxonomy: conflict resolution

(b) **Source dependency**: When one source copies data from another one, wrong data may be propagated to other sources. As a result, a wrong data value can be found as a majority (i.e., as truth).
Example 14: *If the sea surface temperature is measured by 6 sources and 4 sources copy the wrong data from another one, the wrong one will become a true value if voting is applied.*

(c) **Truth computation**: Refers to the result of obtaining the true value from various sources. Disparate data sources may contain single or multiple true values based on the data domain they collect.

   (i) **Single truth**: Computation of single truth is when a data can have only one single true value (e.g., sea surface temperature can have only one value for the same spatial and temporal resolution).
   (ii) **Multi-truth**: When a data item may have more than one single true value (e.g., sea surface temperature of one week has multiple true values or a book can be written by multiple authors).

(d) **Trustworthiness score**: Indicates how much we can rely on a particular data source. In truth discovery, this score is determined by applying weights to the sources (see Example 15). The process of determining the truth involves iteratively initializing and updating these source weights until they converge.

(i) **Initialization**: Involves setting the starting weights for each data source. These weights can either be uniform if all the sources are treated equally trustworthy at the beginning or they can be based on prior knowledge if any pre-existing information about the reliability of the sources is available.
Example 15: *Initially, if we consider both the Climate data source and Copernicus data source to be equally reliable, we might assign them both a uniform weight of 0.90. This means that before any data is analyzed, we assume both sources have the same level of trustworthiness.*

(ii) **Update**: After determining the likely truth for a data item, the weights for each source are updated. This update is based on the average probability of the facts or true values that each source provides. Sources that provide more accurate information receive higher weights, reflecting their greater reliability.

(iii) **Convergence criteria**: Convergence criteria determine when to stop updating the weights. One common criterion is to stop when the cosine similarity between the weights of successive iterations is below a certain threshold. Another criterion is to stop when the computed truth scores fall within a given interval.
Example 16: *If in one iteration, the weights for the Climate and Copernicus sources are 0.96 and 0.84, and in the next iteration, they are 0.90 and 0.78, we might check the cosine similarity between these sets of weights. If the similarity is below a threshold, such as 0.50, the process converges. Another approach is to stop if the truth scores remain within a specific interval, indicating that the source weights are stable and further updates are unnecessary.*

2.2.3 Data Type

Depending on the levels of data fusion, different types of data are utilized. The data type involved in fusion can vary widely. The data type ranges from sensory raw data, such as signals and images, to data from online web portals (Fig. 2.7). Data fusion work incorporates various data types such as *signals*, *images*, *video*, *text*, *categorical*, and *numeric* data.

(a) **Signal**: Refers to merging signals collected from multiple sensors to create a more accurate and comprehensive dataset. This type of data is commonly used for monitoring or tracking purposes.

(b) **Image**: Accumulates all the important data from multiple images into a single comprehensive one. This type of data fusion is widely used in image processing to enhance image quality or extract valuable information.

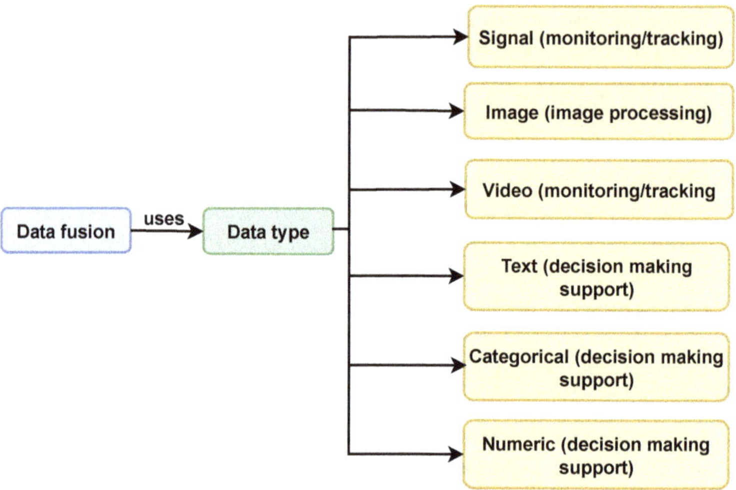

Fig. 2.7 Data fusion taxonomy: data type

(c) **Video**: Fusion of video is closely related to image fusion since the video frames are sequences of images. This data fusion involves integrating multiple video streams or frames to enhance monitoring or tracking.

(d) **Text**: Fuses different information from multiple web portals or any other sources that provide text data. Decision-making systems in healthcare, education, financial market analysis etc. use text data fusion to have more accurate information about any specific matter.

(e) **Categorical**: Refers to the process of fusing data that have different categorical variables, labels, or codes. Consider a scenario where we wish to combine sentiment analysis results classified as *"positive", "negative",* and *"neutral"* with social media post categories such as *"health"* and *"nature".* In this case, fusion entails combining the sentiment categories and subject categorization to enable in-depth analysis.

(f) **Numeric**: Fuses numeric data from multiple sources. Numeric data fusion can take place in monitoring, tracking, and decision-making support.

According to a study by Gutiérrez et al. [2], the focus of data fusion works varies significantly across different data types. The distribution is as follows: image data comprises 33% of fusion works, numeric data 26.98%, text data comprises 14.76%, signal data 11.5%, fusion works involving multiple data types stand at 9.54%, and video data 3.17%. Numeric data finds application at all levels of fusion, signal and picture data are mostly used in low-level fusion, while text, categorical, and video data are primarily used in feature and decision-level fusion.

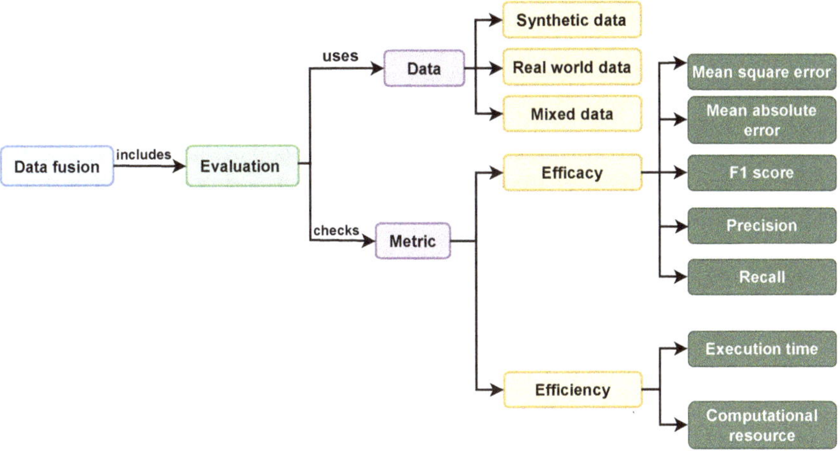

Fig. 2.8 Data fusion taxonomy: evaluation

2.2.4 Evaluation

Evaluation in data fusion assesses the performance and effectiveness of fusion methods. As a result, efficacy and efficiency are used to measure the accuracy and the resource utilization of the fusion process respectively (Fig. 2.8).

(a) **Efficacy**: Refers to the effectiveness of fusion methods to show the quality, reliability and utility of the output. It uses different data and metrics to assess the fusion techniques.

  (i) **Data**: Evaluation of data fusion methods can be performed on three kinds of data: *synthetic, real-world* or *both*.

  (ii) **Metrics**: The common metrics to show the accuracy of the systems are: *mean square error (MSE), mean absolute error (MAE), precision, recall,* and *F1-score*.

(b) **Efficiency**: Checks how much time the fusion system takes based on the available computational resource and size of the dataset.

2.3 State of the Art Research Domains

With the growing volumes of data, many sources provide unusable, improper, inconsistent, and incomplete data, resulting in uncertainty and contradictions that negatively affect data quality [24]. Data fusion enhances data quality by integrating multiple sources to produce more consistent and useful information. In this section, we discuss research work related to various areas of data fusion.

2.3.1   Data Quality

The earliest approaches to data fusion for conflict resolution were generally rule-based. Therefore, they used conflict-handling functions such as minimum, maximum, averaging, or voting. Among these, voting was most popular and used in different scenarios. However, these conflict-handling functions do not evaluate the quality of data sources. In the modern age of big data, when there is an increasing volume of data accessible online, data quality is more important than ever. The publication of several erroneous data points creates veracity issues directly related to data quality problems and inconsistencies [25], which traditional data fusion methods are unable to address. As a result, while using data fusion techniques, data quality is now critically important.

In Sect. 2.2.2, we define data quality simply as how well data serves its intended purpose—its "fit for use" [18]. Data quality is typically evaluated using four dimensions: *accuracy* (how accurate the data is), *completeness* (whether all necessary data is present), *timeliness* (how up-to-date the data is), and *consistency* (how coherent the internal data) [19, 26]. A popular approach to truth evaluation (discovering accurate information [27]) is estimating and initializing the source reliability through the measurement of data quality dimensions. For example, greater *completeness* of a source often indicates higher reliability. Another such method uses the frequency of total trustable values provided by a particular source as compared to other sources [28]. For the degree of sample data trustworthiness, both *confidence score* [18] and *averaged data quality metrics* [29] are shown to be effective. While the former is used as data trustworthiness to exhibit how it impacts and improves data quality metrics, the latter selects a specific data source according to the user query.

2.3.2   Uncertain Data Handling

Uncertain data makes the data sources imperfect and of poor quality. Hence, representing uncertainty is a preprocessing step in truth analysis where uncertain data can be either removed or allowed to be processed further [28]. Uncertainty arises when one source provides a non-null value, but another provides NULL or no information for the same real-world object or when they provide values in a confidence interval. The uncertain data handling approaches are tuple-based, attribute-based, or both [30].

Sundarmurthy et al. [31] propose a technique for representing and query processing over missing values. The authors employed m-tables (m is the symbolic representation of missing values) to represent unknown attribute values, partially known tuples, or completely anonymous tuples. They used positive relational algebra operators (e.g., $\sigma, \pi, \cup, \times, \bowtie, \rho$ excluding set difference and division) over m-tables to obtain certain and possible answers, which requires high computation costs. Additionally, although their technique is applicable for missing data, they are not beneficial for uncertainty with NULL values and aggregation.

Dempster-Shafer is a popular framework to measure uncertainty where the researchers mainly focus on computing the Shannon entropy as the measure of the volume of information to get the degree of uncertainty [32]. In this paper, the authors utilized an exponential factor of the power set of frame of discernment (FOD) in the standard entropy calculation method, which gives an optimal measure for uncertain information within an interval. In another work, Yager et al. [33] propose a measure-based belief function to represent attribute-level uncertainty. In this paper, the authors first selected a set of non-empty values of a particular attribute and used a random variable to determine whether an uncertain value falls into that set or not by calculating the probability score. However, both [32, 33] are limited to measuring the degree of uncertainty within a given set or interval, while in trustworthiness, an actual value can fall outside the estimated interval.

AU-DB [22] focuses on creating an attribute annotated uncertain database where they annotate both tuple and attribute level uncertainty. They annotate with a selected guess from possible uncertain databases applying bound-on attributes and tuple multiplicities. They rely on ETL heuristics (e.g., source trustworthiness) to extract the selected guess value. Bag semantics is employed to match the tuples among possible worlds. AU-DB is the first work that assesses multi-aggregation in computing uncertainty. This approach benefits aggregation and query processing over uncertain data but is limited to ordinal data, while nominal and numerical data should also be considered.

Lenses [34] is an on-demand approach to ETL (extract-transform-load) process. Lenses annotate data uncertainty using a popular uncertainty encoded representation technique called C-table. In general, C-tables encode uncertain data with Boolean expressions and integer valuation with a pre-estimated probability so that users can get certain and possible answers according to their query. While other authors use C-tables for classical relational databases, they introduced virtual C-tables in the transformation part of an ETL process where the VC-tables will be generated based on the query of the users.

Mimir [35] is the extension of the Lenses [34] system, which they improve by optimizing the VC-tables. In this paper, partitioning and inline query evaluation have been applied. During partitioning, the query is normalized to create partitions based on deterministic and non-deterministic clauses. For inline query evaluation, the best-guess relation is pre-materialized to optimize query performance. Besides these, they provide a graphical user interface to the users showing the results.

Kennedy and Glavic [36] discuss several techniques of uncertain data management and how they are encoded, queried, evaluated, and presented to the users. Uncertain data encoding consists of possible world semantics, tuple-independent database, C-table, U-relation, world annotated sample set, and tuple bundles. All of these techniques use probability distribution among the possible worlds and, hence, for the records alongside encoding the uncertainty. Intentional and extensional query evaluation can be applied to these encodings. Techniques like tuple identity, filtering, confidence bounds on attribute values, and record confidence are used to provide the results of a query to the users. The encoding schemes of this paper require adding probability distribution to the relational database, making it a probabilistic database.

Bimonte et al. [23] focus on the management of missing data in the context of data warehouses and OLAP systems instead of representing them with any symbol. They employ horizontal, and vertical functions to estimate missing data and an adjustment method to enhance the imputation process of warehoused multidimensional and multi-granular datasets. They first use horizontal functions (e.g., mean and most frequent) to estimate the missing values and apply an adjustment method (e.g., mixed integer linear programming) if the estimated value does not coincide with a provided aggregated value. Otherwise, a vertical function (e.g., min, max, split) is involved to impute the values. This framework performs better than individually used horizontal or vertical functions in the imputation process.

2.3.3 Conflict Resolution

Resolving the conflicts between two values of the same data item is called *record merging*. Combining the records is needed to preserve the data quality at any level of the data sources. *Record merging* can be divided into two categories: *mediating* and *deciding* (Fig. 2.9).

Deciding: Refers to choosing a value from all the already present values. In this category, minimum, maximum, and voting (as discussed in Sect. 2.3.1) are the most prevalent techniques in conflict resolution. With the increasing volume of data sources, wrong data may be carried out to many sources if they copy from one another. As a result, these techniques can not act properly when the sources are dependent on each other because they publish wrong data after fusing [37]. Therefore, truth discovery methods attained popularity in resolving conflicts with the correct value regardless of the number of sources [38, 39].

Mediating: Refers to choosing a value that does not necessarily exist among the conflicting values [40]. The most popular *mediating* technique is *meet in the middle* which takes the average value. For numerical data, the average is the mean value of all the corresponding attribute values coming from different sources. This approach

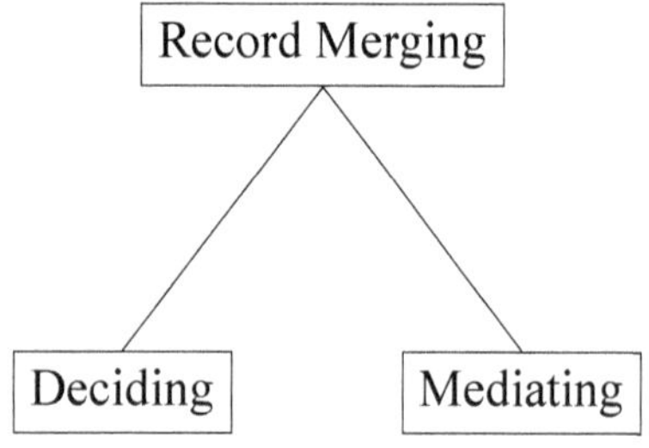

Fig. 2.9 Record merging strategies in conflict resolution

does not provide acceptable outcomes when sources have tremendous record discrepancies.

Weighted mean was incorporated with probability derivation to resolve numerical value conflicts [41]. They split their task into preparation and reconciliation stages. In the preparation stage, they define queries, estimate the cost parameters (i.e., type I, type II, and representation error), obtain attribute retrieval interval (ARI) from query selection intervals, estimate the prior probability, and constructs distortion matrices for all the sources. In the reconciliation stage, they compute the posterior probability, determine candidate values for each interval employing weighted mean and probabilistic median, and pick the best value evaluating the lowest cost. Their system works efficiently for a small number of sources. However, the paper does not address the difficulties of missing value imputation and aggregation operations in the queries. Other *mediating* techniques are the *Bayesian network* [42, 43] which uses posterior probability to get the fused data; *Fuzzy-logic* [44] which applies fuzzy rules and aggregation operators (e.g., fuzzy averaging) to obtain the fused data; and *Dempster-shafer theroy* [32] which uses Dempster's rule of combination to fuse belief masses (i.e., basic probability assignment) from different sources and get the fused data based on highest combined confidence.

2.3.3.1 Truth Discovery Methods

Truth Discovery has become the main component in the general process of data fusion. Truth discovery methods have a general rule: the higher the quality of the source is, the more likely it is that it provides truth, and the more truth it gives, the higher the quality of the source is. Truth discovery algorithms can be divided into two sub-classes: *single truth*—when a data item can have only one single true value and *multi-truth*—when a data item may have more than one single true value.

Yin et al. [45] first propose a single truth-inferring algorithm called Truthfinder for identifying true facts using an iterative method. They first compute the trustworthiness score of the websites from initialized trustworthiness and then compute the confidence score of the facts. The confidence score is adjusted if there is any implication among the facts. After that, the confidence of a fact is calculated using a logistic function, and the trustworthiness of a source is updated. Thresholding has been applied to converge the iterative process. The performance of Truthfinder is significantly better when compared to the results achieved through majority voting.

Lin et al. [46] propose two approaches for discovering multi-truths from conflicting websites. In the first one, they calculate the global domain percentage (i.e., the amount of data provided by the sources in a certain domain or category) and assign weights to the individual domains of the data sources. After that, weighted voting is employed to get the true value. In the second approach (DART), an unsupervised bayesian joint probabilistic model is employed to update the probability of a true value and the trustworthiness of the source. Here, the initialization step consists of domain expertise and domain influence on the data. The iteration converges when the truth computation score remains in an interval.

Fabeo et al. [39] propose a system called STORM, designed to discover multi-truths as an extension of the DART [46] system. They have used a common technique bayesian network to infer the truth. However, they used copy-based source authority to determine the trustworthiness of the sources. Copy-based source authority decides which source is highly copied by other sources making it most reliable. They also incorporated a clustering-based value reconciliation step to remove the discrepancies of the same data item among the sources.

Zhang et al. [47] propose the FTS method that measures the source quality using three metrics: silent rate, true rate, and false rate. In contrast to the previous methods, which do not take into account the impact of null on source quality, this model fully utilizes both claims and null to increase the accuracy of truth finding. The authors evaluate source quality, redesigning the aforementioned metrics and applying the hub authority approach for the truth discovery task.

Jingxue et al. [38] introduce a clustering approach where they first divide the normal and exceptional data by employing a mean shift clustering algorithm and iteratively updating the clusters to get the true value for the conflicting attributes. However, this work can be evaluated when a ground truth is available. In trustworthiness, having a ground truth may not be easy because different sources might provide different levels of data quality [28].

2.4 Open Problems and Future Directions

Despite significant advances in data fusion, several challenges remain unresolved. In this work, we will compare recent truth discovery methods based on the features outlined in the taxonomy (see Sect. 2.2). By analyzing these methods, we aim to identify their key characteristics and highlight open problems that require further research and improvement. This analysis will provide insight into existing gaps and suggest directions for advancing truth discovery in data fusion.

2.4.1 Comparison of Truth Discovery Methods

Truth discovery lies at the core of modern data fusion, systematically assessing both data source reliability and the accuracy of their provided information. These methods fall into three main categories: iterative methods, optimization-based methods, and probabilistic graphical models. Despite their different approaches, they share a fundamental characteristic: the iterative refinement of results through alternating cycles of truth computation and source reliability assessment.

Rather than conducting an exhaustive survey of all fusion techniques, we focus on a selection of recent truth discovery methods and compare them. The comparison is outlined in Tables 2.13 and 2.14 based on the following criteria:

Table 2.13 Comparison of truth discovery methods (part 1)

Method	C1	C2	C3	C4	C5	C6
RTDS [38]	Iterative	No	No	No	No	Single
STORM [39]	Iterative	No	No	Yes	Yes	Multi
DART [46]	Iterative	No	No	Yes	Yes	Multi
FTS [47]	Probabilistic graphical model	No	No	No	No	Single
MTD-VCI [48]	Optimization	No	No	No	No	Multi
RPPTD [49]	Optimization	No	No	No	No	Single
SRTD [50]	Iterative	No	No	No	No	Single
RCHDTD [51]	Optimization	No	No	No	No	Single
Apollo-social [52]	Probabilistic graphical model	No	No	No	Yes	Single
CATD [53]	Optimization	No	No	No	Yes	Single

- **Truth discovery process (C1)**: The specific approach that each study uses to identify and determine the most accurate information from multiple conflicting sources.
- **Symbolic representation for uncertainty (C2)**: Whether the study incorporates symbolic representation to manage and address uncertainties in the data.
- **Handling multi-granular data (C3)**: Whether the research takes into account data at different levels of detail or granularity.
- **Incorporation of domain knowledge (C4)**: Whether the study incorporates domain-specific knowledge or not.
- **Source dependency consideration (C5)**: Whether the research accounts for the possibility that data sources may not be independent and could influence each other, such as when one source copies from another.
- **Single or multiple truths (C6)**: Whether the study finds a single truth or multiple truths.
- **Technique for truth computation (C7)**: The specific technique or algorithm that is used by the system to calculate and determine the truth from the available data.
- **Evaluation techniques (C8)**: The methods or criteria employed to assess and validate the effectiveness of the truth discovery system.
- **Use of gold standard data (C9)**: Whether the research compares its results against a gold standard dataset or not.

Table 2.14 Comparison of truth discovery methods (part 2)

Method	C7	C8	C9	Remarks
RTDS [38]	Mean shift clustering	MAE, MSE, RS	Yes	Removes exceptional data
STORM [39]	Copy-based source authority	P, R, F1	Yes	Introduces a token-based similarity measure that can reconcile different string variants (e.g., abbreviations, typos)
DART [46]	Weighted voting, Unsupervised bayesian joint probabilistic distribution	P, R, F1	Yes	Infers domain specific reliability by examining data richness
FTS [47]	Maximum-likelihood	Recall	Yes	Takes into account the impact of null on source quality
MTD-VCI [48]	Confidence interval estimation	P, R, F1	Yes	Introduces ambiguous claim elimination to deal with ambiguous terms (e.g., "etc.", "et al.")
RPPTD [49]	Majority voting	P, R, F1	Yes	Considers data privacy while extracting reliable information from multiple sources
SRTD [50]	Majority voting based on contribution score of the sources	SPC, MCC, Kappa	Yes	Uses work queue in an HTCondor system to handle large-scale data
RCHDTD [51]	Weighted voting, Weighted median	MNAD	Yes	Incorporates different data types within a unified optimization framework
Apollo-social [52]	Maximum-likelihood	P, R	No	Addresses the challenge of uncertain provenance
CATD [53]	Weighted averaging	MAE, RMSE	No	Handles long-tail phenomenon (i.e., when many sources provide few claims)

P—Precision   **SPC**—Specificity
R—Recall   **MCC**—Matthews Correlation Coefficient
F1—F1 Score   **Kappa**—Cohen's Kappa
MAE—Mean Absolute Error   **RS**—R Squared
MSE—Mean Squared Error
MNAD—Mean Normalized Absolute Distance

2.4.2 Research Challenges and Future Directions

There have been many studies conducted on the topics of truth-finding and data fusion. However, there are still a lot of unresolved issues. This section discusses the research problems identified during the evaluation of the articles included in our study. It also looks at possible future directions for data fusion based on this evaluation.

2.4.2.1 Parameter Initialization

Most of the truth discovery algorithms are required to initialize some input parameters. This initialization has a great impact on the effectiveness of the algorithms [54]. Some parameters such as source trustworthiness score are crucial, because they have a great impact on the result. Current methods usually start with a default reliability for each source and gradually refine it. This process can lead to errors in calculating trustworthiness. Because, if a method starts by assuming certain sources are highly reliable, it may give more weight to their claims. Hence, the result can be biased by the more weighted sources. Conversely, if initial values are too low, even trustworthy sources might not contribute effectively to the final results. Li et al. [55] suggest that starting with an accurate trustworthiness estimation could correct over half of these errors. Hence, if any pre-existing information about the reliability of the sources is available, they can be initialized based on the prior knowledge. However, how to automate parameter initialization is still an open problem.

2.4.2.2 Benchmark Data Utilization

Most of the truth discovery methods utilize some *gold standard* or *benchmarking* data. In the context of truth discovery, *gold standard* or *benchmarking* data are considered highly reliable and accurate. However, in practice, having such a dataset is impossible because of the variations in the degree of accuracy, completeness, and relevancy of data obtained from various sources. Still, there is often a strong emphasis on using *benchmarking* or a *gold standard* data for validating models and algorithms in truth discovery. However, this reliance can be problematic because:

Unavailability: It is uncommon to find perfect datasets. Even if they exist, they may not cover all possible scenarios or domains.

Bias and overfitting: Arises when truth discovery models rely too much on benchmarking data. This can lead to models becoming too specific to a single dataset and not generalizing well to other real-world data.

Fang et al. [22] evaluated several truth discovery methods without using ground truth (i.e., gold standard data). They proposed the *CompTruthHyp* approach to compare the performance of truth discovery methods. In particular, they calculate the probability of observations in a dataset based on the output of different methods. The probability is then ranked to reflect the performance of these methods. They employed *cosine similarity* and *euclidean distance* to measure the distance of two rankings. For *cosine similarity*, a bigger value means better performance, while for *euclidean distance*, a smaller value indicates better performance. Although the *CompTruthHyp* approach could be one possible solution for the new truth discovery methods, still more research is needed to solve this problem.

2.4.2.3 Lack of Error Traceability

Error traceability refers to the ability to track and identify the sources and pathways through which errors or inaccuracies in data propagate within the truth discovery process. This allows researchers and practitioners to understand how errors in initial data inputs affect the final outcome. Traceability also helps to identify specific data sources or variables that contribute to these errors.

In truth discovery, errors can arise from multiple sources, such as unreliable data inputs, incorrect assumptions, or flawed algorithms. Traceability helps in identifying where these errors originate and how they influence the final discovered truths. By understanding this, corrective measures can be applied to minimize errors and improve the accuracy of the truth discovery process.

Provenance techniques can play a crucial role in enhancing error traceability in truth discovery methods. It can provide detailed records of the origin, history, and transformation of data as it moves through the truth discovery process. By keeping track of where data comes from and how it is processed, provenance techniques enable the identification of error sources and the analysis of error propagation. This, in turn, helps improve the accuracy and reliability of the discovered truth.

2.4.2.4 Scalability Issues

Scalability issues in truth discovery occur when methods are applied to large-scale datasets. This challenge emerges as the number of data sources, data points, or data dimensions significantly increases. As the volume, variety, and velocity of data grow, several factors can contribute to scalability problems in truth discovery:

Source diversity: In truth discovery, data often come from multiple sources, each with its own level of reliability. As the number of sources increases, the complexity of assessing and managing these sources also increases. Scalability issues can arise when the truth discovery method struggles to efficiently evaluate and weigh information from a vast number of sources.

Big data challenges: As data volumes increase, the storage and retrieval of data can become a significant challenge. Scalability issues arise when the truth discovery system cannot efficiently handle, store, or access large datasets. This can slow down the entire process and make it difficult to keep up with real-time data demands.

Dimensionality: High-dimensional data (e.g., data with many features or attributes) can also pose scalability challenges, as truth discovery methods may struggle to analyze and extract meaningful truths from such data.

One way to address scalability issues could be to use parallel and distributed processing techniques. By distributing the computation across multiple machines or processors, the truth discovery process can handle larger datasets more quickly.

2.5 Conclusion

In this paper, we explore the foundational concepts within the field of data fusion. We present a comprehensive taxonomy of data fusion procedures, illustrating how truth discovery methods are integrated into the fusion process. The paper provides detailed definitions of fusion levels, data types, truth discovery techniques, evaluation metrics, and data quality dimensions. Furthermore, we examine various research domains that are interconnected within the broader context of data fusion. Based on the features outlined in the taxonomy, we compare recent truth discovery methods. The unresolved challenges associated with these methods are systematically classified into four distinct categories, and potential opportunities for addressing these challenges are also discussed.

References

1. Dong XL, Naumann F (2009) Data fusion—resolving data conflicts for integration. PVLDB 2(2):1654–1655
2. Gutiérrez R, Rampérez V, Paggi H, Lara JA, Soriano J (2022) On the use of information fusion techniques to improve information quality: taxonomy, opportunities and challenges. Inf Fusion 78:102–137
3. Dayal U (1983) Processing queries over generalization hierarchies in a multidatabase system. In: Schkolnick M, Thanos C (eds) 9th international conference on very large data bases, October 31–November 2, 1983, Florence, Italy, Proceedings. Morgan Kaufmann, Italy, pp 342–353
4. Blasch E, Pham T, Chong CY, Koch W, Leung H, Braines D, Abdelzaher T (2021) Machine learning/artificial intelligence for sensor data fusion-opportunities and challenges. IEEE Aerosp Electron Syst Mag 36(7):80–93
5. Senel N, Kefferpütz K, Doycheva K, Elger G (2023) Multi-sensor data fusion for real-time multi-object tracking. Processes 11(2):501
6. Wang M, Perera C, Jayaraman PP, Zhang M, Strazdins P, Shyamsundar RK, Ranjan R (2016) City data fusion: sensor data fusion in the internet of things. Int J Distrib Syst Technol 7(1):15–36
7. Huang X, Liu Y, Huang L, Onstein E, Merschbrock C (2023) BIM and IoT data fusion: the data process model perspective. Autom Constr 149:104792
8. Nahari MK, Ghadiri N, Jafarifard Z, Dastjerdi AB, Sack JR (2017) A framework for linked data fusion and quality assessment. In: Dehghantanha A, Parizi RM (eds) Proceedings of international conference on web research (ICWR'17) . IEEE, Iran, pp 67–72
9. Chen K, Koudas N (2024) Unstructured data fusion for schema and data extraction. Proc ACM Manag Data 2(3):181
10. Berkani N, Bellatreche L, Guittet L (2018) ETL processes in the era of variety. Trans Large Scale Data Knowl Centered Syst 39:98–129
11. Lau BPL, Hasala MS, Zhou Y, Hassan NU, Yuen C, Zhang M, Tan U (2019) A survey of data fusion in smart city applications. Inf Fusion 52:357–374
12. Meng T, Jing X, Yan Z, Pedrycz W (2020) A survey on machine learning for data fusion. Inf Fusion 57:115–129
13. Gao J, Li P, Chen Z, Zhang J (2020) A survey on deep learning for multimodal data fusion. Neural Comput 32(5):829–864
14. Shaik T, Tao X, Li L, Xie H, Velásquez JD (2024) A survey of multimodal information fusion for smart healthcare: mapping the journey from data to wisdom. Inf Fusion 102:102040

15. Canalle GK, Salgado AC, Lóscio BF (2021) A survey on data fusion: what for? in what form? what is next? J Intell Inf Syst 57(1):25–50
16. Lahat D, Adali T, Jutten C (2015) Multimodal data fusion: an overview of methods, challenges, and prospects. Proc IEEE 103(9):1449–1477
17. Li X, Dunkin F, Dezert J (2024) Multi-source information fusion: progress and future. Chin J Aeronaut 37(7):24–58
18. Ardagna D, Cappiello C, Samá W, Vitali M (2018) Context-aware data quality assessment for big data. Future Gener Comput Syst 89:548–562
19. Taleb I, Serhani MA, Dssouli R (2018) Big data quality: a survey. In: Chin FYL, Chen CLP, Khan L, Lee K, Zhang L (eds) Proceedings of international congress on big data (IEEE Big-Data'18). Springer, USA, pp 166–173
20. Imielinski T Jr, WL (1984) Incomplete information in relational databases. JACM 31(4):761–791
21. Abelló A, Cheney J (2024) Eris: efficiently measuring discord in multidimensional sources. PVLDB 33(2):399–423
22. Feng S, Glavic B, Huber A, Kennedy OA (2021) Efficient uncertainty tracking for complex queries with attribute-level bounds. In: Li F, Koutrika G, Tan WC (eds) Proceedings of the 2021 ACM SIGMOD international conference on management of data. ACM, China, pp 528–540
23. Bimonte S, Ren L, Koueya N (2020) A linear programming-based framework for handling missing data in multi-granular data warehouses. Data Knowl Eng 128:101832
24. Tré GD, Dujmović JJ (2021) Dealing with data veracity in multiple criteria handling: an lsp-based sibling approach. In: Andreasen T, Tré GD, Kacprzyk J, Larsen HL, Bordogna G, Zadrożny S (eds) Flexible query answering systems: 14th international conference, FQAS 2021, Bratislava, Slovakia, September 19–24, 2021, Proceedings, Springer, Slovakia, pp 82–96
25. Saha B, Srivastava D (2014) Data quality: the other face of big data. In: Cruz IF, Ferrari E, Tao Y, Bertino E, Trajcevski G (eds) Proceedings of the 30th IEEE international conference on data engineering (ICDE). IEEE, USA, pp 1294–1297
26. Bansal SK (2014) Towards a semantic extract-transform-load (etl) framework for big data integration. In: Chen P, Jain H (eds) 2014 IEEE international congress on big data, Anchorage, AK, USA, June 27 - July 2, 2014. IEEE, Anchorage, AK, USA, pp 522–529
27. Yang Y, Gu L, Zhu X (2019) Conflicts resolving for fusion of multi-source data. In: Jin Z, Mei H, Hu Z, Chen Z (eds) Proceedings of the 2019 IEEE international conference on data science in cyberspace (IEEE DSC 2019), IEEE, China, pp 354–360
28. Li Y, Gao J, Meng C, Li Q, Su L, Zhao B, Fan W, Han J (2015) A survey on truth discovery. SIGKDD Explor 17(2):1–16
29. Safhi HM, Frikh B, Ouhbi B (2019) Data source selection in big data context. In: Indrawan-Santiago M, Pardede E, Salvadori IL, Steinbauer M, Khalil I, Anderst-Kotsis G (eds) Proceedings of the 21st international conference on information integration and web-based applications and services (iiWAS'19). ACM, Germany, pp 611–616
30. Li Y, Chen J, Feng L (2013) Dealing with uncertainty: a survey of theories and practices. IEEE Trans Knowl Data Eng 25(11):2463–2482
31. Sundarmurthy B, Koutris P, Lang W, Naughton JF, Tannen V (2017) m-tables: representing missing data. In: Benedikt M, Orsi G (eds) Proceedings of the 20th international conference on database theory (ICDT 2017). Schloss Dagstuhl–Leibniz-Zentrum für Informatik, Italy, pp 21:1–21:20
32. Mambé MD, Takpé T, Anoh NG, Oumtanaga S (2018) A new uncertainty measure in belief entropy framework. Power 9(11):1–7
33. Yager RR, Alajlan N, Bazi Y (2019) Uncertain database retrieval with measure-based belief function attribute values. Inf Sci 501:761–770
34. Yang Y, Meneghetti N, Fehling R, Liu ZH, Kennedy O (2015) Lenses: an on-demand approach to ETL. PVLDB 8(12):1578–1589
35. Nandi A, Yang Y, Kennedy O, Glavic B, Fehling R, Liu ZH, Gawlick D (2016) Mimir: bringing ctables into practice. CoRR 1–13. arxiv:abs/1601.00073

36. Kennedy O, Glavic B (2019) Analyzing uncertain tabular data. In: Rogova GL (ed) Éloi Bossé. Information quality in information fusion and decision making. Springer, pp 243–277
37. Bakhtouchi A (2022) Data reconciliation and fusion methods: a survey. Appl Comput Inform 18(3/4):182–194
38. Chen J, Yang J, Huang J, Liu Y (2021) Robust truth discovery scheme based on mean shift clustering algorithm. J Internet Technol 22(4):835–842
39. Azzalini F, Piantella D, Rabosio E, Tanca L (2023) Enhancing domain-aware multi-truth data fusion using copy-based source authority and value similarity. PVLDB 32(3):475–500
40. Bleiholder J, Naumann F (2009) Data fusion. ACM Comput Surv 41(1):1–41
41. Jiang Z (2012) A decision-theoretic framework for numerical attribute value reconciliation. IEEE Trans Knowl Data Eng 24(7):1153–1169
42. Kim B, Lee J (2021) A bayesian network-based information fusion combined with DNNs for robust video fire detection. Appl Sci 11(16):7624
43. Wu P, Imbiriba T, Elvira V, Closas P (2022) Bayesian data fusion with shared priors. CoRR 1–12. arxiv:abs/2212.07311
44. Manoharan H, Shitharth S, Sangeetha K, Praveen Kumar B, Hedabou M (2022) Detection of superfluous in channels using data fusion with wireless sensors and fuzzy interface algorithm. Meas Sens 23:100405
45. Yin X, Han J, Yu PS (2008) Truth discovery with multiple conflicting information providers on the web. IEEE Trans Knowl Data Eng 20(6):796–808
46. Lin X, Chen L (2018) Domain-aware multi-truth discovery from conflicting sources. PVLDB 11(5):635–647
47. Zhang J, Wang S, Wu G, Zhang L (2018) An effective truth discovery algorithm with multi-source sparse data. In: Shi Y, Fu H, Tian Y, Krzhizhanovskaya VV, Lees MH, Dongarra J, Sloot PMA (eds) Proceedings of international conference on computational science (ICCS'18). Springer, China, pp 434–442
48. Fang X, Shen C, Sheng QZ, Sun G, Tang Y, Zhuo H (2023) A multi-truth discovery approach based on confidence interval estimation of truths. In: Yang X, Suhartanto H, Wang G, Wang B, Jiang J, Li B, Zhu H, Cui N (eds) Proceedings of international conference on advanced data mining and applications (ADMA'23). Springer, China, pp 599–615
49. Chen J, Liu Y, Xiang Y, Sood K (2022) RPPTD: robust privacy-preserving truth discovery scheme. IEEE Syst J 16(3):4525–4531
50. Zhang D, Wang D, Vance N, Zhang Y, Mike S (2019) On scalable and robust truth discovery in big data social media sensing applications. IEEE Trans Big Data 5(2):195–208
51. Li Y, Li Q, Gao J, Su L, Zhao B, Fan W, Han J (2016) Conflicts to harmony: a framework for resolving conflicts in heterogeneous data by truth discovery. IEEE Trans Knowl Data Eng 28(8):1986–1999
52. Wang D, Amin MTA, Li S, Abdelzaher TF, Kaplan LM, Gu S, Pan C, Liu H, Aggarwal CC, Ganti RK, Wang X, Mohapatra P, Szymanski BK, Le HK (2014) Using humans as sensors: an estimation-theoretic perspective. In: Rowe A, Krishnamachari B, Whitehouse K (eds) Proceedings of international symposium on information processing in sensor networks (IPSN'14). IEEE, Germany, pp 35–46
53. Li Q, Li Y, Gao J, Su L, Zhao B, Demirbas M, Fan W, Han J (2014) A confidence-aware approach for truth discovery on long-tail data. PVLDB 8(4):425–436
54. Waguih DA, Berti-Équille L (2014) Truth discovery algorithms: an experimental evaluation. CoRR 1–13. arxiv:abs/1409.6428
55. Li X, Dong XL, Lyons K, Meng W, Srivastava D (2012) Truth finding on the deep web: is the problem solved? PVLDB 6(2):97–108

Chapter 3
Scalable and Privacy-Aware Relational Data Synthesis

Antheas Kapenekakis, **Daniele Dell'Aglio**, **Martin Bøgsted**,
Minos Garofalakis, and **Katja Hose**

Abstract Sensitive data, especially in the era of GPDR, is under strict regulations and access restrictions. This is a major obstacle for conducting research with the data, as analysts first have to be granted access, requiring a lengthy approval process and often a purpose for accessing the data. Following, they are limited to secure computing resources, which limits their flexibility in choosing the algorithms and tools for their analysis. Data synthesis promises to be a solution to the sharing of sensitive datasets, by acting as a layer of anonymization which aids in reducing access restrictions. Methods for privacy-aware data synthesis of large data under secure with the constraint of limited computing resources is a promising research area, with the goal of liberating sensitive data for research purposes.

Keywords Data synthesis · Privacy preservation · Tabular data · Relational data · GDPR

A. Kapenekakis (✉) · D. Dell'Aglio
Aalborg University, Aalborg, Denmark
e-mail: antheas@cs.aau.dk

D. Dell'Aglio
e-mail: dade@cs.aau.dk

M. Bøgsted
Aalborg University Hospital, Aalborg, Denmark
e-mail: martin.boegsted@rn.dk

M. Garofalakis
Athena Research Center, Athena, Greece
e-mail: minos@athenarc.gr

K. Hose
TU Wien, Vienna, Austria
e-mail: katja.hose@tuwien.ac.at

G. Dejaegere et al. (eds.), *Data Engineering for Data Science*,
https://doi.org/10.1007/978-3-032-18765-9_3

3.1　Introduction

Data synthesis is an emerging field with the aim of promoting responsible data sharing through anonymization of data [1], and correcting for (sampling and societal) bias through the reshaping of datasets [2]. So far, privacy-aware data synthesis has focused on tabular datasets of rudimentary values (numbers and categorical) which fit in memory (up to 5 million rows, <100 MB [3–7]). Nevertheless, most of the interesting synthesis application domains (i.e., multi-modal, hierarchical datasets, and medical data) raise new scalability, performance, and complexity challenges, due to the larger amount and increasing fidelity of the data.

Specifically, medical data is richer and more complex, with multiple tables, each describing a different aspect of a patient (e.g., demographics, lab results, admissions, diagnoses). The relationships between these tables are complex, can be temporal or associative, and contain rich cross-table correlations. As the current state-of-the-art on structured data focuses on simple tabular data, it faces challenges while modeling medical data, and either performs poorly or has negligible privacy guarantees.

Regarding scalability, small datasets (up to 5M rows) avoid most scalability issues, as they can be loaded and processed in memory, with most operations requiring a few gigabytes of RAM, and a few milliseconds, even without the use of parallelization. However, with larger data sets (more than 100M rows), they become unwieldy to load and hold in memory and the processing required for producing synthetic data requires hours to days. Finally, for data above the 1 billion mark, both of these issues are exacerbated, with synthesis execution requiring days to weeks or multiple machines with terabytes of total RAM. This is a prohibitive issue when synthesizing General Data Protection Regulation (GDPR) protected datasets, as data processing requires GDPR-compliant compute, which is more limited and expensive, with compute clusters being out of reach for most use-cases (due to data contamination and access rules).

The research area of this chapter focuses on privacy-aware data synthesis of rich and complex relational medical data, in a scalable and efficient manner. Our goal is to provide avenues for generating accurate and anonymized arbitrary relational data, under the limitations imposed by GDPR-compliance.

3.2　Data Synthesis Landscape

Data synthesis is an upcoming field that promises to lower the barrier to access of sensitive data for research purposes. During generation, a synthesis algorithm learns an embedding of the original data, which is then used to generate a synthetic version. This version bears the same statistical properties as the original data and, when combined with a privacy mechanism, does not leak sensitive information. For data synthesis, we identify five key areas: privacy-awareness, tabular data synthesis (as a prerequisite for relational data synthesis), relational data synthesis, scalability, and evaluation. A taxonomy of these areas is shown in Fig. 3.1.

For privacy-awareness, there are a variety of mechanisms that can be used to ensure privacy prior to synthesis in a formal manner (e.g., differential privacy [8]). Post-synthesis, evaluation methods may be used to gauge the risk of exposure. Unlike formal mechanisms, evaluation methods can only gauge the risk of data exposure today. Both methods are complementary and should be used in conjunction.

Privacy mechanisms embed concrete privacy guarantees into the synthesis process. This means that the output data is guaranteed to have these properties without having to be proven. However, they are hard to convey to the end-user and their effectiveness is not quantifiable in a tangible manner (e.g., a privacy budget of three is not informative about a patient's cancer status being protected). Metrics for evaluation act on synthetic data and gauge the level of privacy. However, this level is derived from metrics that exist today, where a synthetic data release cannot be retroactively revoked. If new metrics or attack methods for synthetic data surface, the level of privacy of released datasets may decrease, in a way that causes harm to individuals years after release.

The foundation of relational data synthesis is tabular data synthesis. As tabular data can be viewed as a simpler case of relational data, most recent work in structured data synthesis has focused on tabular data. We can group algorithms in this area into three categories: Probabilistic Graphical Models (PGMs), Neural Networks (NNs), and the tuning of Large Language Models (LLMs). For Neural Networks, approaches are further divided into Generative Adversarial Networks (GANs) and Variational Autoencoders (VAEs).

PGMs, due to their simplicity, are the most interpretable and easier to add privacy to (e.g., by adding noise to the learned parameters, whose number is smaller). However, they are also the least expressive and have difficulty in capturing complex relationships. Neural Network and LLM approaches can scale their complexity to such an extent where they memorize the whole dataset, which in turn offers excellent performance. But, this memorization leads to an absence of privacy guarantees. When embedding a privacy mechanism to Neural Networks, this prohibits the model from memorization, leading to poor utility. In sum, for a certain privacy level, PGMs offer better utility at high privacy levels with predictable accuracy, but their utility does not scale to the level of NNs and LLMs as the need for privacy decreases.

Relational data synthesis is a very novel research area with high potential. Currently, algorithms for relational data utilize two main approaches: extending tabular data synthesis algorithms to relational data or creating bespoke models for a specific dataset structure (e.g., temporal data).

For synthetic data to be dependable, it needs to feature a certain level of guaranteed privacy and accuracy. This can be achieved with the use of evaluation metrics, a collection of which forms the quality report for a synthetic dataset. We can group these metrics into two main categories: utility and privacy. To measure utility, we can use synthetic metrics (e.g., Kullback-Leibler divergence), real-world use cases (e.g., replicate an analysis performed on the original data), or train a model on synthetic data and evaluate its performance. For privacy, there are two classes of attacks that can

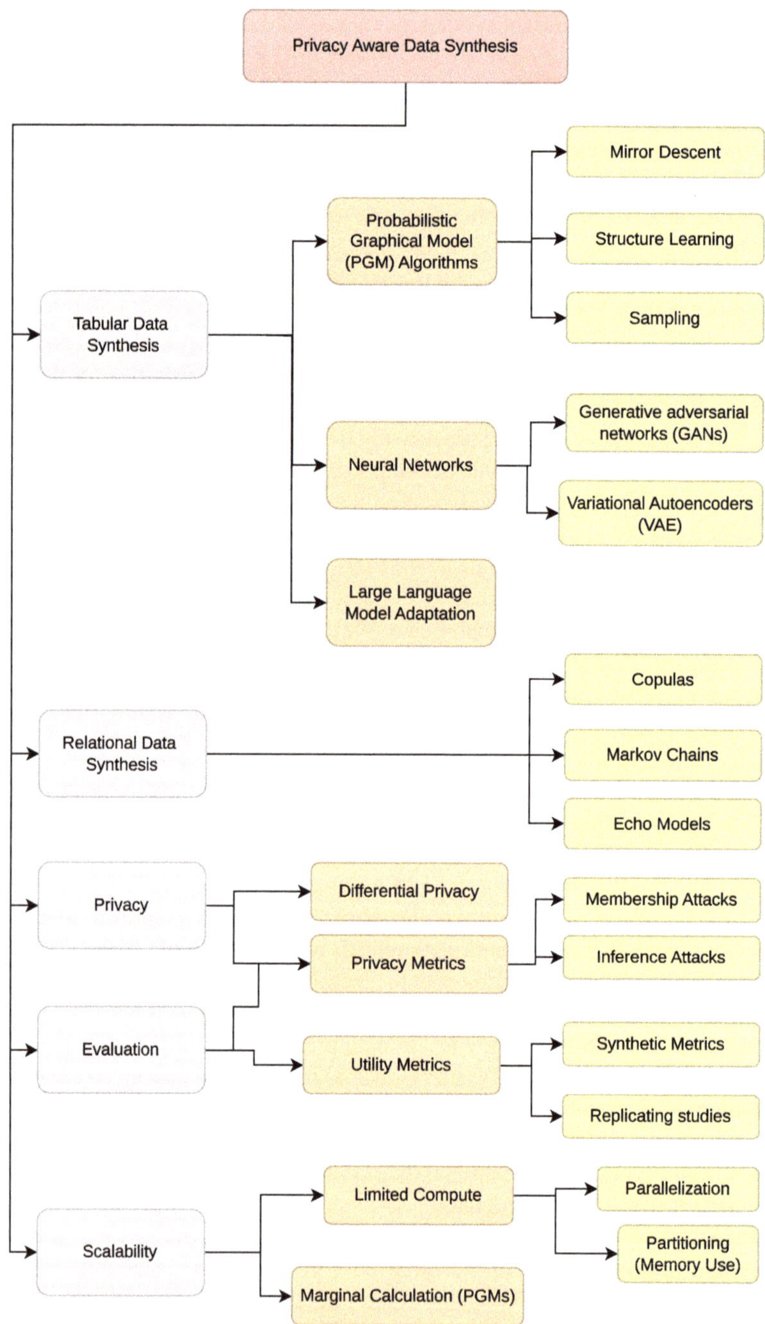

Fig. 3.1 Privacy-aware data synthesis taxonomy

be performed on synthetic data: membership inference and attribute disclosure. First, being able to infer if a person was part of a dataset might be harmful to the individual. This is specific to the nature of a dataset, with the main cause being it containing members of a protected class (e.g., drug addicts). Second, attribute disclosure is the increase in confidence about a protected attribute that concerns an individual (e.g., "has_cancer"), which is higher than the confidence that can be gained from data that excludes the individual. For example, by looking at a study of smokers, we can derive that a person who smokes is more likely to have cancer. Attribute disclosure would occur if we could be measurably more confident about their likelihood of having cancer if they partook in the study.

The final area of interest is scalability. To process and synthesize enterprise data, synthesis algorithms must be able to work at scale, with the ability to process datasets of billions of rows. For sensitive data specifically, this is a challenge, as the data must be processed in a GDPR-compliant manner which limits the scope of available compute resources. Processes and algorithms underlying synthesis must be designed and structured in such a way that ensures they utilize limited compute resources efficiently under a constrained memory envelope.

3.2.1 Research Applications

A key application of data synthesis is in the field of healthcare. Specifically, for the type of data we are investigating in this chapter, Electronic Health Records (EHRs) are a prime candidate. EHR consist of a patient's medical history, and depending on department, can include a variety of data: lab results, diagnoses, treatments, admission information, demographics, and more.

In Fig. 3.2, we see an example of patient trajectory that would be included in a relational EHR dataset. It corresponds to a patient who is undergoing cancer treatments. Cancer treatments consist of macro-blocks, called lines, which determine the type of treatment the patient is undergoing. Then, for each line, there are a number of cycle updates, in which the patient is prescribed a certain set and dosage of drugs.

Such a dataset introduces a variety of challenges when attempting to synthesize it. First, it contains a set of tables with feature correlations between them. Then, it contains temporal data in the tables for lines and cycles, as both the prescribed medicine and the chosen treatment protocol are heavily correlated to the previous ones. Finally, the dataset has a prescriptions table with intercorrelated drugs. The prescriptions table is an associative entity table, with one relationship being the prescription and the other the drugs, which requires special care for preserving the counter-indications between different drugs.

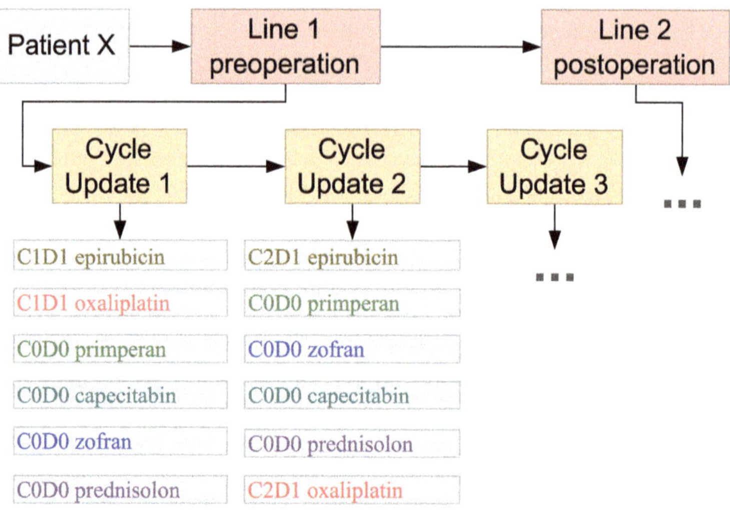

Fig. 3.2 Electronic Health Records example

3.3 State of the Art

3.3.1 Data Science on Medical Data

There is prior work using machine learning for performing diagnosis and prognosis predictions [9–14], private data synthesis [4, 15, 16], and patient similarity [13, 17–19] (as in finding similar patients to an input patient). Potential uses for such data include dynamic prediction [20–22], where using patient history is used to forecast a prognosis, patient similarity [9, 13, 17, 19, 23–25] where a similarity measure is developed and used to cluster similar patients, and anomaly detection. These works signify the need for a privacy-aware version of data that can be shared with researchers and the public with fewer restrictions.

3.3.2 Ensuring Privacy

For ensuring privacy, the most common mechanisms are k-anonymity [26] and Differential Privacy [8], the former being a precursor to the latter, used to ensure that each record in the dataset is indistinguishable from at least $k - 1$ other records. However, in order to achieve this, k-anonymity generalizes by a factor of k, which results in lower dataset fidelity. In addition, due to the lack of concrete privacy guarantees, k-anonymity is subject to re-identification attacks and may leak sensitive information. For example, one issue it does not prevent is private attribute disclosure. If the k individuals that share quasi-identifiers with Bob have the same private attributes, then

Bob's attributes can be inferred. l-diversity [27] addresses this shortcoming by requiring that the k participants have at least l different attributes. Then, t-closeness [28], similar to l-diversity, changes the privacy guarantee to be probabilistic, where an attacker's prior belief about the private attributes of Bob should be equal or close to his posterior.

Differential privacy [8], on the other hand, provides a concrete privacy guarantee that is probabilistic in nature and guards against both inference and membership attacks. Consider a relational dataset D with multiple tables containing information about a set of GDPR protected entities E (e.g., doctors and patients). Each table $T \in D$ may contain foreign keys to other tables (for the collection of tables to be considered a single dataset, there must be a foreign key between any two tables). In data synthesis, the goal is to generate a synthetic dataset S that mirrors the features of D, without privacy exposure.

Definition 3.1 (*Differential Privacy*) A stochastic function K is ϵ-differentially private if for each pair of neighboring datasets $\mathcal{D}$, $\mathcal{D}'$ which differ by only one element and all $V \subseteq \mathrm{Range}(K)$ the following inequality holds:

$$P[K(\mathcal{D}) \in V] \le e^\epsilon \cdot P\left[K(\mathcal{D}') \in V\right]. \tag{3.1}$$

The term ϵ is named the privacy budget and quantifies the privacy exposure of releasing $S = K(\mathcal{D})$.

3.3.3 Tabular Data Synthesis

The foundation of generating lifelike relational data is tabular data synthesis. Without this foundation, handling the intricate relationships between different tables would not yield meaningful results. Data Synthesis algorithms can be separated into three broad categories: Neural Network, Marginal, and NLP based. Notable Neural Network algorithms are PATE-GAN [29], CT-GAN [30], DP-GAN [31], DP-CTGAN [32], and ADS-GAN [4]. For tabular data, a new class of Marginal based algorithms has been shown to outperform NN algorithms when privacy is required [33]. Notable algorithms are: MST [34], PrivBayes [35], AIM [3], PrivMRF [7], which are enhanced by using probabilistic graphical models on inference, through PrivPGM [6]. When privacy is required, most algorithms utilize Differential Privacy [8]. Additionally, Large Language foundational models can be fine-tuned to produce synthetic data.

More specifically, PGM algorithms consist of two steps: structure learning and parameter learning. In structure learning, the model is created, without filling in the parameters. Then, in parameter learning, the dataset is sampled to fill in the model. If privacy is required, both steps should utilize differentially private algorithms.

The algorithm PrivBayes [35] laid the foundation for PGMs, by providing a simple way to fit a Bayesian network to a tabular dataset. Much like other PGM algorithms, it has a structure learning and parameter learning phase. In structure learning, it begins by picking a column randomly, which becomes the base parent. Moving forward, it tests all remaining columns and finds the best parent combination for each of them by choosing a single column each time. Once a column is chosen and its parents are decided, it becomes available to use as a parent. This simple algorithm always results in producing a DAG, which captures a decent level of correlations. Once all columns have parents, the algorithm measures one marginal per column that contains a histogram of the unique value combinations in that column and its parents.

To sample data, PrivBayes iteratively samples each column, by following the ordering it selected them during structure learning. This way, once a column is to be generated, all of its parents have been generated already. This makes it possible to condition its marginal to the parent values and create a new conditioned histogram. This histogram is then sampled to produce a new column value.

PrivBayes has two weaknesses. First, it does not combine the observations between marginals. For example, if two marginals sample the same column, then each contains a noisy observation of its one-way histogram. Combining the two observations could aid to reduce the overall noise. Second, the algorithm it uses to select the parent combinations, "maximal_parents", scales exponentially, which limits PrivBayes to a few dozen columns. To deal with the first issue, PrivPGM [6] was presented as a post-processing step that improved PrivBayes through belief propagation and mirror descent.

Since the publication of PrivPGM, authors have started including it in their algorithms. The algorithm AIM [3], applies mirror descent iteratively while producing the model, which allows for testing which measurements the model fails against and focusing on sampling those. As a workload-aware algorithm, AIM receives a series of queries the data must fit to. If a workload is not available, this may be all two-way marginals. It begins by measuring all one-way marginals and using a model without any correlations. Then, it uses the exponential mechanism to select which two-way marginal is the worstly approximated, and adds it to the measurements. The algorithm repeats until the privacy budget is exhausted.

PrivMRF [7] takes a different approach. It selects a group of viable 2-way marginals based on a scoring function that mirrors utility and measures how much two columns are correlated. Then, it computes the marginals and adds noise using differential privacy. Finally, it uses an approach similar to PrivPGM to combine the resulting measurements into a model.

Neural networks, on the other hand, utilize two main approaches: GANs and VAEs. GANs are used to generate data that is similar to the original data, while VAEs are used to generate data that is similar to the distribution of the original data. For generative approaches, GANs have shown impressive results in producing lifelike data. One of their limitations is that they are not representative of the original joint distribution of the data (e.g., suffer from mode collapse), which renders the resulting data unsuitable for analytics tasks. VAEs, with an appropriate latent representation, perform better in this way but they lack the fidelity of GANs and often fail to capture the underlying distribution as well. In addition, when working with discrete data, all Neural Network approaches require adaptation of the data in order to process it (e.g., one-hot encoding). This preprocessing results in creating a large input layer, which results in the approaches being computationally inefficient. In general, most useful neural networks will require a GPU to at least train, with most modern ones also requiring one for real-time inference.

Moreover, when differential privacy is used, due to NN's large need for data, the amount of noise that needs to be added to the models, e.g., with Stochastic Gradient Descent (SGD) [36], minimizes utility. To counter this, the authors of PATE-GAN [29] adapted the concept of PATE [37] to GANs. PATE is a method for training a model on a teacher-student basis. Essentially, a set of teacher models is trained on partitions of the data. Then, the ensemble of teachers is used to label a public unlabeled dataset through the use of voting with added noise. Finally, the student model is trained on the public dataset. This approach is more information efficient due to only applying differential privacy to the resulting labels, which results in a more accurate model when using the same privacy budget. However, unlike an approach such as adding DP to SGD, the added noise is proportional to the number of teachers instead of the number of individuals in the original data. Therefore, the magnitude of noise added by differential privacy is proportional to the number of individuals the teacher represents (i.e., if a teacher is trained on the data of 100 individuals, it will receive $100\times$ the noise an individual would). This results in a very high amount of noise is added to the voting process. As an unlabeled dataset is required to perform training, approaches such as PATE have been termed as "label-only" differential privacy.

3.3.4 Relational Data Synthesis

Simulacrum[1] is a synthetic dataset by NHS which uses Bayesian networks and first-order Markov Chains (referencing one previous event). The Bayesian networks that form the dataset were handcrafted and, to ensure privacy, a number of histograms, especially those containing a few patients, were shuffled or removed. Simulacrum does not feature differential privacy guarantees and, due to large permutations, has limited downstream utility (e.g., patients can be treated after their death).

The parent company of Simulacrum holds the original data and provides the synthetic data to researchers as a testing ground. Researchers develop their models on the synthetic data, and then submit them to the company, which runs them on the original data and returns the results after verifying they are privacy preserving. While the synthetic data field is maturing, such approaches provide avenues for utilizing synthetic data in scenarios which would otherwise not meet accuracy requirements.

PrivLava [38] is an algorithm that extends the tabular synthesis algorithm PrivMRF to the relational domain. It produces a latent variable through expectation maximization, that is used to provide the ability to condition correlations across tables. The focus is on direct parent-to-child relationships, as the latent variable does not jump

---

[1] https://simulacrum.healthdatainsight.org.uk.

through tables to enable, on a sequence of tables A-B-C, table C to have strong correlations on A. As such, it is not suitable for medical datasets.

HMA1, proposed under the system SDV [39], is a generalized approach for generating hierarchical synthetic data. Underlying it is a statistical approach for generating data using copulas. In a tabular setting, each input column is assumed to be continuous, and the table is fit using copulas. The resulting hyperparameters become a new master table, which is then processed the same way. The process repeats recursively until there is one table left. Then, the models run in reverse, generating the final relational data. Similar to PrivLava, it does not model intra-table correlations such as sequences. Theoretically, the algorithm has the capacity to model inter-table correlations. However, the learning ability of copulas is limited, which results in the model not capturing fine grained correlations.

Finally, Markov-chain Approach for Recursive Events (MARE) is a data synthesis algorithm that specializes on relational and medical data. It orchestrates the execution of an ensemble of tabular synthesis models, each using one of the aforementioned tabular algorithms, in a way that produces semantically correct relational data. By using a framework for modeling relational data, it captures important medical correlations. Specifically, it focuses on three types: parent-child, temporal, and associative. For parent-child relationships, the parent rows that each row references are recursively unrolled and provided to a synthesis algorithm alongside it. Then, the algorithm can generate that row while referencing the parent rows. For temporal relationships, the algorithm uses a Markov Chain to model the a sequence of rows as events. Then, depending on the order n of the Markov Chain, the algorithm can reference n previous rows to generate the next row. Finally, for associative relationships, the algorithm groups the special rows of a table that forms associations into the parent row, which allows them to have arbitrary correlations with each other. For example, if the child rows describe a prescription list, then each medicine can form arbitrary correlations with all other medicines.

3.3.5 Evaluation

We group evaluation metrics into two main categories: utility and privacy. For utility, we can use synthetic metrics (e.g., Kullback-Leibler divergence), real-world use cases (e.g., replicate an analysis performed on the original data), or train a model on synthetic data and evaluate its performance. The main synthetic metrics are histograms [33], target value class distribution [33], Principal Component Analysis (PCA) maps, χ^2, correlation matrices [33], and Kullback-Leibler (KL) Divergence. Modelling metrics [30] use a classifier to train for a target variable and measure the behavior of classifiers trained on synthetic data versus ones trained on original data, with the specific metrics depending on the methodology.

For privacy, there are two classes of attacks that can be performed on synthetic data: membership inference and attribute disclosure. Classifiers may be trained to predict if a record is in the dataset or not (membership inference), or to predict a sensitive attribute of a record (attribute disclosure). Membership inference is damaging if the nature of the dataset is such that an individual is found to be part of it, their reputation may be harmed. Therefore, for certain benign datasets, it might be considered not important. Attribute disclosure, on the other hand, should always be avoided, as it implies that the data of the individual itself are exposed. The extent to which it should be avoided depending on the part of the dataset is still an unexplored question, with current algorithms treating all attributes equally as important.

3.3.6 Scalability

Processing sensitive datasets introduces additional constraints in terms of compute resources. Provisioning GDPR-compliant infrastructure is costly and time-consuming, which often limits the researcher into using a single machine for processing a dataset. Moreover, medical data is often large and complex, which exacerbates the problem. In sum, efficient and scalable algorithms are required to process medical data in a timely manner, while being subject to memory and computing limitations.

The Pasteur system paper introduces methods to enable arbitrary scalability and parallelization in synthesis, through three key contributions. An example of partitioning under Pasteur can be seen in Fig. 3.3. First, it distills and optimizes how data should be stored in memory and disk, such that crossing between the two mediums is efficient and the maximum number of rows can be processed at a time. Then, it introduces a partitioning and parallelization architecture that scales with the available system cores linearly and provides a stable parallelization factor, while limiting the memory usage of synthesis to a predetermined amount. This maximizes compute utilization while preventing crashes due to memory spikes and allowing for processing larger than memory data. Following, Pasteur introduces a novel algorithm for sampling marginals from a dataset. This algorithm is optimized to be faster from a single-core perspective, and then parallelized linearly with the rest of the architecture. Since marginals are the only operation that accesses the whole dataset in

	Non-Partitioned Tables		Partitioned Tables		
PID	Information	Admissions	Prescription	ICU Vitals	
1	1 Row	5 Rows	23 Rows	1523 Rows	**Partition 1**
2	1 Row	9 Rows	102 Rows	5812 Rows	Patients 1-3
3	1 Row	2 Rows	57 Rows	503 Rows	←
4	1 Row	7 Rows	65 Rows	7432 Rows	**Partition 2**
5	1 Row	6 Rows	323 Rows	3244 Rows	Patients 4-6
6	1 Row	2 Rows	34 Rows	321 Rows	←

Fig. 3.3 Partitioning under Pasteur

most Probabilistic Graphical Model (PGM) algorithms, the combination of the three aforementioned contributions allows for end-to-end synthesis of large datasets in a scalable manner while having access to limited compute resources.

3.4 Future Work

In this chapter, we presented a broad overview of the research area of data synthesis, which is an upcoming field focused on privacy preservation and the liberation of data access. In recent years, the field has made great strides in the types of data that can be synthesized, privacy guarantees, and scalability. However, the field still struggles with the accuracy to privacy trade-off, where increasing the privacy guarantees or using algorithms with strong privacy guarantees results in a data fidelity that is unsuitable for many applications. We attribute this lack of accuracy to three main areas: lack of specialization to specific data structures, lack of domain knowledge integration, and a general inefficiency in how privacy guarantees are applied (e.g., all data is equally protected, even if some data is less sensitive than others, privacy guarantees might be too stringent). Then, the lack of specialized metrics concerning the utility and accuracy of resulting data, especially in the relational data synthesis area, makes it hard to quantify the quality and privacy of the synthesized data.

When working with domain specific data (e.g., patient trajectories), certain patterns arise that, given some domain knowledge, can be used to model the data more efficiently, leading to higher accuracy at the same privacy budget. For example, in patient trajectories, when the patients go through treatment plan updates, the dates are monotonically increasing and they do so in a very specific manner. If the algorithms are aware of that, they can simplify the dates to, for example, day span intervals, which removes invalid solutions from the search space and simplifies yielding higher accuracy results. Another example is drug dosages, where for each drug there is only a handful of them. If we treat them as ordinal categorical values versus numerical values, we end up with dosages that look more realistic and are simpler to produce.

In the same vein, domain knowledge can be used to inform the data structures that will be used for the algorithms. Doctors can hand model how dosages are converted to categorical values, removing typos and other errors that might arise from automatic conversion, resulting in values with lower cardinality that yield better accuracy. Moreover, domain experts can specify which aspects of complex columns should be kept, for example, given a date, we might care about the time of day or season, and not the date as an absolute value.

Finally, not all data is equally sensitive. When using approaches such as differential privacy, all data is treated the same, where we end up "wasting" privacy resources or unnecessarily harming data quality. For example, whether a patient has cancer is more important to protect than their height. By strategically analyzing the data, if we are able to make the case that some subset of it is not necessary to protect, or is less sensitive, we can design algorithms that only focus on sensitive data, giving great accuracy to data which is not sensitive, and freeing privacy resources such as the privacy budget to be used on data that is actually sensitive.

For evaluation metrics, there is little work in current research for metrics that can measure temporal and associative qualities, ones that measure the joint-distribution quality. Furthermore, there are too few relational datasets that can be used for synthesis, especially in the medical domain. As a result, there are no widespread and standardized synthesis benchmarks that can be used to cross-compare synthesis algorithms.

3.5 Conclusions

In this chapter, we have presented the landscape of data synthesis, focusing more specifically on privacy-aware relational data synthesis. We have identified five key areas in data synthesis: privacy-awareness, tabular data synthesis, relational data synthesis, scalability, and evaluation. Within those areas, we identified two core challenges: lack of accuracy due to inefficiency and lack of specialization, and lack of specialized metrics, for which we proposed a set of future research directions.

References

1. Liu B, Ding M, Shaham S, Rahayu W, Farokhi F, Lin Z (2021) When machine learning meets privacy: a survey and outlook. ACM Comput Surv 54(2):31:1–31:36
2. Jang T, Zheng F, Wang X (2021) Constructing a fair classifier with generated fair data. Proc AAAI Conf Artif Intell 35(9):7908–7916
3. McKenna R, Mullins B, Sheldon D, Miklau G (2022) AIM: an adaptive and iterative mechanism for differentially private synthetic data. CoRR abs/2201.12677:14
4. Yoon J, Drumright LN, van der Schaar M (2020) Anonymization through data synthesis using generative adversarial networks (ADS-GAN). IEEE J Biomed Health Inform 24(8):2378–2388
5. Zhang Z, Wang T, Li N, Honorio J, Backes M, He S, Chen J, Zhang Y (2021) PrivSyn: differentially private data synthesis, pp 929–946
6. Mckenna R, Sheldon D, Miklau G (2019) Graphical-model based estimation and inference for differential privacy. In: Proceedings of the 36th international conference on machine learning, PMLR, pp 4435–4444
7. Cai K, Wei J, Lei X, Xiao X (2021) Data synthesis via differentially private Markov random fields. VLDB 14:13
8. Dwork C (2006) Differential privacy, languages and programming. In: Bugliesi M, Preneel B, Sassone V, Wegener I (eds) Automata. Springer, Berlin, Heidelberg, pp 1–12
9. Jia Z, Zeng X, Duan H, Lu X, Li H (2020) A patient-similarity-based model for diagnostic prediction. Int J Med Informatics 135:104073
10. Chaudhuri K, Monteleoni C, Sarwate AD (2011) Differentially private empirical risk minimization. J Mach Learn Res JMLR 12:1069–1109
11. Ramos-Pollán R, Guevara-López MA, Suárez-Ortega C, Díaz-Herrero G, Franco-Valiente JM, Rubio-Del-Solar M, González-De-Posada N, Vaz MA, Loureiro J, Ramos I (2012) Discovering mammography-based machine learning classifiers for breast cancer diagnosis. J Med Syst 36(4):2259–2269

12. Ching T, Himmelstein DS, Beaulieu-Jones BK, Kalinin AA, Do BT, Way GP, Ferrero E, Agapow PM, Zietz M, Hoffman MM, Xie W, Rosen GL, Lengerich BJ, Israeli J, Lanchantin J, Woloszynek S, Carpenter AE, Shrikumar A, Xu J, Cofer EM, Lavender CA, Turaga SC, Alexandari AM, Lu Z, Harris DJ, DeCaprio D, Qi Y, Kundaje A, Peng Y, Wiley LK, Segler MHS, Boca SM, Swamidass SJ, Huang A, Gitter A, Greene CS (2018) Opportunities and obstacles for deep learning in biology and medicine. J R Soc Interface 15(141):20170387
13. Sharafoddini A, Dubin JA, Lee J (2017) Patient similarity in prediction models based on health data: A scoping review. JMIR Med Inform 5(1):17
14. Rajkomar A, Oren E, Chen K, Dai AM, Hajaj N, Hardt M, Liu PJ, Liu X, Marcus J, Sun M, Sundberg P, Yee H, Zhang K, Zhang Y, Flores G, Duggan GE, Irvine J, Le Q, Litsch K, Mossin A, Tansuwan J, Wang D, Wexler J, Wilson J, Ludwig D, Volchenboum SL, Chou K, Pearson M, Madabushi S, Shah NH, Butte AJ, Howell MD, Cui C, Corrado GS, Dean J (2018) Scalable and accurate deep learning with electronic health records. npj Digit Med 1(1):1–10
15. Goncalves A, Ray P, Soper B, Stevens J, Coyle L, Sales AP (2020) Generation and evaluation of synthetic patient data. BMC Med Res Methodol 20(1):108
16. Chen RJ, Lu MY, Chen TY, Williamson DFK, Mahmood F (2021) Synthetic data in machine learning for medicine and healthcare. Nature Biomed Eng 5(6):493–497
17. Lee J, Sun J, Wang F, Wang S, Jun CH, Jiang X (2018) Privacy-preserving patient similarity learning in a federated environment: development and analysis. JMIR Med Inform 6(2):e20
18. Zhu Z, Yin C, Qian B, Cheng Y, Wei J, Wang F (2016) Measuring patient similarities via a deep architecture with medical concept embedding. In: 2016 IEEE 16th international conference on data mining (ICDM), pp 749–758
19. Pai S, Bader GD (2018) Patient similarity networks for precision medicine. J Mol Biol 430(18 Pt A):2924–2938
20. Keogh RH, Seaman SR, Barrett JK, Taylor-Robinson D, Szczesniak R (2018) Dynamic prediction of survival in cystic fibrosis: a landmarking analysis using UK patient registry data. Epidemiology 30(1):29–37
21. Andrinopoulou ER, Rizopoulos D, Geleijnse ML, Lesaffre E, Bogers AJJC, Takkenberg JJM (2015) Dynamic prediction of outcome for patients with severe aortic stenosis: application of joint models for longitudinal and time-to-event data. BMC cardiovascular disorders
22. Kim J, Kim K, Callaway CW, Doh K, Choi J, Park J, Jo YH, Lee JH (2017) Dynamic prediction of patient outcomes during ongoing cardiopulmonary resuscitation. Resuscitation 111:127–133
23. Sun J, Wang F, Hu J, Edabollahi S (2012) Supervised patient similarity measure of heterogeneous patient records. SIGKDD Explor Newsl 14(1):16–24
24. Brown SA (2016) Patient similarity: emerging concepts in systems and precision medicine. Front Physiol 7
25. Suo Q, Ma F, Yuan Y, Huai M, Zhong W, Gao J, Zhang A (2018) Deep patient similarity learning for personalized healthcare. IEEE Trans Nanobiosci 17(3):219–227
26. Sweeney L (2002) k-anonymity: a model for protecting privacy. Int J Uncertain Fuzziness Knowl Based Syst 10(5):557–570
27. Machanavajjhala A, Kifer D, Gehrke J, Venkitasubramaniam M (2007) L-diversity: privacy beyond k-anonymity. ACM Trans Knowl Discov Data 1(1):3–es
28. Li N, Li T, Venkatasubramanian S (2007) t-closeness: privacy beyond k-anonymity and l-diversity. In: 2007 IEEE 23rd international conference on data engineering, pp 106–115
29. Yoon J, Jordon J, van der Schaar M (2019) PATE-GAN: generating synthetic data with differential privacy guarantees. In: International conference on learning representations
30. Xu L, Skoularidou M, Cuesta-Infante A, Veeramachaneni K, Wallach H (2019) Modeling tabular data using conditional GAN. In: Larochelle H, Beygelzimer A, d'Alché-Buc F, Fox E, Garnett R (eds) Advances in neural information processing systems. Curran Associates Inc
31. Ho S, Qu Y, Gu B, Gao L, Li J, Xiang Y (2021) DP-GAN: differentially private consecutive data publishing using generative adversarial nets. J Netw Comput Appl 185:103066
32. Fang ML, Dhami DS, Kersting K (2022) DP-CTGAN: differentially private medical Data Generation Using CTGANs. In: Abidi SSR, Abidi S (eds) Artificial intelligence in medicine. Springer International Publishing, pp 178–188

33. Tao Y, McKenna R, Hay M, Machanavajjhala A, Miklau G (2021) Benchmarking differentially private synthetic data generation algorithms. CoRR abs/2112.09238
34. McKenna R, Miklau G, Sheldon D (2021) Winning the NIST Contest: a scalable and general approach to differentially private synthetic data. contest
35. Zhang J, Cormode G, Procopiuc CM, Srivastava D, Xiao X (2017) PrivBayes: private data release via bayesian networks. ACM Trans Database Syst 42(4):25:1–25:41
36. Song S, Chaudhuri K, Sarwate AD (2013) Stochastic gradient descent with differentially private updates. In: 2013 IEEE global conference on signal and information processing, pp 245–248
37. Papernot N, Song S, Mironov I, Raghunathan A, Talwar K, Úlfar Erlingsson (2018) Scalable private learning with pate. arXiv
38. Cai K, Xiao X, Cormode G (2023) Privlava: Synthesizing relational data with foreign keys under differential privacy. Proc ACM Manag Data 1(2):142:1–142:25
39. Patki N, Wedge R, Veeramachaneni K (2016) The Synthetic Data Vault. In: 2016 IEEE international conference on data science and advanced analytics (DSAA), pp 399–410

Part II
Storage and Processing

Chapter 4
Comprehensive Approach to Feature Selection

Uchechukwu Fortune Njoku⬤, Alberto Abelló⬤, Besim Bilalli⬤,
and Gianluca Bontempi⬤

Abstract Feature selection (FS) is essential in machine learning (ML) to improve model accuracy, interpretability, as well as execution time. Existing FS techniques are categorized across multiple perspectives, including data characteristics, label dependency, search strategies, feature interactions, evaluation methods, and design choices. This categorization provides a structured framework for understanding their strengths, limitations, and practical applications. This chapter provides a comprehensive overview of FS, covering foundational techniques, a detailed taxonomy, state-of-the-art methods, and emerging trends. The introduced taxonomy is compared with existing ones, providing novel insights into the field. Also, an example demonstrating how different FS techniques can be applied to ML is presented, highlighting their capabilities in addressing ML challenges, including improving model performance, reducing complexity, and lowering training costs. Open problems such as stability under data perturbation, scalability for high-dimensional datasets, and multi-objective optimization are explored, along with practical discussions of FS tools and libraries to bridge theory and practice. In summary, this chapter combines foundational knowledge with future directions, serving as a guide for researchers and practitioners, equipping them to make informed decisions when selecting and implementing FS strategies to address complex real-world ML challenges.

Keywords Feature selection · Dimensionality reduction · Machine learning · Model interpretability · Optimization

U. F. Njoku · A. Abelló · B. Bilalli
Universitat Politècnica de Catalunya, Barcelona, Spain
e-mail: uchechukwu.fortune.njoku@upc.edu

A. Abelló
e-mail: alberto.abello@upc.edu

B. Bilalli
e-mail: besim.bilalli@upc.edu

U. F. Njoku · G. Bontempi
Université libre de Bruxelles, Brussels, Belgium
e-mail: gianluca.bontempi@ulb.be

© The Author(s) 2026

G. Dejaegere et al. (eds.), *Data Engineering for Data Science*,
https://doi.org/10.1007/978-3-032-18765-9_4

4.1 Overview

Data is the foundation of many technological innovations and advancements, especially in Machine Learning (ML) and Artificial Intelligence (AI). As these fields continue to grow, fueled by the availability of big data, a crucial challenge emerges: determining which data is most relevant for a given task. Not all available data is useful, and distinguishing between what is merely abundant and what is valuable is essential. This transition from *big* to *good* data raises key questions: What defines *good* data? How do we measure *relevance*? These challenges are addressed through Feature Selection (FS)—techniques within ML that identify the most relevant data for a given task based on defined criteria. This challenge is explored in depth in this chapter.

An end-to-end ML pipeline usually has three sub-pipelines: feature, training, and inference pipelines, as shown in Fig. 4.1. FS falls under the feature pipeline, also called data preprocessing. At this stage, the focus is on transforming the raw data into a suitable form for model training. Proper FS before model training provides several advantages, such as improved model performance [1, 2], better data and model understanding [3], faster training times, and reduced computational costs [4]. Therefore, FS is essential to ensure that the most is made of the available data.

However, FS presents several challenges. One of the primary difficulties lies in choosing the appropriate technique for a specific dataset and learning algorithm. Fundamentally, FS involves defining *relevance*, which is the criterion by which features are compared; however, what constitutes relevance is not one-size-fits-all. Different use cases may prioritize different criteria for relevance. For instance, privacy might be a critical factor in healthcare, whereas in the construction industry, cost-efficiency may take precedence. The flexible application of FS methods, allowing for evolving definitions of relevance, remains an ongoing research area in ML. Additionally, as data scales, the complexity of FS methods increases, raising concerns about scalability and performance. Regardless of these challenges, FS remains an indispensable necessity in ML.

We illustrate the importance of FS with an anonymized but real telecommunications company tasked with building ML models to predict two outcomes: whether a user will add a card to their existing contract, and whether a user will upgrade

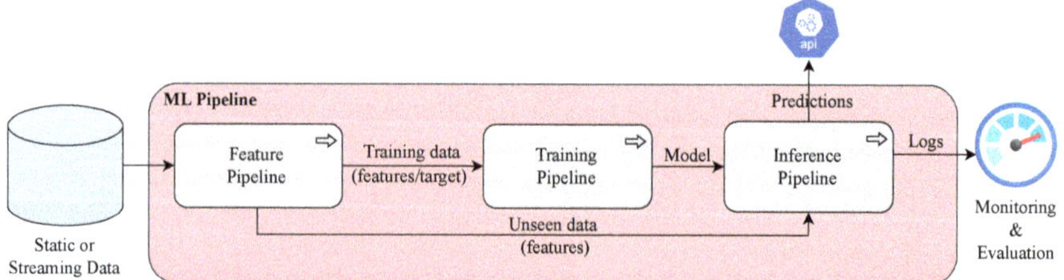

Fig. 4.1 Machine learning end-to-end pipeline

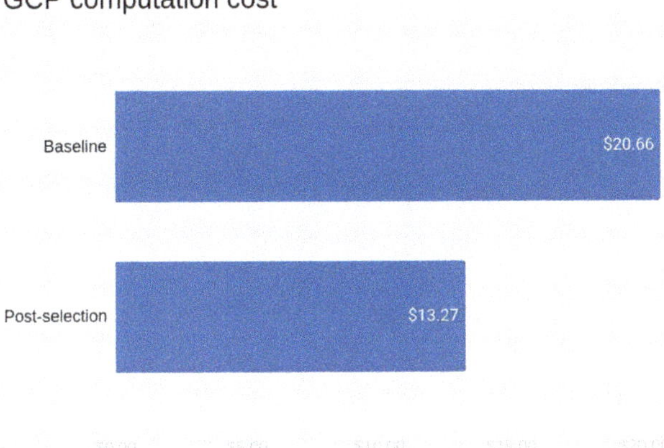

Fig. 4.2 GCP computational cost for "add card" task

to a higher tariff. For both tasks, the company relied on a shared tabular dataset containing 280 features of customer information. Without FS, the feature pipeline through feature engineering (creating new features from existing ones) generated 697 features for the "add card" task and 247 features for the "upscale" task. The models, built using XGBoost [5], achieved Area Under the ROC Curve (AUC) scores of 50.53% and 79.08%, respectively. However, auditing the models[1] revealed that only a fraction of these features were used. This implies that the model training was overly complex and relied on excess features. When we incorporated FS into the pipeline, the number of features was reduced to 150 and 147, respectively, with AUC scores of 51.87% and 79.12% (i.e., slightly better in both cases). This demonstrates that adding more features does not necessarily improve model quality but rather the opposite.

Furthermore, this work was done on the Google Cloud Platform (GCP), so the financial cost of computation was relevant and measured. Focusing on the "add card" and implementing feature set size reduction, we observed a reduction in monetary costs of up to 35% as shown in Fig. 4.2. This use case highlights the practical benefits of FS, particularly in reducing the size of the feature set and, consequently, the computational cost without compromising model performance.

Hence, this chapter is devoted to FS, its advantages, challenges, state-of-the-art, and future directions. Its contributions are:

- An overview of FS, including the various existing techniques, their classifications and supporting implementations. Outlining their advantages, limitations, and applicability to different use cases.

---

[1] The XGBoost class has a *feature_importances_* method for checking each feature's importance.

- A look at the current state-of-the-art in FS, identifying key trends and areas where further research is needed.
- An illustrative example showcasing the practical impact of 41 FS techniques on a binary classification ML task.
- Demonstration of the impact of FS techniques on execution time, feature set size, model accuracy and AUC.

4.2 Feature Selection Landscape

FS involves four key steps: (1) feature subset generation, (2) subset evaluation, (3) stopping criteria, and (4) subset selection. It can be examined from various perspectives both within and beyond these steps, including data characteristics, evaluation approach, number of objectives, and implementation choices. This section presents the full landscape of FS, offering a structured taxonomy that categorizes the field into its key perspectives as shown in Fig. 4.3. Through this taxonomy, we explore the existing methods, challenges, and trade-offs involved, providing a detailed framework that helps navigate the complexities of selecting the most relevant features for a given ML task.

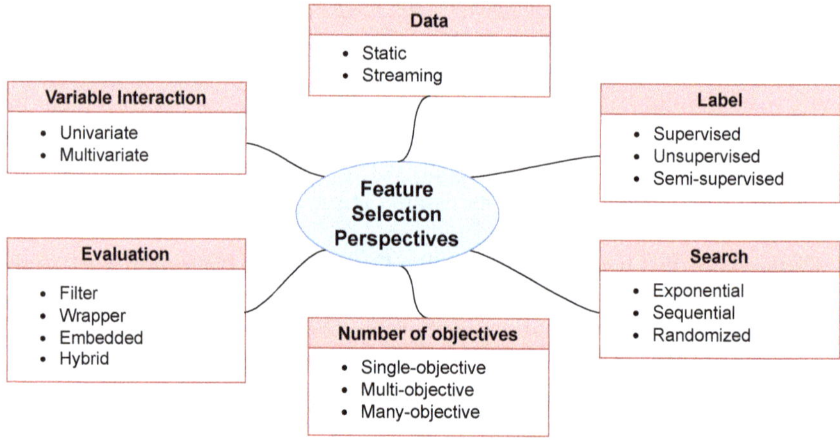

Fig. 4.3 Feature selection taxonomy based on six perspectives

4.2.1 Data Perspective

Data in ML is diverse and can vary significantly in format and structure. Specifically, data can be broadly categorized into static and streaming [6], each posing unique challenges and requirements for FS techniques.

Static. The majority of FS techniques have been designed with static data in mind. Static datasets are typically considered complete at the time of analysis, with a fixed number of features and instances that do not evolve over time [6]. This assumption simplifies the process, as there is no need for continuous choice reconsideration. However, with the growing prevalence of big data, static datasets may no longer suffice for certain applications that require more flexibility and adaptability.

Stream. In many real-world applications, data is continuously generated, leading to evolving datasets with either new instances or new features. Streaming data can thus be subdivided into two scenarios: data streams, where additional data points (instances) are introduced over time, and feature streams, where new features are added or even removed as the dataset evolves [7]. FS techniques that accommodate these changes are referred to as streaming or online FS. The goal is to dynamically identify and update the subset of features relevant to the ML task as new data arrives without retraining from scratch.

Various FS methods have been developed to address the demands of streaming data, such as Grafting, Alpha-Investing [8], Online Streaming Feature Selection (OSFS), Scalable and Accurate Online Approach (SAOLA), Fast-OSFS, OFS-A3M, and OFS-Density [7], each providing distinct approaches to adapting FS to streaming environments.

4.2.2 Label Perspective

The choice of an FS method for a given use case depends largely on the dataset [6]. Some methods require the presence of a target variable, while others do not. Based on this distinction, FS methods are categorized as supervised, unsupervised, or semi-supervised.

Supervised Supervised methods require a target variable in the dataset, as the relevance of a subset of features is evaluated with regard to this target. These are the most commonly used techniques [9]. Examples include Joint Mutual Information (JMI), Minimum Redundancy Maximum Relevance (mRMR), and Correlation-Based Filter (CBF) [6].

Unsupervised Unsupervised methods identify similarities among features without requiring a target variable [10]. These methods are particularly valuable when labelling data for a target variable is either impractical or costly. A common approach is to use distance metrics to find features that preserve the intrinsic data structure or similarity. Unsupervised methods are commonly used alongside unsupervised ML techniques like clustering. Examples include the Laplacian score and Spectral FS (based on spectral graph theory) [11].

Semi-supervised   Labeling data can be difficult or expensive, so in some cases, only part of the dataset is labelled. In this case, both the labelled and unlabeled data are used to evaluate the relevance of features [12]. Some examples of semi-supervised FS are Constraint score [13] and using ensemble learning [14].

4.2.3   Search Perspective

The initial step in the FS process is feature subset generation. In a dataset containing n features, there are $2^n - 2$ possible subsets of features, causing the search space to grow exponentially with the number of features. Consequently, defining an effective strategy for exploring this solution space is essential to identifying all-around efficient solutions. Based on the search strategy employed, FS methods can be classified as exponential, sequential, or randomized approaches.

Exponential   This strategy systematically explores an increasing number of subsets, which grows exponentially with respect to the number of features. It can guarantee the best solution since it traverses most of the solution space. However, it is computationally intensive and infeasible even for a moderate number of features [15]. An example is the branch and bound method, which structures the search space as a tree and explores it top-down with backtracking [16].

Sequential   These simple greedy algorithms traverse the solution space by consecutively updating the feature subset with features that improve or maintain the evaluation criteria. The two variants of this strategy are Sequential Forward Selection (SFS) and Sequential Backward Selection (SBS). The former begins with an empty set and, at each iteration, adds features that improve or preserve the evaluation criteria until a stopping criterion is satisfied, e.g., the maximum number of iterations or feature subset size. The latter conversely begins with the complete set of features and, at each iteration, drops features that decrease the evaluation criterion [15]. Although the overall best solution is not guaranteed, this method is clearly computationally cheaper than the exponential approach.

Randomized   Since traversing the entire solution space is infeasible, this search strategy applies heuristics to obtain near-optimal results. It starts with an initial set of often randomly generated feature subsets called population and iteratively updates it based on the heuristics until a stopping criterion is reached [15]. Genetic algorithms, particle swarm optimization, and bee colony algorithms are examples of randomized search strategies used in FS [17].

4.2.4 Evaluation Perspective

Subset evaluation is the second step in the FS process. Among the numerous FS techniques, the most common categorization is how the subsets of features are evaluated. Depending on whether and how ML algorithms are involved, the techniques are classified into filter, wrapper, embedded, and hybrid methods.

Filter   These methods evaluate and select features independently of any ML algorithm by relying solely on the inherent characteristics and associations among features to determine their relevance. This approach yields a subset of features that is highly generalizable across various ML algorithms but may not be optimal [1]. The main advantages of filter methods, beyond generalizability, are their faster computation time compared to other methods and their robustness to overfitting. However, since they are model-agnostic, the selected features may not be optimal for a specific ML algorithm, and they risk selecting redundant features if inter-feature associations are not considered.

Some filter methods use information gain, correlation, or distance measures to evaluate the association between each predictive feature and the target feature. Other techniques, such as Joint Mutual Information (JMI), account for relationships among the predictive features to reduce redundancy, thereby preventing the selection of highly correlated or redundant features [18].

Wrapper   These methods select features by evaluating generated subsets based on the performance of an ML model (e.g., accuracy, AUC) built using a specific learning algorithm [15]. This iterative approach involves training and testing the model on various subsets of features, ensuring that the selected subset is optimized for the chosen model. However, while this process tailors the selected features to the specific learning model, it lacks generalizability since the selected features may not perform as well with other models. Moreover, wrapper methods can be computationally expensive due to the iterative nature of model-building for each subset, especially when dealing with large datasets or complex models. Despite these challenges, wrapper methods offer a significant advantage by incorporating cross-validation [19], which helps mitigate the risk of overfitting, ensuring that the selected features generalize well with unseen data. Additionally, they tend to produce more accurate results compared to filter methods, as they directly optimize the FS for the learning algorithm used.

Embedded   These are ML algorithms that perform FS intrinsically [20]. They are efficient because they use the target learning algorithm to select features, optimizing them for the algorithm itself. Also, embedded methods do not need to be repeatedly trained. Hence, they offer better computational performance than wrapper methods. However, as with wrapper methods, they lack generalizability since the selected features are tailored to the learning algorithm. Examples include tree-based algorithms like decision trees, which prune irrelevant features using criteria such as the Gini index or Entropy.

Hybrid   These methods aim to leverage the strengths of different FS approaches by combining two or more evaluation perspectives (namely filter, wrapper, and embedded methods) to enhance performance [21]. These methods capitalize on the computational efficiency of filters, the accuracy of wrappers, and the embedded models' inherent FS capability while minimizing the weaknesses of each approach. For example, a hybrid method might begin with a fast filter method to quickly reduce the dimensionality of an extremely large dataset, followed by a more fine-tuned embedded or wrapper method to refine the feature set further. This two-step process balances computational efficiency and accuracy, making hybrid methods particularly useful when dealing with large, complex datasets. By combining techniques, hybrid methods strive to achieve both generalizability and algorithm-specific optimization, offering a more flexible and effective solution.

4.2.5 Variable Interaction Perspective

Still focusing on the subset evaluation step, techniques can be broadly classified based on whether they account for interactions between features or not. This distinction gives rise to two categories: univariate and multivariate methods.

Univariate   Univariate methods evaluate the relevance of each feature individually based on its relationship with the target variable without considering interactions between features. These methods rank features by a statistical measure such as correlation, chi-square, or simple linear regression, selecting those that exhibit the strongest relationship with the target [18]. As a result, univariate techniques are computationally efficient and simple to implement, making them popular for initial filtering in high-dimensional datasets. However, they do not account for redundancy or interactions between features, which may limit their effectiveness in capturing the full complexity of the data. Despite this limitation, univariate methods are useful for identifying the most important individual features early in the process.

Multivariate   Multivariate methods evaluate the relevance of a subset of features by considering interactions between them rather than assessing each feature in isolation [18]. Unlike univariate methods, multivariate approaches deal with redundancy by accounting for interactions between features, offering a more comprehensive analysis of their importance. This makes multivariate methods more suitable for complex datasets where feature interactions play a critical role. These methods tend to be computationally more intensive than univariate techniques but result in a more balanced feature set, reducing redundancy and ensuring that selected features work well together in predicting the target variable. Multivariate methods offer a more holistic approach by capturing the collective contribution of features, often leading to improved model accuracy and interpretability. Techniques such as JMI, mRMR, and CBF are examples of multivariate approaches.

4.2.6 Number of Objectives Perspective

A fundamental aspect of the subset evaluation step of FS is determining what makes a feature relevant. This perspective involves prioritizing the number of objectives and balancing competing objectives. Based on this, FS techniques are categorized into single-objective, multi-objective, and many-objective approaches.

Single-objective The single-objective category is the simplest and most widely used approach in FS. In this approach, the relevance of a subset of features is evaluated based on a single criterion, making it straightforward to implement and interpret. This is found in many filter methods, where a specific metric, such as the Gini index or information gain, is employed to assess feature relevance. With wrapper methods, when only one predictive metric, such as the accuracy score, is used to measure feature relevance, it is classified as single-objective. By focusing on just one criterion, single-objective methods offer simplicity and lower computational cost. Still, they may overlook the trade-offs between other important aspects like model complexity or fairness, which are better captured in more complex methods.

Multi-objective A more advanced and satisfactory approach considers multiple criteria, reflecting the complexity of real-world data analysis tasks. When two or three criteria are used to define and evaluate the relevance of a subset of features, it is referred to as multi-objective FS [22]. This approach acknowledges that no single metric can fully capture all aspects of what makes a feature set useful for a task. The criteria used in multi-objective FS can be combined in several ways. One approach is to merge multiple objectives into a single composite metric. For example, the mRMR method combines relevance, measured by mutual information with the target variable, and redundancy, which seeks to minimize correlations among the selected features. Another way of combining objectives is to assign weights to each criterion, allowing for a flexible trade-off between them.
Alternatively, rather than combining objectives into a single metric, multi-objective optimization techniques can be used to evaluate multiple criteria independently but simultaneously [23]. These techniques generate a set of solutions that offer different trade-offs among the objectives, providing more insight and flexibility in FS. This approach is especially useful in complex scenarios where optimizing for one criterion may negatively impact another.

Many-objective Many-objective FS builds on the principles of the multi-objective FS but considers a larger number of objectives, typically ranging from four to fifteen [24]. While multi-objective approaches traditionally rely on dominance principles, they encounter substantial limitations when applied to scenarios with higher-dimensional objective spaces. These challenges include a large portion of the population becoming non-dominated, computationally expensive diversity evaluations, and inefficient recombination operations, necessitating specialized techniques for handling higher-dimensional objective spaces. To address these issues, specialized algorithms such as NSGA-III [24], which uses a reference-point based non-dominated sorting approach [24] have been developed for many-objective optimization. Although this adds complexity, these methods are able to balance a larger number of competing objectives and allow for a more comprehensive and nuanced definition of feature relevance by realistically accounting for the multiple objectives that may impact a given ML task.

4.2.7 Tools and Libraries

Most data analysis platforms, like WEKA, RapidMiner, and KNIME, offer FS as part of their broader analytics capabilities, allowing users to integrate these methods within their ML pipelines. These tools typically provide a variety of techniques, including filter, wrapper, and embedded methods, along with search strategies like sequential forward or backward selection. Despite their versatility, they tend to focus on general techniques without fully addressing more specialized needs, such as multi-objective or many-objective.

In the Python ecosystem, numerous libraries are available to support FS, each offering distinct functionalities. These libraries allow for greater range and customization, enabling users to implement a wider array of methods. Below is an overview of some commonly used Python libraries and the specific functionalities they provide, highlighting their contributions to various stages of the FS process.

- SkFeature[2]

  - Filter Methods: Implements various statistical techniques to evaluate the relevance of features independently.
  - Sequential Methods (wrapper): Utilizes forward and backward search techniques to identify the optimal subset of features.

- MLXtend[3]

  - Exponential search: Performs a comprehensive search over all possible subsets of features.
  - Sequential Methods (wrapper): Includes forward and backward sequential FS.

- JmetalPy[4]

  - Randomized Methods: Uses evolutionary algorithms and metaheuristics to select features.

---

[2] https://github.com/jundongl/scikit-feature.

[3] https://rasbt.github.io/mlxtend/api_subpackages/mlxtend.feature_selection.

[4] https://github.com/jMetal/jMetalPy.

- SKLearn[5]

  - Filter Methods: Provides techniques such as mutual information and chi-squared tests.
  - Sequential Methods (wrapper): Implements Recursive Feature Elimination (RFE) and other sequential strategies.
  - Embedded Methods: Offers FS as part of models like Decision Trees and Lasso regression.

Although many tools and libraries are available, a dedicated library that comprehensively covers the entire spectrum of methods remains unavailable. Such a library would provide a unified approach to implement and compare different FS techniques, effectively addressing various needs and paradigms.

4.3   State of the Art

As data volumes continue to grow exponentially, FS remains an essential research area for optimizing ML models and enhancing computational efficiency. Over the years, significant progress has been made in addressing the challenges outlined in the previous section, leading to the development of various methods and techniques. This section provides an in-depth review of the state-of-the-art approaches across the six key perspectives, illustrating how these methods have evolved and their impact on modern ML applications.

4.3.1   Data Perspective

The majority of research on FS has focused on static datasets, which is the focus of the rest of this section. However, significant progress has been made in developing FS techniques for streaming data, where all features are not known a priori. One of the earliest approaches, Alpha-investing [8], proposed in 2005, used a dynamic threshold to control the inclusion of new features. Subsequently, the OSFS algorithm [25] was proposed, which utilizes the Markov blanket concept from information theory. OSFS classifies features into four categories: irrelevant, redundant, weakly relevant but non-redundant, and strongly relevant. It operates through a two-step process comprising online relevance analysis and online redundancy analysis. Other proposals include the use of rough set theory, information theory-based techniques, Bayesian Network, symmetrical uncertainty, evolutionary algorithms, windowing, and regularisation [7]. Streaming FS methods have been successfully applied in various domains, including healthcare and bioinformatics, computer vision, social networks, and the Internet of

---

[5] https://scikit-learn.org/stable/modules/feature_selection.html

Things (IoT) [7]. However, a comprehensive benchmark comparing these streaming FS methods is still lacking.

4.3.2 Label Perspective

A significant portion of the work in FS has focused on labelled data, primarily because supervised ML tasks and algorithms are more prevalent compared to semi-supervised or unsupervised approaches, but also because the target variable plays a crucial role in measuring feature relevance and guiding the selection process. A comprehensive tutorial on supervised FS is provided in [9] and presents a taxonomy that classifies methods into three broad categories: feature ranking, subset selection, and embedded methods, which correspond loosely to filter, wrapper, and embedded techniques.

Despite their effectiveness, these methods face limitations when labelling data becomes too costly or impractical. In such cases, semi-supervised and unsupervised methods become more viable alternatives. In a recent survey of semi-supervised FS methods [12], approaches are categorized both by their interaction with learning algorithms and by the specific semi-supervised learning techniques they employ. This dual taxonomy reflects the diversity of methods in this space, yet most semi-supervised FS approaches align with broader categories of semi-supervised ML, such as generative models, self-training, co-training, semi-supervised support vector machines (S^3VM), and graph-based techniques. Among these, graph-based methods have gained the most prominence, largely due to their ability to capture relationships between data points effectively when labels are sparse.

When no labels are available, unsupervised methods come into play, offering distinct advantages: (1) they perform well without prior knowledge of the data, and (2) they reduce the risk of overfitting [11]. Since these methods operate without any prior information about underlying patterns, they rely on intrinsic properties of the data, such as variance, separability, and distribution, to assess feature relevance [10]. They can also be categorized using the common filter, wrapper, hybrid, and embedded evaluation perspectives [10, 11]. Noteworthy examples include the Laplacian Score [26], which evaluates features based on spectral similarity, and Feature Selection using Feature Similarity (FSFS) [27], a multivariate approach grounded in statistical analysis.

4.3.3 Search Perspective

In ML, FS can be framed as a search problem where the objective is to find a subset of features that minimizes model complexity while maximizing model performance or other important criteria. An exhaustive approach, which evaluates every possible feature subset, provides the most thorough solution but is often impractical due to the exponential growth of the search space. For instance, even when limited to only 15 features, there are already 32,766 possible subsets [28]. As a result, various strategies have been developed to explore the search space more efficiently. Techniques like branch-and-bound [29] or beam search [28] restrict the number of subsets evaluated while still covering a substantial part of the search space. However, due to their high computational cost, these methods have seen limited use in practice.

In contrast, sequential search methods offer a more computationally feasible approach, employing greedy algorithms that add or remove one or more features iteratively. These methods, which include SFS, SBS, and their floating variants like SFFS and SBFS, are faster and typically exhibit quadratic complexity [30]. Floating variants are particularly beneficial in cases where nonmonotonic evaluation criteria are applied, as they allow backtracking to revisit previously excluded features. Extensive experimental evaluations have demonstrated the effectiveness of these methods. For example, early work by [31] showed that SFS and SBS improved accuracy for cloud pattern classification, though SBS did not consistently outperform SFS. More recent studies, such as the one focused on customer churn prediction [32], have reinforced these findings, highlighting that SBS and SBFS tend to yield the best accuracy and AUC scores on specific datasets.

Despite their practicality, greedy search methods may miss regions of the search space with better solutions due to their narrow exploration. To address this, randomized search using heuristic techniques such as simulated annealing, particle swarm optimization, and genetic algorithms have been developed [28]. These methods help escape local minima by exploring different parts of the search space, offering a more diverse exploration strategy. Comparative studies, such as the one by [33], have evaluated ten different search algorithms across multiple datasets, concluding that while each method finds high-quality solutions, there is no clear overall winner. Similarly, a review of 21 algorithms by [34] concluded that no single method universally outperforms the others, aligning with the "no free lunch" theorem [35], which suggests that the effectiveness of a search strategy depends on the characteristics of the dataset and the task at hand.

Finally, comparative studies like that by [36] and [37] have shown that heuristic search methods often outperform other strategies in terms of FS quality. These works evaluated multiple search techniques on various datasets using diverse classifiers and consistently found that heuristic methods improved model accuracy by identifying the most relevant features. Despite the challenges in parameter tuning, heuristic algorithms are robust enough to deliver solid performance, even with suboptimal control settings [28].

4.3.4 Evaluation Perspective

The most widely used framework for evaluating FS methods categorizes them into four types: filter, wrapper, embedded, and hybrid methods. Extensive research has been conducted in each category, particularly with benchmarks and comparisons.

Filter methods are the simplest and most computationally efficient, as they evaluate features independently of the learning algorithm. A significant portion of the research in this area involves benchmarking the effectiveness of different filter methods on various domain-specific datasets. For instance, software defect data has been analyzed in [38], object-based imagery data in [39], biomedical data in [40, 41], cancer data in [42, 43], intrusion detection data in [44], and text categorization data in [42]. These studies demonstrate the broad applicability of FS across domains. However, dataset homogeneity in benchmarks can introduce bias, as the performance of FS methods is influenced by the specific characteristics of the datasets [45]. For example, biomedical datasets often have many features but few instances that could favour certain FS methods. A broader benchmark using 32 diverse real-world datasets, including both binary and multiclass classification tasks, showed that filter methods tend to be better suited for binary classification while remaining computationally efficient [1]. The results of filter methods are typically evaluated by measuring their performance with a learning algorithm, such as accuracy or AUC in classification tasks [1, 18] and RMSE or MAE in regression tasks [46, 47], which demonstrated the value of filter-based FS for high-dimensional datasets.

Wrapper methods differ from filter methods in that they incorporate the learning algorithm into the feature evaluation process. While they generally produce higher accuracy, their computational cost is a significant drawback, as multiple subsets must be evaluated through repeated model training. The evaluation criterion (such as accuracy [48], kappa index [49], mean misclassification error [50], or AUC [51]) depends on the specific analysis task (i.e., classification, regression, clustering, etc.) and the choice of learning algorithm. Wrappers often use sequential or randomized search strategies rather than exhaustive search due to the high computational cost. Several benchmarks discussed in Sect. 4.3.3 under both sequential [31, 32] and randomized [33, 34] search have compared wrapper methods.

Embedded methods offer a compromise by embedding FS within the model training process itself, where the model inherently selects relevant features. Common algorithms that perform FS as part of their learning process include decision trees, random forests, and regularization methods. A benchmark study on random forests across 13 datasets demonstrated improved accuracy due to the model's ability to capture nonlinear relationships and interactions between features [52]. However, random forests may introduce bias by favouring features with more categories or higher variance.

Regularization methods are another commonly used embedded FS technique, with L_1 (lasso), L_2 (ridge), and elastic net (a combination of L_1 and L_2) being the most notable. Ridge regression penalizes less relevant features by shrinking their coefficients towards zero but never to absolute zero. Lasso, which shrinks irrelevant feature coefficients to zero, is particularly effective for high-dimensional datasets where feature selection is critical to avoid overfitting [53]. However, it struggles when features are correlated, often selecting only one from a group and ignoring the rest. Elastic net, which combines both L_1 and L_2 regularization, addresses this issue by selecting groups of correlated features together [54] and is effective for problems where the number of features is greater than the number of instances.

Table 4.1 Summary of the literature on hybrid FS methods, detailing the filter and wrapper (i.e., learning algorithms and search strategies) techniques used

Literature	Filter	Learning algorithm	Search strategy
[55]	BW ratio	KNN	Randomized
[56]	Mutual information maximization	Extreme learning machine	Randomized
[57]	Correlation-based FS	KNN	Randomized
[60]	Chi-square	KNN	Randomized
[61]	Document frequency, mutual information, chi-square, information gain	Decision tree, SVM	Randomized
[63]	Information gain	Naive Bayes	Randomized
[64]	F-score, information gain	SVM	Sequential
[58]	ReliefF	Bayesian Net, Naive Bayes, KNN, SVM	Sequential
[65]	Seven methods including: FCBF, mRMR, JMIM	Multi-Layer perceptron	Sequential
[66]	Fisher, T-test, Entropy	Bayesian classifier	Exponential
[67]	Gini index, mRMR, Maximum information coefficient	Random forest	Sequential
[59]	F-score	SVM	Sequential

A benchmark comparing elastic net to lasso and ridge on several gene expression datasets showed improved accuracy and stability for the elastic net, particularly in cases of multicollinearity [54].

Hybrid methods combine the strengths of filter, wrapper, and embedded methods, aiming to mitigate their respective weaknesses. This approach has been applied across various domains, including microarray analysis [55–58], stock trend prediction [59], and text classification [60, 61]. Table 4.1 summarizes recent works on hybrid FS methods. The most common combination involves using filter methods to pre-select features, followed by wrapper methods with randomized search strategies. These hybrid methods have demonstrated superior performance on high-dimensional data, achieving higher classification accuracy, better precision and recall balance, and greater efficiency compared to using only the original or filtered datasets [58, 62].

4.3.5 Variable Interaction Perspective

Univariate and multivariate FS methods have not been studied independently. The literature examines them under the label perspective, which has been mentioned several times, or under the evaluation perspective, as filter methods. In general, however, univariate methods are faster but prone to redundancy, while multivariate methods handle redundancy at a computational cost.

4.3.6 Number of Objective Perspective

Traditionally, FS has been driven by single-objective approaches. In filter methods, for instance, common metrics like information gain, correlation measures, or data similarity are used to evaluate feature relevance. For wrapper methods, evaluation is often centred on model performance measures such as accuracy [18], ROC [42], AUC [40], F-Score [42], precision [42], recall [42], or the kappa index [39]. While optimizing a single metric is simpler and widely used, this approach fails to capture the complexity of feature relevance. There is a growing realization that focusing on one metric does not fully reflect the interplay between features and model performance, making it necessary to adopt a multi-objective perspective.

Thus, multi-objective FS has gained popularity by considering multiple, often conflicting, objectives. For example, in [2], two model performance metrics (namely accuracy and AUC) were combined to provide a more nuanced evaluation of feature subsets. Other works, such as [68] and [69], optimized both classification performance and feature reduction. Similarly, in [70], the authors proposed a multi-objective function that optimizes both the training and validation kappa index to minimize classification error and overfitting. Evolutionary algorithms, such as NSGA-II [71], have been widely used, as seen in [72, 73], where objectives like minimizing the number of features and classification error, or maximizing Expected Maximum Profit (EMP) while minimizing the number of features, are considered.

Most multi-objective approaches focus on balancing two criteria: one that minimizes the number of selected features and another that optimizes a model performance metric, such as accuracy, AUC, or kappa index. However, in recent years, the emphasis on fairness in ML models has added a new layer of complexity. FS has been instrumental in balancing predictive performance with fairness, particularly in scenarios where ML models must reduce bias towards individuals or groups. For example, [74] used NSGA-II to select features that maximized both the predictive performance (F1-score) and fairness (Statistical Parity Difference) of the models. In [75], multiple fairness metrics (Demographic Parity, Equality of Odds, and Equal Opportunity) were averaged and combined with the F1-score in a scalarized utility function to balance performance and fairness. Another example is the work in [76], where the authors employed a two-phase approach that first maximizes predictive performance using Sequential SFS and then reduces bias using SBS to eliminate features that lower fairness. While most works focus on two or three objectives, the need for optimizing more than three objectives has become increasingly important in creating feature subsets that balance these many, often conflicting, criteria.

Many-objective FS, which considers four or more objectives, represents the most complex approach and holds promise for addressing increasingly complex problems in ML and data analysis. Although this has received less attention, it has been applied in various ML problems, including multi-class classification [77–79], anomaly detection [80], motor imagery action prediction [81], software defect prediction [82], and gesture recognition [83]. In these studies, the number of objectives optimized ranges from four [78, 79, 82, 84, 85], five [80, 81, 83, 86], to as many as seven objectives [77]. The most commonly optimized objectives include the number of selected features [77–80, 82–87], ranking loss [77, 79], precision [77, 80, 81], coverage [77], hamming loss [77, 79], F1-score [77, 81–83], and accuracy [80, 81, 84, 86, 87]. Other objectives considered in many-objective FS include inter-class and intra-class distance [84, 87], classification error rate [78, 85], correlation [78, 82], feature complexity [78, 79], and relevance [86, 87]. In all these works, randomized search strategies have been employed to balance the trade-offs between objectives, with the number of selected features often being a constant and prominent objective.

Choosing an appropriate FS method requires careful consideration of several key perspectives. The dataset's characteristics—such as whether labels are available, the relationships among features, and the size of the dataset in terms of both features and instances—are crucial factors. The nature of the task, whether classification, regression, or clustering, along with the specific objectives or criteria that are most important for the task, also play a significant role in this decision. As discussed, there is no universal FS method that suits all scenarios, so understanding these factors is essential to selecting the most effective approach for a given problem.

4.4 Illustrative Example

Combining all the unique dimensions of the taxonomy defined in Fig. 4.3 leads to 432 FS techniques. However, our illustrative example involves a static dataset where the label is present (supervised), significantly reducing the techniques that need to be considered. Also, univariate methods are sequential by definition and so cannot be combined with exponential or randomized search. Lastly, hybrid evaluation involves at least two criteria and cannot be a single-objective design. Hence, we focus only on 41 FS techniques by combining the different interaction, evaluation, design, and search perspectives. Thus, our case study focuses on a binary classification task, predicting whether customers of a telecommunications company will churn or remain based on 17 features and 7043 instances. The dataset used is the IBM Telecommunication Industry Sample Data.[6]

---

[6] https://bit.ly/ibm_telco_data.

Table 4.2 FS techniques that are considered in the illustrative example

Perspective	Category	Method
Interaction	Univariate	Gini score, Mutual information, Chi2, Symmetric uncertainty
	Multivariate	Relevance/Redundancy, Conditional Mutual Information, Joint Mutual Information
Evaluation	Filter	mRMR, JMI, CMIM, MIFS
	Wrapper	Naive Bayes, Simple Logistic Regression
	Embedded	Decision Tree (Entropy), Lasso Logistic Regression
	Hybrid	mRMR + Naive Bayes, Gini + Simple Logistic Regression
Search	Exponential	Exponential
	Sequential	SFS, Ranking
	Randomized	Genetic algorithm, NSGA-II, NSGA-III
Objectives	Single	One
	Multi	Two
	Many	Four

Table 4.2 shows the methods considered under each category in the various perspectives of the FS taxonomy. A combination of these different methods led to the 41 FS techniques examined in this illustrative example. For scenarios involving the use of ML algorithms, the following model performance metrics were considered: Accuracy, AUC, Precision, Recall, and F1-score. All source code used for this example is available on GitHub.[7]

The results of each FS technique are evaluated based on four key metrics: execution time, feature subset size, and the predictive accuracy and AUC of the final classifier, XGBoost, a powerful algorithm renowned for its ability to achieve state-of-the-art performance across various ML tasks [88]. Execution times ranged from 0.02 seconds to 7170.42 seconds, while subset sizes varied between 1 and 17 features. For univariate methods, which typically rank features based on an evaluation criterion and return the full (17) set of features, the top 7 features were selected to measure accuracy and AUC. Also, compared to the baseline model performance, FS improved the accuracy of the final XGBoost classifier by up to 1.07%, while the AUC improved by up to 7.51%.

To detect the relationship between the FS techniques and the considered metrics (i.e., execution time, subset size, model accuracy, and AUC), the time and size results

---

[7] https://github.com/F-U-Njoku/illustrative-example.

Table 4.3 Discretization of the four metrics based on results from the FS techniques

Metric	Cluster	Value(s)
Execution time	1-Fast	0–10
	0-Moderate	15–500
	2-Slow	>600
Subset size	1-Small	1–5
	2-Moderate	6–10
	0-Large	>10
Accuracy	1-Improved	>0.7492
	0-Unimproved	≤ 0.7492
AUC	1-Improved	>0.7071
	0-Unimproved	≤ 0.7071

are discretized using a k-means clustering approach while the XGBoost model accuracy and AUC are set to two clusters (1 or 0) with respect to if there is an improvement or not compared to the baseline. The baseline is the XGBoost classifier trained on the complete dataset with all 17 features. The clusters for each metric are shown in Table 4.3, and a snapshot of the final dataset used to analyse relationships between the perspectives and the metrics is shown in Table 4.4.

Improvements in model performance were observed: accuracy increased in 11% of the cases, while AUC improvements were more frequent, occurring in 89% of the cases.

Using the processed results (Table 4.4) from all FS techniques, a different decision tree with a depth of 4 is constructed to analyze the relationship between the FS dimensions (i.e., interaction, evaluation, design, search) and each metric (i.e., execution time, subset size, model accuracy, and AUC). These trees provide insights into how different dimensions influence each metric. Hence, we derive numerous relationships depicted as flowcharts to serve as an illustrative guideline for selecting and using FS techniques.

Beginning with execution time, the search strategy has the most significant influence. As shown in Fig. 4.4, choosing sequential search is preferable for a fast execution time. Randomized search often leads to moderate execution time except when the design is multi- or many-objectives. Also, the exponential search without a hybrid evaluation leads to slow execution time.

Considering the subset size of the results, the relationships are visualized in Fig. 4.5. When employing a univariate interaction criterion, the method returns a ranking of the entire feature set based on their individual scores. Consequently, an additional step is required to determine the appropriate number of features to select for the subset. Next, filter evaluation methods generally yield small subsets when using multivariate criteria. Similarly, other evaluation methods combined with

Table 4.4 Results from 41 FS techniques uniquely combining variable interaction, evaluation, number of objectives, and search perspectives. Performance is assessed using Time, Size, Accuracy (ACC), and AUC (see Table 4.3 for metric details)

Interaction	Evaluation	Design	Search	Time	Size	ACC	AUC
Multivariate	Embedded	Many	Exponential	2	2	0	1
Multivariate	Embedded	Many	Randomized	0	2	0	1
Multivariate	Embedded	Many	Sequential	1	1	0	1
Multivariate	Embedded	Multi	Exponential	2	2	0	1
Multivariate	Embedded	Multi	Randomized	0	2	0	1
Multivariate	Embedded	Multi	Sequential	1	2	1	1
Multivariate	Embedded	Single	Exponential	2	2	0	1
Multivariate	Embedded	Single	Randomized	0	0	1	1
Multivariate	Embedded	Single	Sequential	1	1	0	0
Multivariate	Filter	Many	Randomized	2	1	0	1
Multivariate	Filter	Many	Sequential	1	1	0	1
Multivariate	Filter	Multi	Randomized	2	1	0	1
Multivariate	Filter	Multi	Sequential	1	1	0	1
Multivariate	Filter	Single	Randomized	0	1	0	1
Multivariate	Filter	Single	Sequential	1	1	0	1
Multivariate	Hybrid	Many	Exponential	0	2	0	1
Multivariate	Hybrid	Many	Randomized	0	2	0	1
Multivariate	Hybrid	Many	Sequential	1	2	0	1
Multivariate	Hybrid	Multi	Exponential	0	1	0	1
Multivariate	Hybrid	Multi	Randomized	0	2	0	1
Multivariate	Hybrid	Multi	Sequential	1	1	0	0
Multivariate	Wrapper	Many	Exponential	2	2	0	1
Multivariate	Wrapper	Many	Randomized	0	2	0	1
Multivariate	Wrapper	Many	Sequential	1	2	0	1
Multivariate	Wrapper	Multi	Exponential	2	2	0	1
Multivariate	Wrapper	Multi	Randomized	0	2	0	1
Multivariate	Wrapper	Multi	Sequential	1	1	0	1
Multivariate	Wrapper	Single	Exponential	2	1	0	1
Multivariate	Wrapper	Single	Randomized	0	2	0	1
Multivariate	Wrapper	Single	Sequential	1	1	0	0
Univariate	Embedded	Many	Sequential	1	0	0	1
Univariate	Embedded	Multi	Sequential	1	0	0	1
Univariate	Embedded	Single	Sequential	1	0	1	1
Univariate	Filter	Many	Sequential	1	0	0	1
Univariate	Filter	Multi	Sequential	1	0	0	1
Univariate	Filter	Single	Sequential	1	0	0	1
Univariate	Hybrid	Many	Sequential	1	0	0	1
Univariate	Hybrid	Multi	Sequential	1	0	0	1
Univariate	Wrapper	Many	Sequential	1	0	0	1
Univariate	Wrapper	Multi	Sequential	1	0	0	1
Univariate	Wrapper	Single	Sequential	1	0	0	1

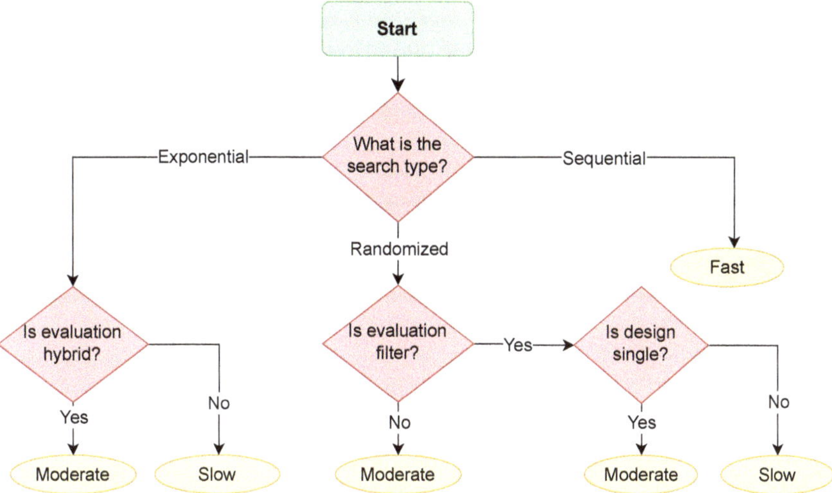

Fig. 4.4 Relationship between FS perspectives and execution time

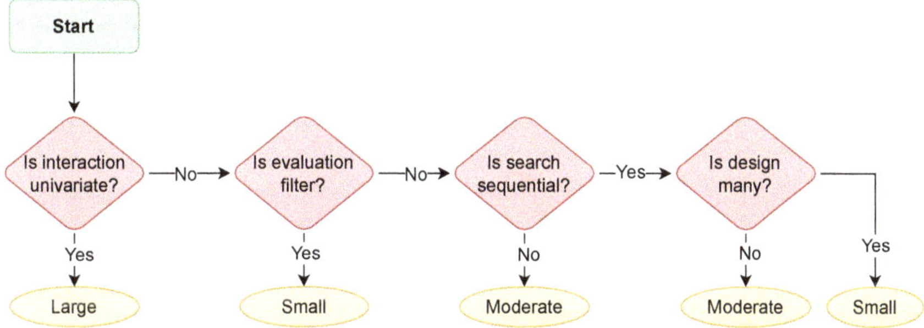

Fig. 4.5 Relationship between FS perspectives and subset size

sequential search and a single design often yield small subset sizes. Otherwise, in other configurations, the resulting subset size is frequently moderate.

Moving on to the model performance, we begin with accuracy. Given the relatively fewer cases of accuracy improvement, it is insightful to identify the configurations that result in better accuracy results. The relationships for the accuracy metric are shown in Fig. 4.6. Accuracy improvement is observed when the evaluation method is embedded, the design is single objective, and the search is randomized. Additionally, accuracy also improves when the evaluation method is embedded, the design is single, and the search is not randomized but instead uses a univariate interaction criterion. For the other configurations, reduced accuracy is observed..

Finally, for the model AUC, improvement is the most frequent outcome. Therefore, it is more insightful to identify the configurations to avoid (i.e., those that lead to reduced AUC). Figure 4.7 illustrates the associations between the FS dimensions and

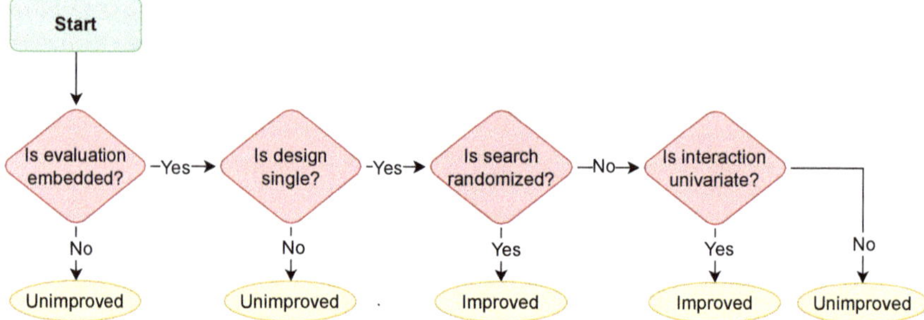

Fig. 4.6 Relationship between FS perspectives and model accuracy

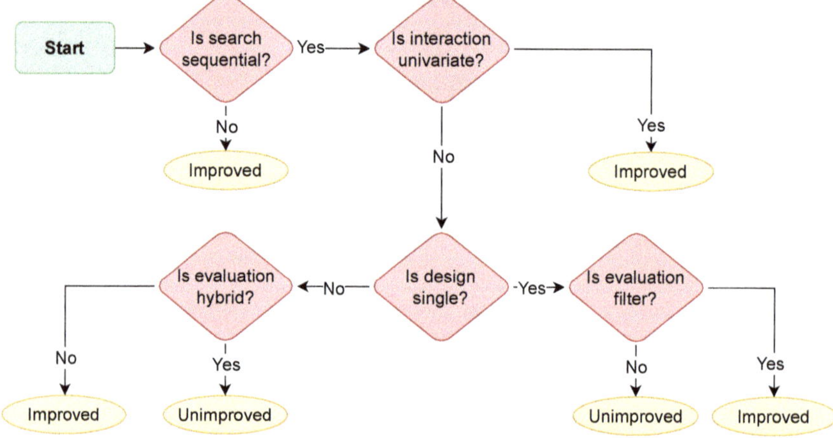

Fig. 4.7 Relationship between FS perspectives and model AUC

model AUC. A reduction in AUC is observed when the search strategy is sequential, the interaction is multivariate, and the design is single objective with any evaluation method other than filter. Additionally, a decrease in AUC is also seen when the search is sequential, the interaction is multivariate, and the design is not single, but the evaluation method is hybrid.

This illustrative example demonstrates how different perspectives influence the quality of results obtained in FS. It highlights the importance of understanding each perspective and carefully selecting them based on the specific requirements of a given use case. In FS, there is no universally optimal technique, as the effectiveness of a method depends on the problem context, dataset characteristics, and evaluation criteria.

4.5 Open Problems and Future Directions

Although the field of FS has made significant advancements, several challenges remain. In this section, we present these challenges to outline potential directions for future research.

4.5.1 Open Problems

The open problems in FS include the following:

1. *Scalability*: Similar to most ML tasks, big data impacts the time to train models and perform FS. Generally, filter methods are more scalable than wrapper and embedded methods. Also, the time complexity increases from single-objective to many-objective design. However, the focus has been on making filter methods more scalable using distribution/partitioning [89–91], parallel processing [90, 92], GPUs [93], MPI [94, 95], and early dropping heuristics [91, 96]. Conversely, even though the computational cost of wrapper methods worsens with the growing size of data, they have not received much attention regarding scalability. Indeed, distribution, partitioning [69, 97], and parallel processing [68, 69] have been used to improve some wrapper methods too. Nevertheless, as demonstrated in [51], the good performance of wrapper methods motivates the need to propose more scalable versions.
2. *Stability*: Beyond the internal evaluation of FS models through evaluation criteria, the selected features must be reliable and robust, i.e., little changes in the data should not lead to high variability in the selected features. Stability measures the robustness of FS methods to slight changes in the data. The instability of feature subsets from FS remains an issue, especially in domains such as bioinformatics, with high-dimensional datasets and low sample sizes.
   The stability of feature rankers (a subset of filter methods) on four microarray data is studied in [98] by running 100 Monte Carlo simulations and bootstrapping the original dataset each time to create a new, slightly altered dataset. They demonstrate the sensitivity of FS methods to changes in data, particularly microarray data. In [99], the stability of filter and wrapper methods was evaluated by randomly sampling a dataset without replacement 30 times for four overlapping percentages. Results show that filter methods are more stable. Causal FS [100] offers a promising approach to stability by going beyond statistical associations and focusing on features with causal relationships with the target, thereby enhancing robustness to data distribution shifts.
   Although stability is typically measured by data perturbation, cross-validation, or partitioning, the impact of data shuffling (i.e., a re-ordering of features and instances) on the stability of FS methods has yet to be studied. Practically, current tools and libraries for FS need to readily provide this supplementary information

on the stability of FS methods necessary to strengthen confidence in the results of the methods.

3. *Many-objective FS*: Many-objective FS presents an interesting yet underexplored challenge. As the number of objectives increases, the complexity of the search space grows, making it more difficult to efficiently identify optimal subsets of features. This problem becomes particularly pressing in scenarios involving a between the competing objectives. Designing scalable and efficient algorithms for many-objective FS, which can balance competing criteria without overly increasing computational demand, is a promising avenue for future research.

4. *Visualization and Interpretability of FS Results*: Despite the advances in FS, interpreting the relationship between the selected features and the underlying models remains a significant challenge. The ability to visualize FS results, especially when dealing with multiple solutions, is crucial for users to understand how FS impacts model behaviour and performance. Further work is needed in developing intuitive visualization techniques that help to communicate the importance of features, their interactions with models, and the trade-offs involved in many-objective FS [101].

5. *FS for Unconventional Data*: FS methods must also evolve to accommodate more complex and unconventional data properties. For example, unbalanced or multi-label data classification and real-time data processing scenarios present unique challenges. Existing FS methods may not generalize well to these contexts, requiring new approaches that can handle the intricacies of these data types while maintaining computational efficiency and accuracy.

4.5.2 Future Directions

Building upon an understanding of the current challenges faced by existing FS methods, we outline the following potential directions for future research:

1. *Diversification of objectives*: While traditional FS often focuses on accuracy, new objectives such as fairness, privacy, interpretability, and model robustness are gaining attention. The next wave of FS research will likely involve designing methods that can consider these diverse objectives without compromising computational efficiency or stability.

2. *Scalability of many-objective FS*: As the demand for models that can optimize multiple conflicting objectives continues to grow, designing scalable many-objective FS methods will become increasingly important. Hence, research should focus on developing algorithms capable of efficiently navigating the vast search space of many-objective problems, potentially leveraging ML techniques like reinforcement learning or deep neural networks to guide the search process.

3. *Visualization of final results*: With the growing complexity of FS techniques, especially in many-objective settings, there is a need for more advanced visualization tools that can effectively communicate results to users. These tools

should be able to illustrate the trade-offs between different objectives, the relationships between features and models, and the overall interpretability of the selected feature subsets.
4. *Robustness of FS tools*: Finally, future FS tools and libraries should provide robustness. This means providing not only feature subsets that improve model performance but also accompanying information about the stability and reliability of those selections. Such improvements would enhance the usability and trustworthiness of FS methods, especially in critical domains like healthcare and finance, where the costs of erroneous FS can be especially high.

4.6 Conclusions

In this chapter, we have explored the vast landscape of FS, covering six perspectives: data, label, search, number of objectives, evaluation, and variable interaction. We began with an overview of FS, supported by a use case that highlighted its advantages in enhancing model performance and reducing computational and monetary costs.

Next, we surveyed the state-of-the-art through the lens of the presented taxonomy. The majority of research has focused on static data with labels, using single-objective approaches—univariate or multivariate—commonly falling under filter or wrapper methods. In contrast, methods tailored for streaming data or having multiple objectives have received comparatively less attention.

Despite the significant strides made, challenges remain, particularly in scaling these methods to handle very large datasets and adapting them to continuously evolving applications. Future research must focus on refining these methods to enhance efficiency and flexibility to incorporate more objectives as the need arises.

Finally, we presented an example demonstrating how choices across the six perspectives can influence key metrics such as execution time, feature subset size, model accuracy, or AUC.

Ultimately, the evolution of FS methods will play a pivotal role in the future of AI, as models become more intricate and their societal impact continues to grow. We hope this chapter provides both a foundational understanding and a forward-looking perspective on the opportunities and challenges ahead.

References

1. Njoku UF, Abelló Gamazo A, Bilalli B, Bontempi G (2022) Impact of filter feature selection on classification: an empirical study. In: Proceedings of the 24rd international workshop on design, optimization, languages and analytical processing of big data (DOLAP): co-located with the 24th international conference on extending database technology and the 24th international conference on database theory (EDBT/ICDT 2022), CEUR-WS.org, UK, pp 71–80

2. Njoku UF, Abelló Gamazo A, Bilalli B, Bontempi G (2023) Wrapper methods for multi-objective feature selection. 26th international conference on extending database technology (EDBT 2023). CEUR-WS.org, Greece, pp 697–709

3. Cai J, Luo J, Wang S, Yang S (2018) Feature selection in machine learning: a new perspective. Neurocomputing 300:70–79

4. Cheng X (2024) A comprehensive study of feature selection techniques in machine learning models. Insights Comput Signals Syst 1(1):10–70088

5. Chen T, Guestrin C (2016) Xgboost: a scalable tree boosting system. In: Proceedings of the 22nd ACM SIGKDD international conference on knowledge discovery and data mining. ACM, USA, pp 785–794

6. Li J, Cheng K, Wang S, Morstatter F, Trevino RP, Tang J, Liu H (2017) Feature selection: a data perspective. ACM Comput Surv (CSUR) 50(6):1–45

7. Zaman EAK, Mohamed A, Ahmad A (2022) Feature selection for online streaming high-dimensional data: a state-of-the-art review. Appl Soft Comput 127:109355

8. Zhou J, Foster D, Stine R, Ungar L (2005) Streaming feature selection using alpha-investing. In: Proceedings of the eleventh ACM SIGKDD international conference on Knowledge discovery in data mining, ACM, USA, pp 384–393

9. Huang SH (2015) Supervised feature selection: a tutorial. Artif Intell Res 4(2):22–37

10. Dwivedi R, Tiwari A, Bharill N, Ratnaparkhe M, Tiwari AK (2024) A taxonomy of unsupervised feature selection methods including their pros, cons, and challenges. J Supercomput 80(16):1–29

11. Solorio-Fernández S, Carrasco-Ochoa JA, Martínez-Trinidad JF (2020) A review of unsupervised feature selection methods. Artif Intell Rev 53(2):907–948

12. Sheikhpour R, Sarram MA, Gharaghani S, Chahooki MAZ (2017) A survey on semi-supervised feature selection methods. Pattern Recogn 64:141–158

13. Kalakech M, Biela P, Macaire L, Hamad D (2011) Constraint scores for semi-supervised feature selection: a comparative study. Pattern Recognit Lett 32(5):656–665

14. Bellal F, Elghazel H, Aussem A (2012) A semi-supervised feature ranking method with ensemble learning. Pattern Recognit Lett 33(10):1426–1433

15. El Aboudi N, Benhlima L (2016) Review on wrapper feature selection approaches. 2016 international conference on engineering & MIS (ICEMIS). IEEE, Morocco, pp 1–5

16. Narendra F (1977) A branch and bound algorithm for feature subset selection. IEEE Trans Comput 100(9):917–922

17. Agrawal P, Abutarboush HF, Ganesh T, Mohamed AW (2021) Metaheuristic algorithms on feature selection: a survey of one decade of research (2009–2019). IEEE Access 9:26766–26791

18. Bommert A, Sun X, Bischl B, Rahnenführer J, Lang M (2020) Benchmark for filter methods for feature selection in high-dimensional classification data. Comput Stat Data Anal 143:106839

19. Berrar D, et al. (2019) Cross-validation. https://dberrar.github.io/#Publications

20. Guyon I, Elisseeff A (2003) An introduction to variable and feature selection. J Mach Learn Res 3(Mar):1157–1182

21. Almugren N, Alshamlan H (2019) A survey on hybrid feature selection methods in microarray gene expression data for cancer classification. IEEE Access 7:78533–78548

22. Al-Tashi Q, Abdulkadir SJ, Rais HM, Mirjalili S, Alhussian H (2020) Approaches to multi-objective feature selection: a systematic literature review. IEEE Access 8:125076–125096

23. Jiao R, Nguyen BH, Xue B, Zhang M (2023) A survey on evolutionary multiobjective feature selection in classification: approaches, applications, and challenges. IEEE Trans Evol Comput 28(4):1156–1176

24. Deb K, Jain H (2013) An evolutionary many-objective optimization algorithm using reference-point-based nondominated sorting approach, part i: solving problems with box constraints. IEEE Trans Evol Comput 18(4):577–601

25. Wu X, Yu K, Wang H, Ding W (2010) Online streaming feature selection. In: Proceedings of the 27th international conference on machine learning (ICML-10), Omni Press, USA, pp 1159–1166

26. He X, Cai D, Niyogi P (2005) Laplacian score for feature selection. Adv Neural Inf Process Syst 18(8):507–514
27. Mitra P, Murthy C, Pal SK (2002) Unsupervised feature selection using feature similarity. IEEE Trans Pattern Anal Mach Intell 24(3):301–312
28. Doak J (1992) An evaluation of feature selection methods and their application to computer security. Techninal Report CSE-92-18
29. Somol P, Pudil P, Kittler J (2004) Fast branch & bound algorithms for optimal feature selection. IEEE Trans Pattern Anal Mach Intell 26(7):900–912
30. Pudil P, Novovičová J, Kittler J (1994) Floating search methods in feature selection. Pattern Recognit Lett 15(11):1119–1125
31. Aha DW, Bankert RL (1995) A comparative evaluation of sequential feature selection algorithms. In: Pre-proceedings of the Fifth International Workshop on Artificial Intelligence and Statistics, PMLR, USA, pp 1–7
32. Yulianti Y, Saifudin A (2020) Sequential feature selection in customer churn prediction based on naive bayes. In: IOP conference series: materials science and engineering, IOP, Indonesia, p 012090
33. Diao R, Shen Q (2015) Nature inspired feature selection meta-heuristics. Artif Intell Rev 44:311–340
34. Sadeghian Z, Akbari E, Nematzadeh H, Motameni H (2023) A review of feature selection methods based on meta-heuristic algorithms. J Exp Theor Artif Intell 37(1):1–51
35. Adam SP, Alexandropoulos SAN, Pardalos PM, Vrahatis MN (2019) No free lunch theorem: a review. Approx Optim Algorithms Complex Appl 145:57–82
36. Bansal A, Jain A (2021) Comparison of meta-heuristic with evolutionary and local search methods for feature selection. Metaheurist Evol Comput Algorithms Appl 916:529–554
37. Ay Ş, Ekinci E, Garip Z (2023) A comparative analysis of meta-heuristic optimization algorithms for feature selection on ml-based classification of heart-related diseases. J Supercomput 79(11):11797–11826
38. Xu Z, Liu J, Yang Z, An G, Jia X (2016) The impact of feature selection on defect prediction performance: an empirical comparison. 2016 IEEE 27th international symposium on software reliability engineering (ISSRE). IEEE, Canada, pp 309–320
39. Laliberte AS, Browning D, Rango A (2012) A comparison of three feature selection methods for object-based classification of sub-decimeter resolution ultracam-l imagery. Int J Appl Earth Obs Geoinf 15:70–78
40. Drotár P, Gazda J, Smékal Z (2015) An experimental comparison of feature selection methods on two-class biomedical datasets. Comput Biol Med 66:1–10
41. Meyer PE, Schretter C, Bontempi G (2008) Information-theoretic feature selection in microarray data using variable complementarity. IEEE J Sel Topics Signal Process 2(3):261–274
42. Aphinyanaphongs Y, Fu LD, Li Z, Peskin ER, Efstathiadis E, Aliferis CF, Statnikov A (2014) A comprehensive empirical comparison of modern supervised classification and feature selection methods for text categorization. J Assoc Inf Sci Technol 65(10)
43. Meyer PE, Bontempi G (2006) On the use of variable complementarity for feature selection in cancer classification. Workshops on applications of evolutionary computation. Springer, Germany, pp 91–102
44. Nguyen HT, Petrović S, Franke K (2010) A comparison of feature-selection methods for intrusion detection. International conference on mathematical methods, models, and architectures for computer network security. Springer, Germany, pp 242–255
45. Oreski D, Oreski S, Klicek B (2017) Effects of dataset characteristics on the performance of feature selection techniques. Appl Soft Comput 52:109–119
46. Ren K, Fang W, Qu J, Zhang X, Shi X (2020) Comparison of eight filter-based feature selection methods for monthly streamflow forecasting-three case studies on camels data sets. J Hydrol 586:124897
47. Bommert A, Welchowski T, Schmid M, Rahnenführer J (2022) Benchmark of filter methods for feature selection in high-dimensional gene expression survival data. Briefings Bioinf 23(1), bbab354

48. Chong J, Tjurin P, Niemelä M, Jämsä T, Farrahi V (2021) Machine-learning models for activity class prediction: a comparative study of feature selection and classification algorithms. Gait & Posture 89:45–53

49. Vieira SM, Kaymak U, Sousa JM (2010) Cohen's kappa coefficient as a performance measure for feature selection. International conference on fuzzy systems. IEEE, Spain, pp 1–8

50. Rodriguez-Galiano VF, Luque-Espinar JA, Chica-Olmo M, Mendes MP (2018) Feature selection approaches for predictive modelling of groundwater nitrate pollution: an evaluation of filters, embedded and wrapper methods. Sci Total Environ 624:661–672

51. Bagherzadeh-Khiabani F, Ramezankhani A, Azizi F, Hadaegh F, Steyerberg EW, Khalili D (2016) A tutorial on variable selection for clinical prediction models: feature selection methods in data mining could improve the results. J Clin Epidemiol 71:76–85

52. Breiman L (2001) Random forests. Mach Learn 45:5–32

53. Ng AY (2004) Feature selection, l 1 vs. l 2 regularization, and rotational invariance. In: Proceedings of the twenty-first international conference on Machine learning, ACM, USA, p 78

54. Zou H, Hastie T (2005) Regularization and variable selection via the elastic net. J R Stat Soc Ser B Stat Methodol 67(2):301–320

55. Lee CP, Leu Y (2011) A novel hybrid feature selection method for microarray data analysis. Appl Soft Comput 11(1):208–213

56. Lu H, Chen J, Yan K, Jin Q, Xue Y, Gao Z (2017) A hybrid feature selection algorithm for gene expression data classification. Neurocomputing 256:56–62

57. Chuang LY, Yang CH, Wu KC, Yang CH (2011) A hybrid feature selection method for DNA microarray data. Comput Biol Med 41(4):228–237

58. Hameed SS, Petinrina OO, Hashi AO, Saeed F (2018) Filter-wrapper combination and embedded feature selection for gene expression data. Int J Adv Soft Comput Appl 10(1)

59. Lee MC (2009) Using support vector machine with a hybrid feature selection method to the stock trend prediction. Expert Syst Appl 36(8):10896–10904

60. Imani MB, Keyvanpour MR, Azmi R (2013) A novel embedded feature selection method: a comparative study in the application of text categorization. Appl Artif Intell 27(5):408–427

61. Gunal S (2012) Hybrid feature selection for text classification. Turk J Electr Eng Comput Sci 20:1296–1311

62. Chen CW, Tsai YH, Chang FR, Lin WC (2020) Ensemble feature selection in medical datasets: combining filter, wrapper, and embedded feature selection results. Expert Syst 37(5):e12553

63. Naseriparsa M, Bidgoli AM, Varaee T (2013) A hybrid feature selection method to improve performance of a group of classification algorithms. Int J Comput Appl 69(17):28–35

64. Hsu HH, Hsieh CW, Lu MD (2011) Hybrid feature selection by combining filters and wrappers. Expert Syst Appl 38(7):8144–8150

65. Viharos ZJ, Kis KB, Fodor Á, Büki MI (2021) Adaptive, hybrid feature selection (AHFS). Pattern Recognit 116:107932

66. Cateni S, Colla V, Vannucci M (2014) A hybrid feature selection method for classification purposes. 2014 European modelling symposium. IEEE, Italy, pp 39–44

67. Wei G, Zhao J, Feng Y, He A, Yu J (2020) A novel hybrid feature selection method based on dynamic feature importance. Appl Soft Comput 93:106337

68. Galar M, Triguero I, Bustince H, Herrera F (2018) A preliminary study of the feasibility of global evolutionary feature selection for big datasets under apache spark. In: 2018 IEEE congress on evolutionary computation (CEC), IEEE, Brazil, pp 1–8

69. Peralta D, Del Río S, Ramírez-Gallego S, Triguero I, Benitez JM (2015) Herrera F (2015) Evolutionary feature selection for big data classification: a mapreduce approach. Math Probl Eng 1:246139

70. González J, Ortega J, Damas M, Martín-Smith P, Gan JQ (2019) A new multi-objective wrapper method for feature selection-accuracy and stability analysis for BCI. Neurocomputing 333:407–418

71. Deb K, Pratap A, Agarwal S, Meyarivan T (2002) A fast and elitist multiobjective genetic algorithm: Nsga-ii. IEEE Trans Evol Comput 6(2):182–197

72. Hamdani TM, Won JM, Alimi AM, Karray F (2007) Multi-objective feature selection with nsga ii. Adaptive and natural computing algorithms: 8th international conference, ICANNGA 2007, Warsaw, Poland, April 11–14, 2007, Proceedings, Part I 8. Springer, Germany, pp 240–247

73. Kozodoi N, Lessmann S, Papakonstantinou K, Gatsoulis Y, Baesens B (2019) A multi-objective approach for profit-driven feature selection in credit scoring. Decis Support Syst 120:106–117

74. Rehman AU, Nadeem A, Malik MZ (2022) Fair feature subset selection using multiobjective genetic algorithm. In: Proceedings of the Genetic and Evolutionary Computation Conference Companion, ACM, USA, p 360–363

75. Dorleon G, Megdiche I, Bricon-Souf N, Teste O (2022) Feature selection under fairness constraints. In: Proceedings of the 37th ACM/SIGAPP symposium on applied computing, ACM, USA, pp 1125–1127

76. Salazar R, Neutatz F, Abedjan Z (2021) Automated feature engineering for algorithmic fairness. Proc VLDB Endow 14(9):1694–1702

77. Dong H, Sun J, Sun X, Ding R (2020) A many-objective feature selection for multi-label classification. Knowl-Based Syst 208:106456

78. Asilian Bidgoli A, Ebrahimpour-Komleh H, Rahnamayan S (2021) A novel binary many-objective feature selection algorithm for multi-label data classification. Int J Mach Learn Cybern 12(7):2041–2057

79. Bidgoli AA, Ebrahimpour-Komleh H, Rahnamayan S (2019) A many-objective feature selection algorithm for multi-label classification based on computational complexity of features. 2019 14th international conference on computer science & education (ICCSE). IEEE, Canada, pp 85–91

80. Zhang Z, Wen J, Zhang J, Cai X, Xie L (2020) A many objective-based feature selection model for anomaly detection in cloud environment. IEEE Access 8:60218–60231

81. Pal M, Bandyopadhyay S (2016) Many-objective feature selection for motor imagery EEG signals using differential evolution and support vector machine. 2016 international conference on microelectronics, computing and communications (MicroCom). IEEE, India, pp 1–6

82. Mao Q, Zhang J, Zhao T, Cai X (2024) A many objective based feature selection model for software defect prediction. Concurr Comput Pract Exp 36(19):e8153

83. Otis MJD, Vandewynckel J (2021) A many-objective simultaneous feature selection and discretization for LCS-based gesture recognition. Appl Sci 11(21):9787

84. Li H, He F, Liang Y, Quan Q (2020) A dividing-based many-objective evolutionary algorithm for large-scale feature selection. Soft Comput 24:6851–6870

85. González J, Ortega J, Damas M, Martín-Smith P (2019) Many-objective cooperative co-evolutionary feature selection: a lexicographic approach. International work-conference on artificial neural networks. Springer, Switzerland, pp 463–474

86. Shu L, He F, Hu X, Li H (2021) A novel feature selection with many-objective optimization and learning mechanism. 2021 IEEE 24th international conference on computer supported cooperative work in design (CSCWD). IEEE, China, pp 684–689

87. Li Y, Sun Z, Liu X, Chen WT, Horng DJ, Lai K (2021) Feature selection based on a large-scale many-objective evolutionary algorithm. Comput Intell Neurosci 1:9961727

88. Hakkal S, LAHCEN AA, (2024) Xgboost to enhance learner performance prediction. Comput Educ Artif Intell 7:100254

89. Bolon-Canedo V, Sechidis K, Sanchez-Marono N, Alonso-Betanzos A, Brown G (2019) Insights into distributed feature ranking. Inf Sci 496:378–398

90. Palma-Mendoza RJ, de Marcos L, Rodriguez D, Alonso-Betanzos A (2019) Distributed correlation-based feature selection in spark. Inf Sci 496:287–299

91. Reggiani C, Le Borgne YA, Bontempi G (2017) Feature selection in high-dimensional dataset using mapreduce. Benelux conference on artificial intelligence. Springer, Switzerland, pp 101–115

92. Khaire UM, Dhanalakshmi R (2022) Stability of feature selection algorithm: A review. J King Saud University-Comput Inf Sci 34(4):1060–1073

93. Ramírez-Gallego S, Lastra I, Martínez-Rego D, Bolón-Canedo V, Benítez JM, Herrera F, Alonso-Betanzos A (2017) Fast-mrmr: fast minimum redundancy maximum relevance algorithm for high-dimensional big data. Int J Intell Syst 32(2):134–152

94. Beceiro B, González-Domínguez J, Touriño J (2022) Parallel-fst: a feature selection library for multicore clusters. J Parallel Distrib Comput 169:106–116

95. González-Domínguez J, Bolón-Canedo V, Freire B, Touriño J (2019) Parallel feature selection for distributed-memory clusters. Inf Sci 496:399–409

96. Nguyen T, Phan N, Nguyen N, Nguyen BT, Halvorsen P, Riegler MA (2022) Parallel feature selection based on the trace ratio criterion. 2022 international joint conference on neural networks (IJCNN). IEEE, Italy, pp 1–8

97. Bolón-Canedo V, Sanchez-Marono N, Alonso-Betanzos A (2013) A distributed wrapper approach for feature selection. In: ESANN, i6doc.com, Belgium

98. Moulos P, Kanaris I, Bontempi G (2013) Stability of feature selection algorithms for classification in high-throughput genomics datasets. 13th IEEE international conference on bioinformatics and bioengineering. IEEE, Greece, pp 1–4

99. Wald R, Khoshgoftaar TM, Napolitano A (2013) Stability of filter-and wrapper-based feature subset selection. 2013 IEEE 25th international conference on tools with artificial intelligence. IEEE, USA, pp 374–380

100. Bontempi G, Meyer PE (2010) Causal filter selection in microarray data. In: Proceedings of the 27th international conference on machine learning (icml-10), OmniPress, USA, pp 95–102

101. Njoku UF, Abelló Gamazo A, Bilalli B, Bontempi G (2024) A data-science pipeline to enable the interpretability of many-objective feature selection. In: Proceedings of the 26th international workshop on design, optimization, languages and analytical processing of big data (DOLAP 2024): co-located with the 27th international conference on extending database technology and the 27th international conference on database theory (EDBT/ICDT 2024), CEUR-WS.org, Italy, pp 83–87

Chapter 5
Current Systems for Managing Massive High Frequency Time Series

Abduvoris Abduvakhobov⬭, Søren Kejser Jensen⬭, Christian Thomsen⬭, and Esteban Zimányi⬭

Abstract Manufacturers and owners use high-frequency sensor data to optimize energy production. Data is collected on the edge (i.e., wind turbines) and transferred to the cloud for analytics. General-purpose RDBMSs are unable to handle the volume and velocity of sensor data. As a remedy, Time Series Management Systems (TSMSs) have been offered to manage sensor data across the entire pipeline efficiently. This chapter surveys TSMSs developed through academic or industrial research and documented through peer-reviewed papers. The chapter uses classification criteria for surveying systems by architecture, year, primary purpose, deployment, maturity, scale shown, data processing engine, API, approximation, latency, data store and storage layout. A collection of open research problems is provided based on the surveyed systems.

Keywords Time series management systems · Distributed systems · Distributed databases · Sensor data · Approximation

5.1 Overview

Data from high-quality sensors is used to optimize the production of renewable energy from devices such as wind turbines, solar panels and hydroelectric dams. They are equipped with many sensors that produce vast amounts of *time series* at

A. Abduvakhobov (✉) · S. K. Jensen · C. Thomsen
Aalborg University, Aalborg Øst, Denmark
e-mail: abduvorisa@cs.aau.dk

S. K. Jensen
e-mail: skj@cs.aau.dk

C. Thomsen
e-mail: chr@cs.aau.dk

E. Zimányi
Université Libre de Bruxelles, Bruxelles, Belgium
e-mail: esteban.zimanyi@ulb.be

© The Author(s) 2026

G. Dejaegere et al. (eds.), *Data Engineering for Data Science*,
https://doi.org/10.1007/978-3-032-18765-9_5

high frequencies. A time series is a sequence of *data points* ordered by time. A wind turbine generating 2500 data points every 10 ms produces more than 321 GiB of time series daily, assuming timestamps and values require 8 bytes each [1].

General-purpose RDBMSs manage sensor data inefficiently and fail to address high volume and velocity of data [2–5]. For example, for managing wind turbine data, this further limits the frequency at which time series can be collected due to limited hardware on edge (e.g., wind turbines, etc.), limited bandwidth to the cloud, high storage cost in the cloud and low data quality after compression [6]. Instead, time series are stored in big data formats such as Apache Parquet [7]. However, extra engineering work is required to store data in big data formats. To remedy these issues, TSMSs have been developed [8, 9]. Thus, this chapter surveys TSMSs developed through academic or industrial research and documented through peer-reviewed papers. In addition, a set of problems is provided based on the surveyed TSMSs. The focus of the survey is TSMSs for managing vast high frequency time series. Thus, it omits systems focused on spatiotemporal data management systems, such as MobilityDB [10], and systems designed for resource-constrained devices, such as RibesDB [11].

Papers were collected using the following method. For each paper cited in [8, 9], the following structured search was performed:

- Newer publications citing the paper (16 systems were found).
- The publication history of each author of the paper since 2022.

Another iteration with the same steps was performed for each new paper. Publications in the following conference proceedings and journals since 2022 were also checked:

- ACM SIGMOD conference (1 system was found)
- ACM SIGMOD Record journal
- ACM TODS conference
- EDBT conference
- IEEE ICDE conference
- IEEE Big Data conference
- The VLDB journal (1 system was found)
- DASFAA conference (1 system was found).

The chapter omits TSMS **Mach** [12, 13] which is described in the 2022 TSMS survey [8]. Systems having a new publication that describes a new contribution are included. The rest of the chapter is structured as follows. Section 5.2 defines the classification criteria for the systems and presents methods commonly used by TSMSs. Sect. 5.3 describes the TSMSs. Sect. 5.4 describes open problems and Sect. 5.5 concludes.

5.2 Landscape of TSMSs

This section provides an overview of the different characteristics that will be used to classify the different TSMSs. Furthermore, it also provides a brief introduction to concepts widely used in TSMSs.

5.2.1 Classification Criteria

This section describes the classification criteria for a TSMS. TSMSs in Sects. 5.3.1 and 5.3.2 are sorted by year.

Primary Purpose: The primary purpose for which the TSMS was created: "Monitoring" and "Data Analytics". Monitoring systems are designed to monitor and detect emergencies, system errors and faulty behavior. Data analytics systems are designed for large-scale time series analytics in either real-time or batch mode.

Deployment: The TSMS's intended type of deployment: "Centralized" and "Distributed". Centralized systems are deployed on a single node, and distributed systems allow scaling out through distributed computing and storage. The deployment type also significantly impacts the system's architecture and its functionality.

Maturity: The TSMS's maturity: "Proof-of-concept", "Demonstration" and "Mature". Proof-of-concept systems implement only enough functionalities to evaluate new methods. Demonstration systems have enough functionalities to be tested in real-life scenarios and provide a user interface. Mature systems are deployed to solve real-life problems and have an open-source community or commercial support.

Scale Shown: The measure of scalability that is defined by the "size of the largest data set" and the "highest number of physical nodes". If the size of the largest dataset is not specified in the paper, the "name of used benchmark suite" is written instead. "Unknown" is written if the information is not available.

Data Processing Engine: The data processing engine used for query processing and supported analytical tasks. If an existing system is used, the "name of the system" is written, and if a new data processing engine was implemented, "proprietary" and the implementation language are written.

API: The primary API provided by the TSMS: "unknown", an "existing query language", an "extended version of an existing query language", a "client library and its implementation language" or a "web service".

Approximation: Approximation can be used when storing data or executing queries. During ingestion, approximation can be performed by *lossy compression* methods like polynomial functions, sketches or aggregation using the aggregate function. When executing queries, "Approximate Query Processing (AQP)" can be performed on lossy compressed data using approximated representations of data like polynomial functions. AQP can also be performed using pre-computed aggregates that match the requested time interval or by sampling. Another approximation method is "Timestamp Approximation (TA)" that is applied to store timestamps in a lossy manner and thus answer queries approximately. We write "Not Supported" if a TSMS does not support AQP.

Latency: The latency from data ingestion and processing. "Batch" systems must ingest large batches of time series before they can execute queries. "Near Real-Time" systems can ingest and process ad-hoc queries simultaneously. "Real-Time" systems can execute user-defined continuous queries.

Data Store: The data store used by the TSMS for storing data. If an existing system is used, then the "name of the system" is shown. If a TSMS implements a new data store, then "proprietary" and the implementation language are written.

Storage Layout: The internal data representation used in TSMS's data store. This impacts query execution time and the amount of storage required.

Sections 5.3.1 and 5.3.2 summarize the contributions of each system with regard to the classification criteria. Tables 5.1 and 5.2 provide a summary of systems using the classification criteria. A system is marked with a ($\star$) if it is open-source.

5.2.2 Preliminaries

To use uniform terminology for describing TSMSs, we define the following terms and use them throughout the chapter. A *data point* represents a tuple consisting of a timestamp and one or more values and, if available, tags. *Tags* are metadata about a data point stored as strings and *values* are measurements stored as floats. A sequence of data points ordered by time is a *time series*. A *univariate time series* has one value per data point and a *multivariate time series* has multiple values per data point. Time series is identified using a time series ID and, if available, tags.

5.2.3 Design of Figures

All figures are redrawn from the papers to make their style uniform across the chapter. The same style as in [8] is used: *boxes with complete lines* for system components; *lines* between components for an undefined connection; *boxes with rounded corners* for data; *arrows* for data flow between components; *boxes with dotted lines* for logically related components; *cylinders* for data stores; *boxes with dashed lines* for nodes; *boxes with dashed lines* for nodes as components; *text labels* for annotations.

A component can represent different concepts depending on the system. For a large distributed system, it can be a node in a cluster. For a centralized system, it can be a query parser. Boxes with complete lines are used for the in-memory data layout and triangles are used for trees. Other constructs used in a single figure are described using labels in the figure.

5.2.4 Background

This section describes common methods, architectures and data formats used by TSMSs in Sect. 5.3. We hope this will help readers understand the description of systems.

5.2.4.1 Log Structured Merge Tree

A Log Structured Merge (LSM) tree was first introduced by O'Neil et al. [14] as an indexing technique that was inspired by the log-structured file system proposed by Rosenblum and Ousterhout [15]. The use of LSM trees has gained traction after Google's Bigtable paper [16], which introduced the concepts of Memtable and Sorted String Table (SSTable). LSM tree is an append-only data structure for use cases where writes are far more frequent than reads. During ingestion, data points are first written to a Memtable. A Memtable is a collection of sorted key-value pairs stored in a tree. When its size reaches a threshold, it is flushed to disk as an SSTable. As LSM trees are append-only storage structures, updates are performed by appending data points. They also support deletion by inserting into an SSTable a special flag called a *tombstone* that marks a specific key or list of keys to be discarded during the reading of SSTables. Periodical *compaction* of SSTables is performed to remove deleted records and duplicate keys from updates and merge them into larger SSTables.

TSMSs use a leveled compaction method that stores SSTables in multiple layers. The maximum number and size of SSTables are configured for each layer. The lowest layer (i.e., Level 0) holds the most recent SSTables and after reaching a certain threshold, these SSTables are compacted and moved to the next layer (i.e., Level 1). The threshold and size of SSTables increase after moving to each layer [17]. To avoid scanning all SSTables for query processing, in-memory indexes are created that store the offsets to the keys in SSTables. More details about LSM tree indexes are provided in Sect. 5.2.4.2.

LSM trees are used by TSMSs such as TimeUnion [18], ForestTI [19] and Apache IoTDB [20, 21] due to their ability to sustain high write throughput. These systems implement time-partitioned LSM trees that partition the SSTables in each level using a time range and apply leveled compaction, which ensures low latency for time-range queries. In time-partitioned LSM trees, each partition maintains metadata that indicates the time range in a given partition. SSTables within each partition also contain an index file that stores information such as time period, time series IDs and time series tags in each SSTable.

5.2.4.2　Log Structured Merge Tree Indexing Techniques

Many indexing structures have been proposed to ensure low latency of queries for systems using LSM trees.

One of the simplest strategies is to maintain an in-memory skip list of time series IDs to their SSTable offsets. Given that SSTables are sorted by key, the offset of the nearest key that is in the skip list can be accessed and further scanned until the specific key is found [22].

A more memory-efficient probabilistic index structure is a Bloom filter [23] that is used for checking whether an element is "probably" or "definitely not" a member of a set (i.e., false positives are possible, but false negatives are not). A Bloom filter consists of a bit array and multiple hash functions. Bloom filters can be used to check if specific time points exist in the database (e.g., when writing new data points or before answering queries [17]).

Forward and *inverted* indexes are used by TSMSs to search for time series using time series tags. Forward indexes are used to create and locate time series IDs that map to one or more time series tags. Inverted indexes map each time series tag to a set of time series IDs. For example, forward indexes can be used during inserts and inverted indexes can be used to answer queries with specific time series tags. Due to the vast amounts of time series, the memory footprint of both indexes can be very large. Thus, TSMSs use different caching and compression strategies to accommodate the increasing size of indexes. InfluxDB [24] and Apache Druid [25] use Concise [26] or Roaring Bitmap [27] for compressing time series IDs of inverted indexes. See [28] for a survey on compression for inverted indexes. Tree data structures are also commonly used for building indexes. ByteSeries [29] uses Cedar Trie [30] for building an inverted index. Chronos [31] uses B-tree and Timon [32] uses a time-partitioned tree index to locate SSTables using a time range.

5.2.4.3　Big Data Formats

General-purpose big data formats, are commonly used by TSMSs. These formats can be used by multiple systems and provide a competitive performance compared to system-specific formats [33]. Apache Parquet (Parquet) [34] and Apache ORC (ORC) [35] are popular big data formats that use a columnar layout with compression for each column. Both formats partition data into smaller groups and each group contains metadata such as simple statistics, null maps and encoding information. This helps to answer some queries directly from metadata and avoid scanning entire groups. Apache Arrow [36] is an in-memory columnar format that is used for computations with on-disk formats like Parquet or ORC.

5.2.4.4   Cloud Storage

Most TSMSs provide support for cloud storage such as block store that is accessed using the local filesystem and object store that is accessed using a web-based API. For example, Amazon offers Amazon EBS for a block store [37] and Amazon S3 [38] for an object store. Block store is a more costly and faster cloud storage option, while the object store is a much cheaper and slower alternative. TimescaleDB [39] supports both types of cloud storage and InfluxDB supports the object store.

5.2.4.5   Distributed Architectures

Distributed TSMSs generally use one of two architectures: shared-disk and shared-nothing. In a shared-disk architecture, all compute nodes globally share the disk storage area [40]. In a shared-nothing architecture, each compute node maintains exclusive access to its memory and disk storage [40]. Both architectures can be implemented on the cloud.

5.3   State of the Art

This section provides a description of TSMSs. Each system description follows the classification criteria and describes how the system ingests data and processes queries. The amount of details in each description may vary from system to system. This is due to the amount of information, such as the level of detail of the paper, source code and documentation that was available.

5.3.1   Internal Data Store

Karlsletter et al. [41] implements a decentralized prototype system in C++ for the evaluation of a method for processing range queries over distributed nodes without maintaining a shared state. The system stores multivariate time series and uses Parquet for storage. The system also stores additional levels of pre-computed aggregates using tumbling windows. The system only supports range queries that select a subset of sensors and specify the maximum required time resolution. Edge nodes, the closest to the sensors, store raw time series and pre-computed aggregates of all levels but only cover a limited time span. Intermediate nodes contain a longer time span than the edge nodes and all levels of aggregates but the raw time series. Cloud nodes store longer time spans, contain aggregates with only higher granularity levels and miss the most recent time series. The user side handles the logic for distributed data retrieval, and thus, no shared state between distributed nodes is required. Queries are first sent to the closest node in terms of latency over a limited

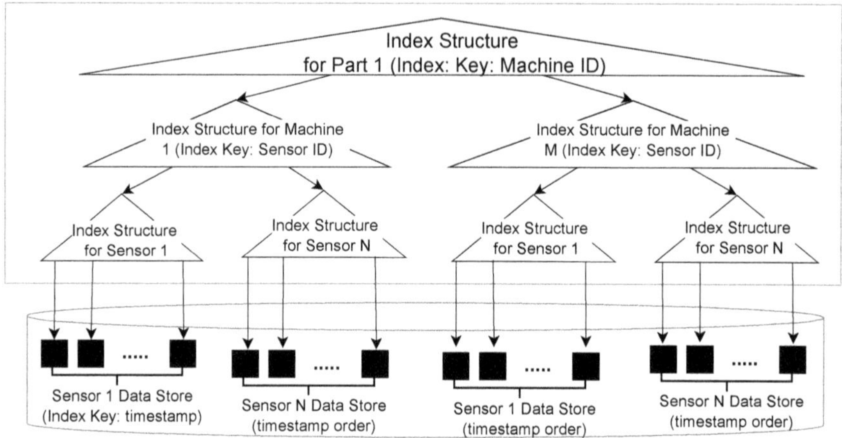

Fig. 5.1 Sen-Store data organization as SDS blocks and k-ary index. Redrawn from [43]

network connection. Upon receiving the query, the node quickly computes the metadata that includes information about the other existing nodes and the time range that can be covered by the current node and sends it to the user before sending the actual response. This allows the user to modify the query and send it to the other respective nodes to retrieve the remaining parts of the queried time series. This eliminates the need to maintain constant communication between nodes, which is an overhead in network-constrained environments. However, the paper does not specify the overhead of computing queries by the user. For its API, the system implements a set of Python functions based on Apache Arrow Flight [42] that is used for client-server communication. AQP is not supported.

Choi et al. [43] propose **Sen-Store** as a prototype key-value data store implemented for the management of data streams generated by industrial IoT systems. It implements a novel indexing structure called IoI (indexing of indexing) and uses LevelDB. IoI assumes that in industrial IoT environments, every time series has a fixed schema that consists of *part ID, machine ID, sensor ID, timestamp* and *sensor value*. Thus, IoI uses a hierarchical in-memory k-ary tree index (k-ary tree) shown in Fig. 5.1 and partitions data based on Part ID, Machine ID and Sensor ID. The leaf nodes contain sensor data blocks that store pointers to Memtables called PSbuf or SSTables called per-Sensor Data Store (SDS) for each sensor. SDS is a dedicated block on disk that stores an append-only sequential sensor file (SSF), SSTables and SDS index. During the ingestion, Sen-Store either assigns a new ID for the sensor or retrieves an existing ID from the k-ary tree. Then, values are inserted into the PSbuf that is created for each sensor ID. When the PSbuf is full, it is flushed to the SSF and Sen-Store creates or updates SSF's in-memory index. When the size of the SSF reaches a threshold, Sen-Store converts the SSF and its in-memory index into an SSTable and updates the SDS index. Subsequently, the k-ary tree index is also updated. That ensures that the time series for a specific part, machine and sensor are stored in the same location, allowing the avoidance of further compaction of SSTables. Sen-Store can be integrated into Apache IoTDB [21] as its data processing engine. Sen-Store does not support AQP.

Cai et al. [44] propose **NBTSMS**, a centralized TSMS that is implemented on raw Non-Volatile Memory (NVM) storage. NBTSMS modifies the driver for Intel Optane DC Persistent Memory (PMEM) and implements modules for creating logical blocks using multiple default PMEM blocks and computing queries on them. NBTSMS can only store time series with regular sampling intervals. The system assumes that the most recent data for each time series is accessed more often and proposes three levels of data block organization strategy. Level 0 uses PMEM's default block size of 256 B and level 1 uses 1 KB logical blocks that combine four level 0 blocks. Level 2 consists of four level 1 blocks, respectively. For each time series, NBTSMS stores the most recent data in level 0 (i.e., the fastest block) and uses a threshold to move the time series to the next data block. To construct level 1 and level 2 blocks, NBTSMS uses a metadata file to store the addresses of reserved physical blocks. For each ingested time series, NBTSMS maintains the metadata file to maintain the following information: device ID, tag name, sampling interval, value type, threshold used for converting data blocks to the next data block type, pointers to the first data block in each data block level, pointers to median value in each level block, the first timestamp of each data block level and average compression rate of all data blocks. Values are compressed using delta encoding. A time series metadata file is used to process queries. NBTSMS uses a binary search for querying level 0 and 1 data blocks and queries on level 2 data blocks are computed using a custom prediction algorithm that computes the pointer for the required level 2 blocks. In this manner, NBTSMS eliminates the need to build custom indexes.

Wang et al. [20, 21] propose **Apache IoTDB** (IoTDB), a distributed and open-source TSMS for IoT data analytics in edge and cloud environments. The architecture of Apache IoTDB is represented in Fig. 5.2. Apache IoTDB uses a tree structure data model that aims to leverage the hierarchical nature of time series from IoT systems. Apache IoTDB treats each time series as a univariate time series with a tag. The system uses a native columnar file format called TsFile that can be stored in the local file system or in an HDFS cluster. Values in TsFile are compressed using Facebook's Gorilla [45]. Other compression methods like Run Length Encoding or custom-built precision erasing method (TS_2DIFF) are also supported. When ingesting multivariate time series where several values share the same timestamps, TsFile allows storing the timestamp only once. TsFile stores the file-level index and simple statistics for each page to accelerate queries. Apache IoTDB supports the synchronization of TsFiles produced on the edge to cloud nodes. The system supports automatic schema identification where inserts are allowed without defining the schema. Data points are first written to a WAL. Apache IoTDB uses an LSM tree, and thus, newly inserted data points are stored in Memtables before being flushed to disk as TsFiles. Out-of-order data points are inserted into another Memtable and stored in different TsFiles. Periodic compaction is performed to merge TsFiles storing out-of-order inserts with regular TsFiles. Apache IoTDB also supports deletion using tombstones. For the cloud nodes, Apache IoTDB allows replication using the non-blocking replication

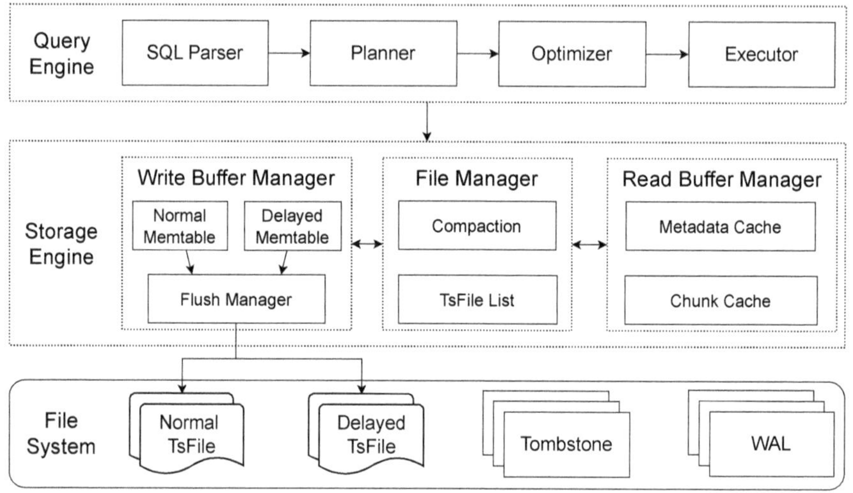

Fig. 5.2 An architecture of Apache IoTDB. Redrawn from [21]

protocol NB-Raft [46]. Apache IoTDB provides built-in connectors for interfacing with downstream applications such as Apache Flink, Apache Spark, Grafana and MATLAB. Apache IoTDB is implemented in Java and provides an SQL-like interface TQSL and native API in multiple languages. Apache IoTDB does not support AQP.

Liu et al. [47] propose a real-time indexing framework for a distributed TSMS **Khronos** which is deployed in Alibaba for performance monitoring. Khronos implements a time-partitioned LSM tree for data storage and creates inverted and k-ary tree indexes using key-value tags of time series. Indexes store a postings list of time series IDs. Data points are first ingested into a Memtable using a build thread and indexes are created. When data points in the Memtable reach a threshold, the Memtable is labeled as immutable. Then, the dump thread performs index supplementing by reusing indexes from the previous two SSTables that are preserved in memory. This allows to avoid periodical visible delays caused by building a local index for each newly started Memtable, which halts the ingestion process of a time series including many tags. Each time series is assigned a time series ID using a function that keeps track of obsolete IDs from the flushed Memtables. Before flushing, the dump thread checks if any of the time series IDs from the immutable Memtable are present in the indexes of the previous two SSTables. If they are, the thread will supplement the indexes of the immutable Memtable and store only the reference to the indexes of previous SSTables and add a pointer to the data points in the immutable Memtable. Then, the flush thread flushes the immutable Memtable to an SSTable that contains compressed data points and local indexes. Inverted indexes are compressed using SIMDNewPfor [48] and ART [49] is used for the k-ary tree index, respectively. Data points are compressed using Facebook's Gorilla compression [45] method.

Faisal et al. [50] propose **TVA**, a distributed open-source TSMS that supports the encrypted processing of time series during all steps of query execution with support for tumbling window aggregate queries and storage of unordered irregular time series. TVA stores time series using columnar in-memory tables and supports vectorized operations on them. TVA is implemented in C++ and provides a C++ library for API with similar syntax to the Pandas library [51]. The time series is distributed among multiple TVA servers residing in different geographical regions (e.g., different cloud providers). TVA supports secure multi-party computations and constantly monitors the trustworthiness of each TVA server's computations. TVA maintains a public schema of time series that is used by all users to query the system. TVA assumes that TVA servers are informed about the type of operators that are required for computation and ensures that the user only sends the query predicate values in an encrypted way. TVA supports tumbling window aggregates and aggregate queries. TVA does not support AQP.

Ding et al. [52] propose **NexusDB**, a distributed TSMS developed to manage time series generated in industrial environments. NexusDB specializes in industrial environments and assumes that time series are produced at regular intervals and that timestamps can be stored in a lossy manner by sacrificing certain precision. NexusDB uses a custom file format, NexusFile (NF), that consists of a fixed-size header stored in 16 bits: 2 bits for the NF version, 8 bits for the start timestamp, 4 bits for the sampling interval and 2 bits for the data length. The values are appended to the NF file after the header. Timestamps are not stored explicitly but are approximated by rounding to the sampling interval in the NF file header. Thus, NexusDB only supports the storage of univariate time series with regular sampling intervals. NexusDB uses a component called Index Manager for building the in-memory k-ary tree index from time series tags where leaf nodes are pointers to corresponding NF files. Thus, all queries are managed by the Index Manager to locate the physical location of NF files. The Index Manager also manages the creation, merging and deletion of NF files.

5.3.1.1 Discussion

Table 5.1 shows a summary of systems with internal data stores. Seven systems were presented. The summary shows that four out of seven TSMSs use LSM trees. Apache IoTDB [20, 21] and Khronos [47] are the only mature systems, and Apache IoTDB is an open-source system. Apache IoTDB also proposes an open-source columnar file format specifically tailored for storing time series, while no information about Khronos' on-disk format was provided. Half of the systems are distributed. Five out of seven systems provide a client library interface and only Apache IoTDB provides an SQL-like interface. Only Khronos is a real-time system. Sen-Store [43] uses embeddable LevelDB as a data store, while other systems implement proprietary data stores. Systems using LSM trees are generally the preferred option when high ingestion speed is required. Apache IoTDB, being a mature system, can be used in cases when edge-to-cloud deployment is required along with fast data ingestion. The use of compression methods like Gorilla helps to use network bandwidth efficiently

Table 5.1 Summary of systems with internal data store

	Year	Primary purpose	Deployment	Maturity	Scale shown	Data processing engine	API	Approximation	Latency	Data store	Storage layout
Karlsletter et al.	2022	Data analytics	Distributed	Proof-of-concept	4.3 TB 3 nodes	Proprietary	Client library (Python)	Not supported	Near real-time	Proprietary (C++)	Original time series and pre-computed aggregates with 13 levels of aggregates stored in Parquet
Sen-store⋆	2022	Monitoring	Centralized	Demonstration	TPCx-IoT 1 node	IoTDB	Client library (C)	Not supported	Near real-time	LevelDB	A dedicated disk block (SDS) for each time series. SDS consists of append-only log file (SSF), SSTables and index files
NBTSMS	2022	Data analytics	Centralized	Proof-of-concept	YCSB-TS 1 node	Proprietary	Unknown	Not supported	Batch	Proprietary	Three levels of data blocks on NVM disk. Each time series is distributed across levels. Level 0 uses the default disk block (256B). Level 1 (1KB) and level 2 (4KB) are logical blocks
Apache IoTDB⋆	2023	Data analytics	Distributed	Mature	IoT benchmark 20 nodes	Proprietary (Java)	SQL-like	Not supported	Near real-time	Proprietary (Java), HDFS	Columnar TsFile consisting of pages and each page stores compressed data points using lossy/lossless compression, metadata and statistics
Khronos	2023	Monitoring	Distributed	Mature	196 TB 1 node	Proprietary	Unknown	Not supported	Real-time	Proprietary	Time-partitioned LSM tree with optimized indexing that reuses the indexes from previous SSTables to supplement the index for the new SSTable
TVA	2023	Data analytics	Distributed	Proof-of-concept	Unknown 3 nodes	Proprietary (C++)	Client library (C++)	Not supported	Near real-time	Proprietary (C++)	Time-partitioned LSM and SSTables contain compressed data points and local indexes
NexusDB	2024	Data analytics	Centralized	Proof-of-concept	Unknown	Proprietary	Unknown	Timestamps	Batch	Proprietary	NexusFile consisting of a 16-bit header and values for each univariate time series with regular sampling intervals

and increase the amount of data transferred from the edge to the cloud. As the target use case of Sen-Store is industrial IoT, it confines its users to a fixed schema to offer fast ingestion and good data locality. However, the system might not suit broader use-cases. In situations where monitoring of a large number of diverse devices is required, Khronos can be used to address the high storage costs arising from the index creation. TVA can be used in cases when data security of query processing and data storage is a paramount requirement; however, ensuring this brings additional overhead like decreased speed of ingestion and query execution. NexusDB [52] is the only system that offers a timestamp approximation and thus can provide additional gains in compression. However, the system might not suit cases where exact times-tamps are required for each data point, such as in scientific data storage. NBTSMS [44] stands out from the other systems as it exclusively specializes in using modern hardware, thus offering very high ingestion and query speed. However, the system's maturity might be a bottleneck for deployment in real-life scenarios.

5.3.2 *External Data Store*

Halder et al. [53] propose **SmartCrypt**, a centralized TSMS for securely storing and access sharing for industrial IoT data. SmartCrypt uses a symmetric homomor-phic encryption method to encrypt data before storing it in Apache Cassandra [54]. SmartCrypt provides controlled access to computational queries by (1) restricting the granularity at which data consumers can run analytical queries (e.g., per-minute, per-hour) (2) restricting the time interval over which analytical queries are computed (3) restricting the data consumers to run only allowed type analytical queries. The data owner determines selective access policies for each data consumer before granting access permission. Every data consumer is provided a unique decryption key using the encrypted authorization token. Access permissions are cryptographically bound to a particular data owner's public key. A cloud server can run analytical queries using ciphertext (i.e., without decrypting the data) and send the response to a data consumer as ciphertext. Only the data consumer who is granted access permission can decrypt the ciphertext to obtain the final result. SmartCrypt uses an in-memory encrypted k-ary tree-based index to improve latency for analytical queries. SmartCrypt does not support AQP and provides a Java client library for API.

Yan et al [55]. propose CnosDB as a distributed TSMS for data analytics with support for cloud storage. CnosDB aims to provide high scalability, high ingestion speed and data compression for storage efficiency. CnosDB uses Apache DataFu-sion as its query engine. The system uses master-worker architecture and requires the deployment of at least one management node and one data node in a cluster. The management node performs query parsing, query execution planning, and job scheduling. It also stores all metadata required to maintain the databases and cluster of data nodes. It also partitions time series into so-called "bucket-vnode" groups based on timestamps. Each bucket-vnode group is then managed by a data node. CnosDB stands out from the other systems with wide support for compression of each data type it supports. It uses Delta-of-Delta Encoding for timestamp compression, a combination of ZigZag Encoding and Variable-Length Encoding for integers, Gorilla compression method for floats, Bit Packing for booleans and Snappy compression for strings. Each data node consists of a WAL module, an indexing module and a data module. The WAL module ensures data is written to WAL and the indexing module is responsible for maintaining the index cache and building forward and inverted indexes. The data module is responsible for ingesting data into the LSM tree and further management of Memtables and SSTables. CnosDB provides an SQL interface and supports AQP by downsampling time series.

Wang et al. [19] propose **ForestTI** as an open-source centralized TSMS for high ingestion performance and efficient memory use for building in-memory indexes. ForestTI optimizes inverted index implementation and uses a time-partitioned LSM tree by extending the embeddable key-value store LevelDB [56]. For ingestion, ForestTI keeps an in-memory hash table of time series tags and their physical IDs (Fig. 5.3). If a new time series is not found in the hash table, the time series is assigned a physical and logical ID using an indirection layer and adds both IDs to the hash table and inverted index. The physical ID uniquely identifies time series and is also used as a key when ingesting time series into a Memtable and SSTable. Logical ID is for the indirection layer that stores a pointer to the physical time series ID to identify if the time series object is in memory or on disk. Then, data points of a time series are ingested into a small fixed-size memory buffer with 32 values called a time series object that contains a physical time series ID, time series tags and a compressed array for timestamps and value fields. For compressing timestamps and values, Facebook's Gorilla [45] compression method is used. The time series object tags and logical ID are inserted into the two-level inverted index (i.e., distinct key-value pairs of a tag key and tag values) shown in Fig. 5.4. The first index level only contains tag keys stored in an Adaptive Radix Tree (ART) [49, 57]. Tag keys in the first index level are mapped to the second index level that manages tag values and corresponding posting lists (i.e., matching logical time series IDs).

ForestTI manages tag values and respective posting lists in three types of nodes: (1) a hash table is used for tag values with small posting lists; (2) when the size of a posting list exceeds a threshold (i.e., 10), the tag values are stored in ART with posting lists in the leaf nodes; (3) when memory space is minimum, ForestTI selects cold trie nodes according to the last accessed timestamp and evicts them to a Memtable as a key-value pair where the key is a physical time series ID concatenated with a timestamp of the first data point and the value is compressed data points. Memtables are periodically flushed to level 0 SSTables. SSTables in level 0 are stored in 30 min partitions and older partitions are flushed to level 1 once the level 0 partitions exceed the one-hour threshold. ForestTI performs compaction to store partitions with the same time series IDs in a single SSTable in level 1. This provides better data locality. The system supports storing SSTables in Amazon EBS and S3. ForestTI does not support AQP and provides an HTTP API.

Shen et al. [58] propose **Lindorm TSDB**, a distributed TSMS developed for large-scale monitoring of systems at Alibaba. Lindorm TSDB is a part of Alibaba's Cloud database service Lindorm [59]. The system uses shared-disk architecture and

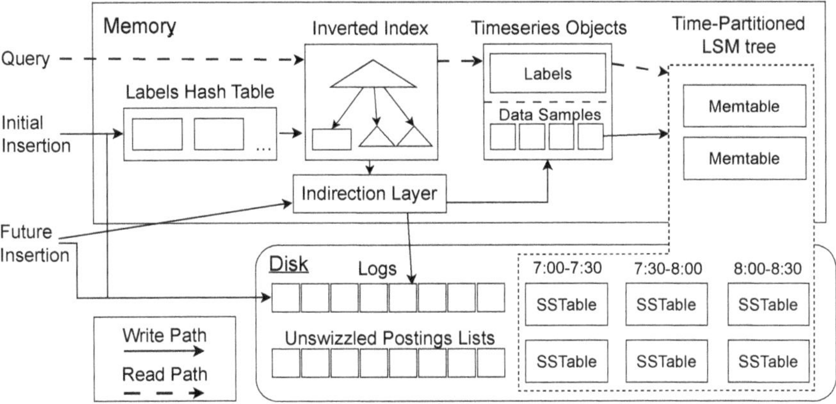

Fig. 5.3 Architecture of ForestTI. Redrawn from [19]

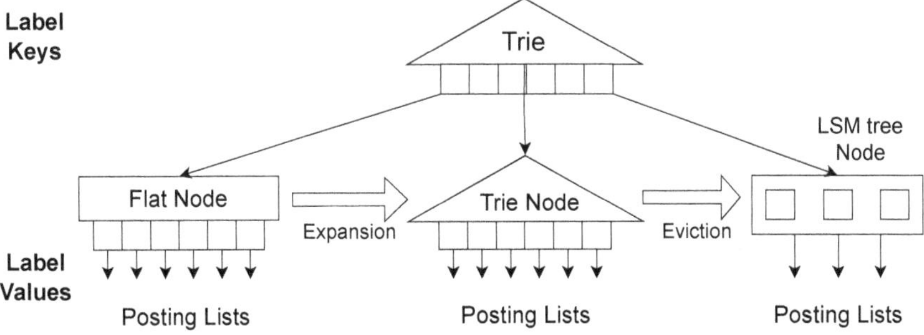

Fig. 5.4 Two-level inverted index. Redrawn from [19]

implements a Machine Learning (ML) component for ML-related computations. As Fig. 5.5 shows, Lindorm TSDB consists of four major components: TSProxy, TSCore, Lindorm DFS and Lindorm ML. TSProxy partitions data into different shards according to two dimensions: (1) *Time series id* that is used to uniquely identify one value field and a set of all associated tags of the time series; (2) *Time*. Each shard is managed by a single TSCore node. Shards reside in Lindorm DFS, a distributed file system that offers an HDFS-compatible interface. TSCore uses a time-partitioned LSM tree. Ingested data points are first written to the WAL, compressed using dictionary compression, and then stored in a Memtable in TSCore. Memtable contains the FwdIdx (i.e., forward indexes), InvIdx (i.e., inverted indexes), and TSD object for each time series. The TSD is a compressed representation of data points using Delta-of-Delta, ZigZag, XOR and Run Length Encoding. The FwdIdx and InvIdx are created for efficient querying. Periodically, TSCore flushes the Memtable to its own shards as SSTables. To check index files stored on disk, TSCore creates in-memory bloom filters. Lindorm DFS performs periodical compaction of index files and TSD files on disk. During the flush and compaction process, Lindorm TSDB

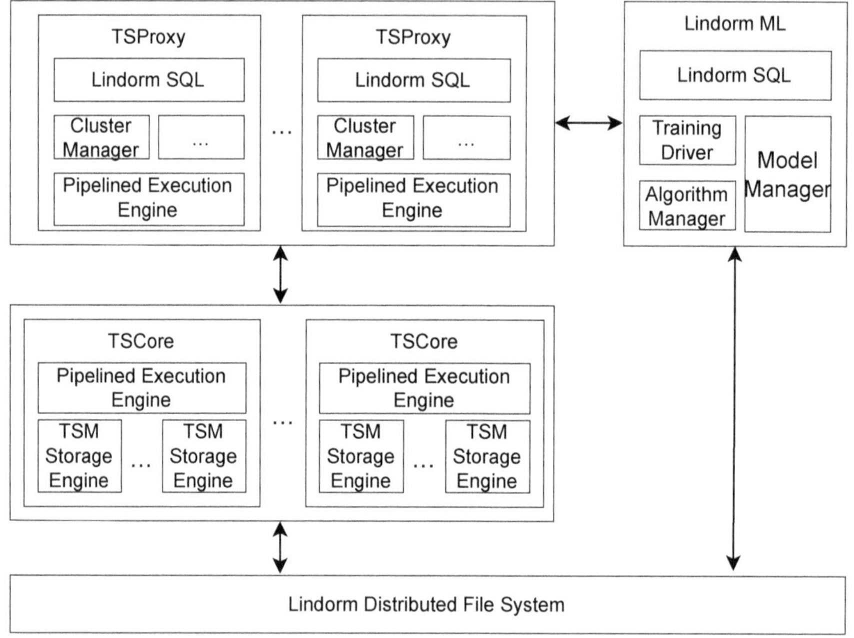

Fig. 5.5 Lindorm TSDB architecture. Redrawn from [58]

also pre-computes aggregates for different time windows and stores them in separate unspecified files. Lindorm TSDB [58] uses Apache Calcite [60] for SQL API by adding special keywords for time series aggregate queries and ML tasks. For insert and aggregate queries, Lindorm TSDB implements its own parser and executor to bypass Apache Calcite's query parsing and optimization. Queries (except for insert) are first processed by Apache Calcite's syntax parsing and then handled by a custom executor and finally passed to TSCore's execution engine to quickly scan specified time series from the data store. Lindorm ML is a machine learning component of Lindorm TSDB that provides time series analytics such as anomaly detection and time series forecasting in a distributed manner for each time series. Trained models are stored in Lindorm DFS using an unspecified format and training metadata are stored in Apache ZooKeeper [61]. Lindorm TSDB does not support approximate query processing (AQP).

Yang et al. [62] propose **MostDB** as a centralized TSMS that supports pointwise error-bounded compression of time series with a separate storage method for outliers. MostDB uses the novel MOST (model-based compression with outlier storage) com- pression method for pointwise error-bounded compression and InfluxDB [24] as an external data store (Fig. 5.6). MostDB's implementation using Apache IoTDB [20, 21] as a data store is also available. MOST compression method extends a Shrinking Cone [63] algorithm to extract outliers and split each value field into segments that can be approximated using a linear function. Then segments are approximated using a

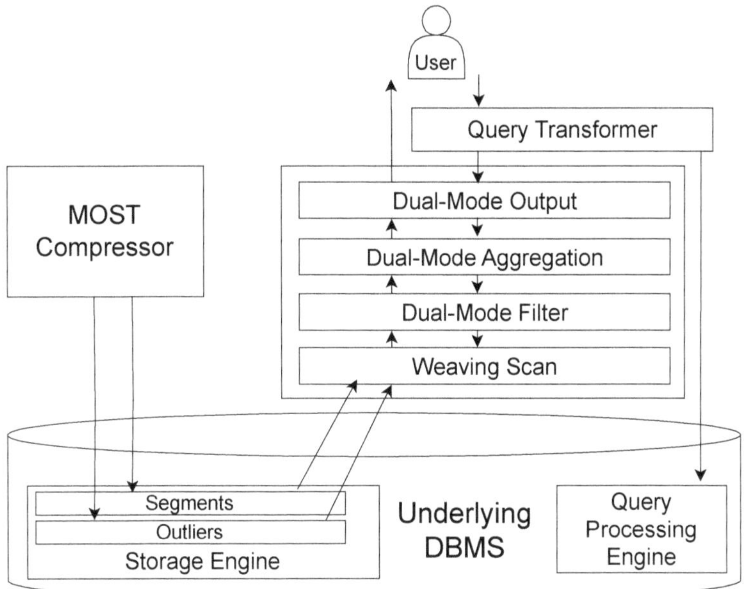

Fig. 5.6 MostDB architecture. Redrawn from [62]

first-degree polynomial function. To compress outliers, MOST creates a quantization parameter that can reconstruct the outlier value using the linear function of the adjacent segment. The quantization parameter for each outlier is further compressed using a ZigZag [64] and Variable-Length Integer Encoding. In InfluxDB, a segment's slope and intercept are compressed using Facebook's Gorilla [45] compression method. An outlier's quantization parameter is stored using Simple8b. Timestamps are encoded using Delta-of-Delta compression [45]. MostDB uses an LSM tree and stores time series tags as a key and compressed segments and outliers as a value in a Memtable before flushing Memtables to disk as SSTables. MostDB implements a dual-mode query processing method that allows handling each query operator concurrently in segment mode and outlier mode on MOST compressed data. By using volcano query execution, on every *next()* function call, MostDB returns a segment and a list of outliers associated with the segment. MostDB supports storing multivariate time series. In the case of a query including multiple columns, the so-called weave operator is applied to align the segments and outliers of all relevant value fields to create global segments and outliers. MostDB provides different query computation modes using AQP: (1) Default style computes the results using the global segments and outliers and results in lower query accuracy; (2) Style-E provides the expected error rate of the +; (3) Style-EB generates both the expected error rate and the deterministic bounds of the aggregation result (4) Style-EBS computes standard deviation in addition to the output of Style-EB. MostDB offers a proprietary Java API.

Shuo et al. [65] propose a distributed TSMS for the management of massive univariate time series from payload devices installed in a Chinese space station. The system uses Apache Cassandra [54] as its storage and data processing engine. A dynamic partitioner is implemented to intercept and rewrite each write request into index information and data entity information. The index information and data entity information are stored in the index and data entity tables, respectively. Upon a read request, the system first queries the index table to extract the needed key for extracting the data records from the data entity table. To accommodate aggregate queries, pre-computed aggregates are computed for the data in Cassandra Memtable using a 32s time window. Each time window is assigned a serial number and node ID using a tiered binary tree algorithm. Pre-computed aggregates are stored in a separate table so that each aggregation query is rewritten using a custom aggregation query algorithm that computes the set of serial numbers and node IDs required to retrieve pre-computed aggregates to compute the final aggregation result. The system uses Cassandra's CQL query language and does not support AQP.

Ouyang et al. [66] design **Lindorm-UWC** as a distributed TSMS proposed for the monitoring of time series produced by Internet of Vehicle (IoV) systems such as cars at Alibaba. The system uses a shared-disk architecture and proposes a cold-hot data separation mechanism and ultra-wide-column data store called UWCSE that combines a row-store with a column-store using an LSM tree for the row-store and Parquet for the column-store, respectively. As part of Alibaba Cloud's Lindorm [59], Lindorm-UWC shares a common component with Lindorm TSDB [58] (i.e., the underlying shared distributed file system Lindorm DFS). As Fig. 5.7 shows, Lindorm-UWC consists of three major components: LDServer, LDMaster and Lindorm DFS. LDMaster is responsible for load balancing and failover recovery for the cluster of LDServer nodes. LDMaster is implemented in an active-standby mode and uses Apache Zookeeper. Time series are distributed across different shards and each LDServer node manages a disjoint subset of shards. Shards physically reside in Lindorm DFS. During the ingestion, a data point is first written to the WAL for durability and then to a Memtable. Lindorm UWC supports a document data model where insert statements only need to specify a primary key, e.g., a combination of vehicle ID and timestamp for the key, and append any type and amount of key-value pairs (i.e., measurements) for the value (Fig. 5.8). This provides a high data ingestion rate and flexible schema that is suitable for storing IoV data from different cars with many types of components producing data in irregular sampling intervals. Periodically, LDServer flushes the Memtable to disk as SSTables. Each SSTable consists of fixed-size blocks (e.g. 32 KB) of four types: data blocks (compressed using LZ4, ZSTD) for storing value fields as key-value pairs, index blocks for storing starting primary key and an offset to sequence of data blocks, bloom block for storing the bloom filter for the primary keys in the SSTable and root block that stores the starting primary key and an offset for each index block and root block in the SSTable. Lindorm DFS performs compaction and merging of SSTables. During the flushing, UWCSE also creates a Parquet file for each value field addition to SSTables, however, users are required to specify the type of storage layout (i.e., row-store or column-store) for each query. Lindorm-UWC does not support AQP.

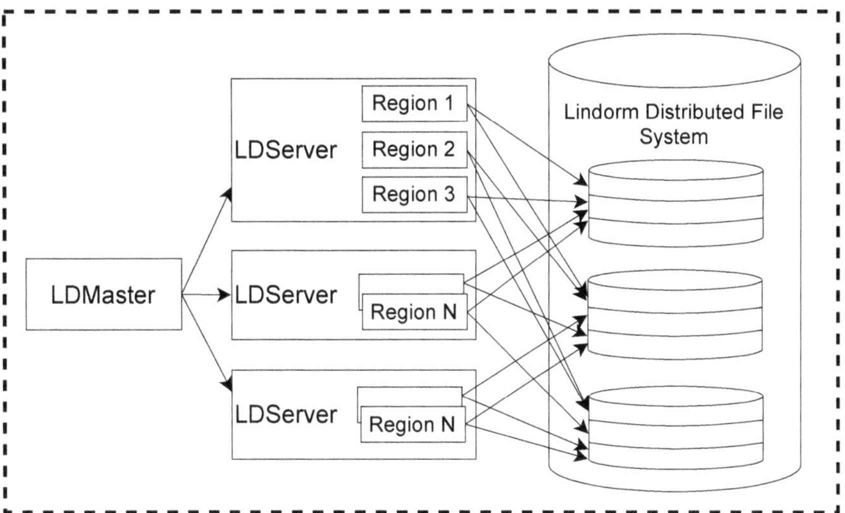

Fig. 5.7 Lindorm-UWC architecture. Redrawn from [66]

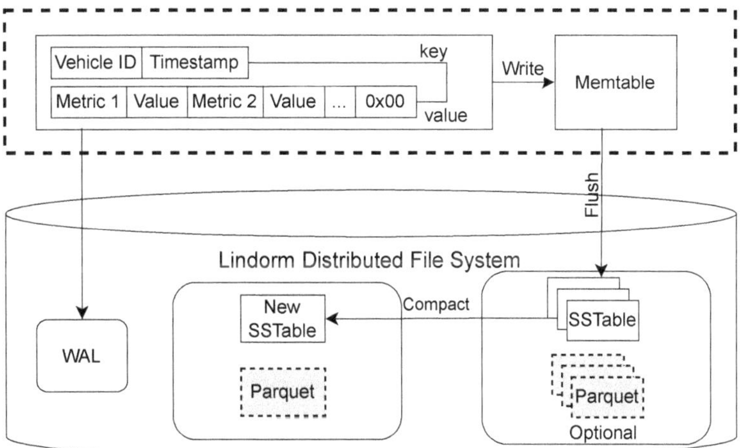

Fig. 5.8 Architecture of Lindorm-UWC's ultra-wide-column data store. Redrawn from [66]

Wang et al. [18] propose **TimeUnion** as an open-source centralized prototype TSMS developed for managing time series in hybrid cloud storage environments by combining cloud block stores and object stores. TimeUnion extends the embeddable key-value store LevelDB and uses a time-partitioned LSM tree with an optimized compaction mechanism and support for out-of-order inserts. TimeUnion proposes a grouping mechanism of time series that share common tags. For grouping time series with common tags during the ingestion, TimeUnion allocates a fixed-size multi-dimensional in-memory array of 32 values called group chunk that consists of the following arrays: (1) time series offsets; (2) timestamp array of all unique timestamps of time series in the group; (3) multiple value arrays to store values of each time series in the group. Value arrays are compressed using an extended version of Facebook Gorilla's [45] compression method with support for null values. The group ID is created from the common tags of time series in the group. For example, time series from the same region are stored in the same group. This also eliminates the need to store the same tags for each time series separately. Group IDs are used as posting lists for an inverted index. The rest of the time series tags are used to allocate a time series ID for each time series within the group. TimeUnion manages a single global inverted memory-mapped index using a double array trie [30]. The group chunk is inserted into a Memtable as a key-value pair where the key stores the group ID (64 bits) and starting timestamp for the group (64 bits) and the value stores timestamps and values in compressed byte arrays. In this way, TimeUnion ensures that the time series from the same group end up in the same Memtable in sorted order. When the Memtable is full, it is converted into an immutable Memtable and then flushed to level 0 disk storage as an SSTable. Compaction is performed to move SSTables from level 0 to level 1. Both levels are partitioned using 30-minute partitions and use fast cloud storage Amazon EBS. Level 1 has a threshold of two hours, after which the older time partitions are compacted to level 2, which uses Amazon S3. Level 2 uses one-hour partitions. In this way, TimeUnion stores more recent data in faster cloud storage. TimeUnion provides a C++ library and an HTTP interface. TimeUnion does not support AQP.

5.3.2.1 Discussion

Table 5.2 shows a summary of systems with external data stores. Eight systems with external data stores were surveyed. TimeUnion [18], CnosDB [55], ForestTI [19] and MostDB [62] are open-source systems. TimeUnion, ForestTI, MostDB and SmartCrypt [53] are the only centralized and not mature systems. For a fast and possibly lossy compression of time series, MostDB can be used. The rest of the systems are mature and support distributed deployment: CnosDB, Lindorm TSDB [58], Lindorm-UWC [66] and the system by Shuo et al. [65]. CnosDB uses reusable components such as Apache DataFusion which improves the system's compatibility with third-party tool ecosystems. Using CnosDB is applicable for cases when distributed TSMS is required that offers good data compression and high ingestion speed. The system by Shuo et al. offers fast ingestion. However, the system uses an indirection layer, which may increase the query processing time compared to using an inverted index. Lindorm TSDB stands out from the other systems as it targets a novel analytical use-case of training ML models for classification, forecasting and anomaly detection, and performing model inference directly within the system. Due to their large-scale operations, Lindorm TSDB and Lindorm-UWC use distributed cloud storage. As Lindorm-UWC targets real-time monitoring of a large fleet of vehicles, it offers a flexible schema to store sensor data from any part of the vehicle. However, both Lindorm TSDB and Lindorm-UWC are tightly coupled with a single

Table 5.2 Summary of systems with external data store

	Year	Primary purpose	Deployment	Maturity	Scale shown	Data processing engine	API	Approximation	Latency	Data store	Storage layout
SmartCrypt	2022	Data analytics	Centralized	Proof-of-concept	1.7 GB 1 node	Proprietary (Java)	Client library (Java)	Not supported	Batch	Apache cassandra	Encrypted data tables stored in Cassandra tables
TimeUnion ★	2022	Monitoring	Centralized	Demonstration	TSBS (DevOps) 1 node	Proprietary (C++)	Web interface, Client library (C++)	Not supported	Near real-time	LevelDB	Time-partitioned LSM trees with three levels of storage. Levels 0 and 1 use AWS EBS with 30 min partitions. Level 2 uses AWS S3 with 1 h partitions
CnosDB ★	2023	Data analytics	Distributed	Mature	Unknown	Apache dataFusion	SQL	AQP	Near real-time	Cloud storage	Time-partitioned LSM with two levels of TSM files (SSTables)
ForestTI ★	2023	Monitoring	Centralized	Demonstration	TSBS (DevOps) 1 node	Proprietary (C++)	Web interface, Client library (C++)	Not supported	Near real-time	LevelDB	Two-level time-partitioned LSM tree. Both levels use 30 min time partitions. Periodical compaction for merging SSTables with same time series ID and move them to next level
Lindorm TSDB	2023	Monitoring	Distributed	Mature	TSBS (DevOps) 3 nodes	Apache calcite	SQL-like	Not supported	Real-time	Lindorm DFS	Time-partitioned LSM where time series are stored as TSD (SSTables) files in a distributed cloud storage Lindorm DFS. A separate TSD file is created for each time series ID and time period

(continued)

Table 5.3 (continued)

	Year	Primary purpose	Deployment	Maturity	Scale shown	Data processing engine	API	Approximation	Latency	Data store	Storage layout
MostDB ★	2023	Data analytics	Centralized	Demonstration	1.7 GB 1 node	Proprietary (Java)	Client library (Java)	AQP	Near real-time	InfluxDB	Compressed timestamps, value segments and outliers stored in SSTables
Shuo et al.	2024	Monitoring	Distributed	Mature	768 TB 30 nodes	Proprietary (Java)	CQL	Not supported	Near real-time	Apache cassandra	Inserts written into an index and data entity information and stored index and data entity tables using Cassandra, respectively
Lindorm-UWC	2024	Monitoring	Distributed	Mature	3.2 TB 3 nodes	Proprietary	Client library (Java)	Not supported	Real-time	Lindorm DFS	Row-store combined with column-store. Time-partitioned SSTables for the row-store and Parquet files for the column-store. SSTable consisting of four fixed-size block types: data blocks for storing values, index blocks for storing pointers to a sequence of data blocks, bloom blocks for storing the Bloom filter for SSTable and root blocks that store the starting primary key and a pointer for each index block and root block

vendor's distributed cloud storage system (i.e., LDFS). TimeUnion can be used for storing data in different cloud storage options (i.e., block store and object store). Only Lindorm TSDB and the system by Shuo et al. support an SQL-like interface, while the other systems provide a client library interface. Only MostDB supports AQP using linear functions and offers different query computation modes. Only CnosDB uses extensible data processing engines and Lindorm TSDB uses Apache Calcite for query parsing and optimization. TimeUnion and ForestTI use embeddable LevelDB as a data store. ForestTI can be used as an open-source system for real-time monitoring of a large number of diverse devices with an efficient index creation strategy.

5.4   Open Problems and Future Directions

The analysis of TSMSs in Sect. 5.3 shows that the open problems in the 2022 survey [8] remain. Thus, this section relates them to the systems described in Sect. 5.3.

Support for Data Analytics
The surveyed TSMSs need better support for data analytics and user-defined methods. Most systems provide a client library that is mainly confined to a specific language and supports basic aggregate operations such as Max, Min, Avg, Count, and Last. Lindorm TSDB, a mature system, provides support for OLAP queries and novel analytical use-cases such as anomaly detection, forecasting and clustering using ML models. Regarding user-defined methods, Apache IoTDB [20, 21] is the only system that supports user-defined methods.

Integration of Edge and Cloud
Most of the surveyed TSMSs are intended for only one type of deployment. However, for efficiently transferring high frequency time series over a low network bandwidth, it is beneficial to deploy TSMSs both on the edge and in the cloud so that ingestion and compression of time series are performed on the edge, after which compressed data points are transferred to the cloud for large-scale analytics [1, 8]. From the surveyed systems, only Apache IoTDB and the system by Karlsletter et al. [41] support edge-to-cloud deployment. However, the latter is specifically designed to evaluate a novel method for computing only range queries among distributed nodes.

Standardization and Portability
Comprehensive evaluation of a TSMS becomes a challenging task due to a lack of standard query interfaces, query languages, data models and benchmarks [8]. For example, both Lindorm TSDB and Apache IoTDB provide SQL-like query languages, but they still exhibit distinct features. Although many benchmarks have been offered for TSMS, such as TS-Benchmark [67], SciTS [68] and TSM-Bench [69], they were not used by any surveyed systems. New systems are evaluated using either system-specific benchmarks or those from the TPC. The need for commonly accepted standard benchmark tools is still an open problem.

Reusable Components for Implementing TSMS
Implementing a robust TSMS is a complicated task due to the lack of reusable components, and developers are required to design most system components from scratch. However, many reusable open-source components have been developed for OLAP RDBMSs. Parquet [34] and ORC [35] have been offered for open-source big data file formats. The accumulation of big data file formats that normally results in data lakes requires efficient management. For that purpose, Delta Lake [70], Apache Hudi [71] and Apache Iceberg [72] were proposed to support transactions, indexing and other DBMS functionality on top of data lakes. To uniform the manipulation of those big data formats in memory, Apache Arrow [73] is proposed as an in-memory format. Apache DataFusion [73] and Velox [74] are used to build specialized data processing engines. In addition, query parser and optimizer Apache Calcite [60] and dataframe library Polars [75] have been proposed. For developers willing to base their system on embeddable key-value stores, LevelDB [56] and RocksDB [76] have been proposed. By reusing those components, developers can invest more time in novel distinct features of their systems.

Limits of Time Series Compression
High velocity and volume of time series result in a massive amount of data that must be stored for many years (e.g., for 25 years [6]). As [6] states, saving storage space through aggregating or simply removing old time series is a missed opportunity for the business. In recent years, there has been a significant advancement in the domain of floating-point lossless compression methods [45, 77–80]. However, the surveyed systems only use Facebook Gorilla's compression method, which is considered to be less effective compared to the methods that have been proposed lately [77–80]. In addition, the storage reduction achieved by lossless compression is limited [81] compared to lossy methods [82].

As a remedy, error-bounded lossy compression methods were proposed [82–84]. These methods are implemented in multiple systems such as Bakkalian et al. [85], TVStore [86] and ModelarDB [4, 87–90]. Several works evaluated downstream analytics with lossily compressed data. The impact of error-bounded lossy compression on the accuracy of state-of-the-art time series forecasting models is studied in [91]. The use of lossily compressed data for incremental ML tasks is studied in [6, 92] evaluates the impact of lossy compressed data on OLAP queries. The early research proves the feasibility of error-bounded lossy compression methods for downstream analytics. Thus, these methods might become a key storage option for next-generation TSMSs.

5.5   Conclusion

This chapter thoroughly surveys the state-of-the-art TSMSs designed for managing massive high frequency time series. The survey's classification criteria aim to aid readers in finding the right system based on their requirements. The majority of systems prioritize high write throughput and memory-efficient indexing structures. Some systems emphasize integration with cloud storage, which is becoming a key component of large-scale distributed systems. The systems with internal data stores primarily use LSM trees and implement proprietary data processing engines. They have fewer mature implementations compared to systems with external data stores. The systems with external data stores are primarily mature and distributed systems.

The surveyed systems generally lack good data analytics support with standard interfaces such as SQL. Most systems support one type of deployment and lack support for edge-to-cloud integration. The systems use different benchmark tools, which adds an extra layer of complexity for users to evaluate them. Reusable components for system development can significantly boost the implementation process and novelty in the domain of TSMS. The systems use generic methods for compressed storage of time series. Increasing the velocity and volume of sensor data requires next-generation TSMSs to implement more efficient compression methods to reduce storage space.

References

1. Jensen SK, Thomsen C (2023) Holistic analytics of sensor data from renewable energy sources: a vision paper. In: 27th European conference on advances in databases and information systems. Springer Cham, Barcelona, Spain, pp 360–366
2. Palpanas T, Beckmann V (2019) Report on the first and second interdisciplinary time series analysis workshop (itisa). ACM SIGMOD Rec 48(3):36–40
3. Bagnall A, Cole RL, Palpanas T, Zoumpatianos K (2019) Data series management (Dagstuhl Seminar 19282). Dagstuhl Rep 9(7):24–39
4. Jensen SK, Pedersen TB, Thomsen C (2018) Modelardb: modular model-based time series management with spark and cassandra. Proc VLDB Endow 11(11):1688–1701
5. Zoumpatianos K, Palpanas T (2018) Data series management: fulfilling the need for big sequence analytics. In: 2018 IEEE 34th international conference on data engineering (ICDE). IEEE, Paris, France, pp 1677–1678
6. Abduvakhobov A, Jensen SK, Pedersen TB, Thomsen C (2025) Scalable model-based management of massive high frequency wind turbine data with modclardb. In: Proceedings of the 51st international conference on very large data bases. VLDB Endowment, London, United Kingdom, pp 4723–4732
7. Rangaraj AG, ShobanaDevi A, Srinath Y, Boopathi K, Balaraman K (2022) Efficient and secure storage for renewable energy resource data using Parquet for data analytics. In: Sharma N, Chakrabarti A, Balas VE, Bruckstein AM (eds) Data management. Analytics and innovation, Springer, Singapore, Kolkata, India, pp 263–292
8. Jensen SK, Pedersen TB, Thomsen C (2022) Time series management systems: a 2022 Survey. In: Palpanas T, Zoumpatianos K (eds) Data series management and analytics (Forthcoming). ACM, p (In press)
9. Jensen SK, Pedersen TB, Thomsen C (2017) Time series management systems: a survey. IEEE Trans Knowl Data Eng 29(11):2581–2600
10. Zimányi E, Sakr M, Lesuisse A (2020) Mobilitydb: a mobility database based on postgresql and postgis. ACM Trans Database Syst (TODS) 45(4):1–42
11. Guo D, Wang JK (2023) Ribesdb: A time-series database at edge for the industrial internet of things. In: 2023 IEEE 16th Malaysia international conference on communication (MICC). IEEE, Kuala Lumpur, Malaysia, pp 146–150

12. Solleza F, Crotty A, Karumuri S, Tatbul N, Zdonik S (2022) Mach: a pluggable metrics storage engine for the age of observability. In: 12th conference on innovative data systems research, CIDR 2022, Chaminade, CA, USA, January 9-12, 2022, Santa Cruz, California, USA, pp 8–15. www.cidrdb.org
13. Solleza F, Li S, Sun W, Tang R, Schwarzkopf M, Tatbul N, Crotty A, Cohen D, Zdonik S (2024) Mach: Firefighting time-critical issues in complex systems using high-frequency telemetry. In: Proceedings of the VLDB endowment. VLDB Endowment, Guangzhou, China, pp 4425–4428
14. O'Neil P, Cheng E, Gawlick D, O'Neil E (1996) The log-structured merge-tree (lsm-tree). Acta Informatica 33:351–385
15. Rosenblum M, Ousterhout JK (1992) The design and implementation of a log-structured file system. ACM Trans Comput Syst (TOCS) 10(1):26–52
16. Chang F, Dean J, Ghemawat S, Hsieh WC, Wallach DA, Burrows M, Chandra T, Fikes A, Gruber RE (2008) Bigtable: a distributed storage system for structured data. ACM Trans Comput Syst (TOCS) 26(4):1–26
17. Petrov A (2019) Database internals: a deep dive into how distributed data systems work. O'Reilly Media, pp 129–152
18. Wang Z, Shao Z (2022) TimeUnion: an efficient architecture with unified data model for timeseries management systems on hybrid cloud storage. In: Proceedings of the 2022 international conference on management of data (SIGMOD). Association for Computing Machinery, Philadelphia, PA, USA, pp 1418–1432
19. Wang Z, Shao Z (2023) Forestti: a scalable inverted-index-oriented timeseries management system with flexible memory efficiency. Proc ACM Manag Data 1(2):1–25
20. Wang C, Huang X, Qiao J, Jiang T, Rui L, Zhang J, Kang R, Feinauer J, McGrail KA, Wang P, Luo D, Yuan J, Wang J, Sun J (2020) Apache IoTDB: time-series database for internet of things. Proc VLDB Endow 13(12):2901–2904
21. Wang C, Qiao J, Huang X, Song S, Hou H, Jiang T, Rui L, Wang J, Sun J (2023) Apache IoTDB: a time series database for IoT applications. Proc ACM Manag Data 1(195):1–27
22. Kleppmann M (2017) Designing data-intensive applications: the big ideas behind reliable, scalable, and maintainable systems. O'Reilly Media, pp 69–111
23. Bloom BH (1970) Space/time trade-offs in hash coding with allowable errors. Commun ACM 13(7):422–426
24. InfluxData (2024) Get started with influxdb v2. https://docs.influxdata.com/influxdb/v2/
25. Apache Druid (2024) Apache Druid. https://druid.apache.org/
26. Colantonio A, Di Pietro R (2010) Concise: compressed 'n' composable integer set. Inf Process Lett 110(16):644–650
27. Lemire D, Kaser O, Kurz N, Deri L, O'Hara C, Saint-Jacques F, Ssi-Yan-Kai G (2018) Roaring bitmaps: implementation of an optimized software library. Softw Practice Exp 48(4):867–895
28. Pibiri GE, Venturini R (2020) Techniques for inverted index compression. ACM Comput Surv (CSUR) 53(125):1–36
29. Shi X, Feng Z, Li K, Zhou Y, Jin H, Jiang Y, He B, Ling Z, Li X (2020) Byteseries: an in-memory time series database for large-scale monitoring systems. In: SoCC '20: proceedings of the 11th ACM symposium on cloud computing. Association for Computing Machinery, USA, pp 60–73
30. Yoshinaga N (2021) cedar—c++ implementation of efficiently-updatable double-array trie. http://www.tkl.iis.u-tokyo.ac.jp/~ynaga/cedar/
31. Chardin B, Lacombe JM, Petit JM (2015) Chronos: a nosql system on flash memory for industrial process data. Distrib Parallel Databases 34(1):293–319
32. Cao W, Gao Y, Li F, Wang S, Lin B, Xu K, Feng X, Wang Y, Liu Z, Zhang G (2020) Timon: a timestamped event database for efficient telemetry data processing and analytics. In: Proceedings of the 2020 ACM SIGMOD international conference on management of data. Association for Computing Machinery, Portland, USA, pp 739–753
33. Tan J, Ghanem T, Perron M, Yu X, Stonebraker M, DeWitt D, Serafini M, Aboulnaga A, Kraska T (2019) Choosing a cloud dbms: architectures and tradeoffs. Proc VLDB Endow 12(12):2170–2182

34. Apache Parquet (2024) Apache Parquet. https://parquet.apache.org/
35. Apache ORC (2024) Apache orc, the smallest, fastest columnar storage for hadoop workloads. https://orc.apache.org/
36. Apache Arrow (2024) The universal columnar format and multi-language toolbox for fast data interchange and in-memory analytics. https://arrow.apache.org/
37. AWS (2024a) Amazon Elastic Block Store. https://aws.amazon.com/ebs/
38. AWS (2024b) Amazon S3. https://aws.amazon.com/s3/
39. Timescale (2024) Get started with timescale. https://docs.timescale.com/self-hosted/latest/
40. Özsu MT, Valduriez P (2020) Principles of distributed database systems, 4th edn, Springer Cham, Chap. Parallel Database Systems
41. Karlstetter R, Widhopf-Fenk R, Schulz M (2022) Querying distributed sensor streams in the edge-to-cloud continuum. In: 2022 IEEE international conference on edge computing and communications (EDGE). IEEE, Barcelona, Spain, pp 192–197
42. Apache Arrow Flight (2024) Arrow flight RPC. https://arrow.apache.org/docs/format/Flight.html
43. Choi K, Han H, Jung H, Kang S (2022) Workload-optimized sensor data store for industrial IoT gateways. Futur Gener Comput Syst 135:394–408
44. Cai T, Ma Y, Niu D, Gao P, Lei T, Dai J (2022) A new IoT storage system based on raw NVM. In: 2022 IEEE international conference on big data (Big Data). IEEE, Osaka, Japan, pp 367–372
45. Pelkonen T, Franklin S, Teller J, Cavallaro P, Huang Q, Meza J, Veeraraghavan K (2015) Gorilla: a fast, scalable, in-memory time series database. Proc VLDB Endow 8(12):1816–1827
46. Jiang T, Huang X, Song S, Wang C, Wang J, Li R, Sun J (2023) Non-blocking raft for high throughput IoT data. In: 2023 IEEE 39th international conference on data engineering (ICDE). IEEE, San Diego, CA, USA, pp 1140–1152
47. Liu X, Wei Z, Yu W, Liu S, Wang G, Liu X, Li Y (2023) Khronos: a real-time indexing framework for time series databases on large-scale performance monitoring systems. In: Proceedings of the 32nd ACM international conference on information and knowledge management (CIKM). Association for Computing Machinery, Birmingham, United Kingdom, pp 1607–1616
48. Lemire D, Boytsov L (2015) Decoding billions of integers per second through vectorization. Softw Practice Exp 45(1):1–29
49. Leis V, Kemper A, Neumann T (2013) The adaptive radix tree: artful indexing for main-memory databases. In: 2013 IEEE 29th international conference on data engineering (ICDE). IEEE, Brisbane, Australia, pp 38–49
50. Faisal M, Zhang J, Liagouris J, Kalavri V, Varia M (2023) TVA: a multi-party computation system for secure and expressive time series analytics. In: Proceedings of the 32nd USENIX security symposium. USENIX, Anaheim, CA, USA, pp 5395–5412
51. Pandas (2024) pandas is a fast, powerful, flexible and easy to use open source data analysis and manipulation tool, built on top of the python programming language. https://pandas.pydata.org/
52. Ding L, Chzhen DY, Li Y, Zhang Z, Xie Z, Li M (2024) Nexusdb: a large-scale distributed time-series database for industrial scenarios. In: Zhang W, Tung A, Zheng Z, Yang Z, Wang X, Guo H (eds) Web and big data. Springer Nature Singapore, Jinhua, China, pp 408–412
53. Halder S, Newe T (2022) Secure time series data sharing with fine-grained access control in cloud-enabled IIoT. In: NOMS 2022–2022 IEEE/IFIP network operations and management symposium. IEEE, Budapest, Hungary, pp 1–9
54. Apache Cassandra (2024) Open source NoSQL database. https://cassandra.apache.org/_/index.html
55. Yan Y, Zheng B, Wang H, Zhang J, Wang Y (2023) CnosDB: a flexible distributed time-series database for large-scale data. In: Wang X, Sapino ML, Han WS, El Abbadi A, Dobbie G, Feng Z, Shao Y, Yin H (eds) Database systems for advanced applications. Springer Nature Switzerland, Cham, pp 696–700
56. Ghemawat S, Dean J (2021) Leveldb. https://github.com/google/leveldb

57. Leis V, Scheibner F, Kemper A, Neumann T (2016) The art of practical synchronization. In: DaMoN '16: Proceedings of the 12th international workshop on data management on new hardware, vol 3. Association for Computing Machinery, San Francisco, California, USA, pp 1–8
58. Shen C, Ouyang Q, Li F, Liu Z, Zhu L, Zou Y, Su Q, Yu T, Yi Y, Hu J, Zheng C, Wen B, Zheng H, Xu L, Pan S, Wu B, He X, Li Y, Tan J, Wang S, Pei D, Zhang W, Li F (2023) Lindorm TSDB: a cloud-native time-series database for large-scale monitoring systems. Proc VLDB Endow 16(12):3715–3727
59. Alibaba Cloud (2024) Lindorm. https://www.alibabacloud.com/en/product/lindorm?_p_lc=1
60. Begoli E, Camacho-Rodríguez J, Hyde J, Mior MJ, Lemire D (2018) Apache calcite: a foundational framework for optimized query processing over heterogeneous data sources. In: SIGMOD '18: Proceedings of the 2018 international conference on management of data. Association for Computing Machinery, Houston, TX, USA, pp 221–230
61. Apache ZooKeeper (2020) Welcome to Apache ZooKeeper. https://zookeeper.apache.org/
62. Yang Z, Chen S (2023) MOST: model-based compression with outlier storage for time series data. Proc ACM Manag Data 1(250):1–29
63. Galakatos A, Markovitch M, Binnig C, Fonseca R, Kraska T (2019) Fiting-tree: a data-aware index structure. In: Proceedings of the 2019 international conference on management of data. Association for Computing Machinery, Amsterdam, Netherlands, pp 1189–1206
64. Google (2001) Protocol buffers encoding. https://developers.google.com/protocol-buffers/docs/encoding#types
65. Yan S, Wang J, Liang J (2024) Big data storage and analysis system for space application. In: 2024 27th international conference on computer supported cooperative work in design (CSCWD). IEEE, Tianjin, China, pp 1764–1769
66. Ouyang Q, Shen C, Yang W, Yu P, Xiao Q, Lei J, Chen Y, Zhong Q, Wang X, Lin Y, Meng Q, Ji Z, Meng W, Zheng C, Wang S, Pei D, Zhang W, Li F, Zhou J (2024) Lindorm-UWC: an ultra-wide-column database for internet of vehicles. Proc VLDB Endow 17(12):4117–4129
67. Hao Y, Qin X, Chen Y, Li Y, Sun X, Tao Y, Zhang X, Du X (2021) Ts-benchmark: a benchmark for time series databases. In: 2021 IEEE 37th international conference on data engineering (ICDE). IEEE, Chania, Greece, pp 588–599
68. Mostafa J, Wehbi S, Chilingaryan S, Kopmann A (2022) SciTS: a benchmark for time-series databases in scientific experiments and industrial internet of things. In: SSDBM '22: Proceedings of the 34th international conference on scientific and statistical database management. Association for Computing Machinery, Copenhagen Denmark, pp 1–11
69. Khelifati A, Khayati M, Dignös A, Difallah D, Cudré-Mauroux P (2023) TSM-Bench: benchmarking time series database systems for monitoring applications. Proc VLDB Endow 16(11):3363–3376
70. Armbrust M, Das T, Sun L, Yavuz B, Zhu S, Murthy M, Torres J, van Hovell H, Ionescu A, Łuszczak A, undefinedwitakowski M, Szafrański M, Li X, Ueshin T, Mokhtar M, Boncz P, Ghodsi A, Paranjpye S, Senster P, Xin R, Zaharia M, (2020) Delta lake: high-performance ACID table storage over cloud object stores. Proc VLDB Endow 13(12):3411–3424
71. Apache Hudi (2024) Apache Hudi. https://hudi.apache.org/
72. Apache Iceberg (2024) The open table format for analytic datasets. https://iceberg.apache.org/
73. Lamb A, Shen Y, Heres D, Chakraborty J, Kabak MO, Hsieh LC, Sun C (2024) Apache arrow datafusion: a fast, embeddable, modular analytic query engine. In: SIGMOD/PODS '24: companion of the 2024 international conference on management of data. Association for Computing Machinery, Santiago, Chile, pp 5–17
74. Pedreira P, Erling O, Basmanova M, Wilfong K, Sakka L, Pai K, He W, Chattopadhyay B (2022) Velox: meta's unified execution engine. Proc VLDB Endow 15(12):3372–3384
75. Polars (2024) Dataframes for the new era. https://pola.rs/
76. RocksDB (2024) A persistent key-value store for fast storage environments. https://rocksdb.org/
77. Liakos P, Papakonstantinopoulou K, Kotidis Y (2022) Chimp: efficient lossless floating point compression for time series databases. Proc VLDB Endow 15(11):3058–3070

78. Liakos P, Papakonstantinopoulou K, Bruineman T, Raasveldt M, Kotidis Y (2024) How to make your duck fly: advanced floating point compression to the rescue. In: Proceedings 27th international conference on extending database technology, EDBT 2024, Paestum, Italy, March 25–March 28, OpenProceedings.org, Paestum, Italy, pp 826–829

79. Li R, Li Z, Wu Y, Chen C, Zheng Y (2023) Elf: erasing-based lossless floating-point compression. Proc VLDB Endow 16(7):1763–1776

80. Afroozeh A, Kuffo LX, Boncz P (2023) ALP: adaptive lossless floating-point compression. Proc ACM Manag Data 1(230):1–26

81. Ringwelski M, Renner C, Reinhardt A, Weigel A, Turau V (2012) The hitchhiker's guide to choosing the compression algorithm for your smart meter data. In: 2012 IEEE international energy conference and exhibition (ENERGYCON). IEEE, Florence, Italy, pp 935–940

82. Kitsios X, Liakos P, Papakonstantinopoulou K, Kotidis Y (2023) Sim-piece: highly accurate piecewise linear approximation through similar segment merging. Proc VLDB Endow 16(8):1910–1922

83. Lazaridis I, Mehrotra S (2003) Capturing sensor-generated time series with quality guarantees. In: Proceedings 19th international conference on data engineering (Cat. No.03CH37405). IEEE, Bangalore, India, pp 429–440

84. Elmeleegy H, Elmagarmid AK, Cecchet E, Aref WG, Zwaenepoel W (2009) Online piecewise linear approximation of numerical streams with precision guarantees. Proc VLDB Endow 2(1):145–156

85. Bakkalian G, Koncilia C, Wrembel R (2016) On representing interval measures by means of functions. In: Bellatreche L, Pastor Ó, Almendros Jiménez JM, Aït-Ameur Y (eds) Model and data engineering. Springer International Publishing, Almería, Spain, pp 180–193

86. An Y, Su Y, Zhu Y, Wang J (2022) TVStore: automatically bounding time series storage via Time-Varying compression. In: 20th USENIX conference on file and storage technologies (FAST 22). USENIX Association, Santa Clara, CA, pp 83–100

87. Jensen SK, Pedersen TB, Thomsen C (2019) Demonstration of modelardb: Model-based management of dimensional time series. In: Proceedings of the 2019 ACM SIGMOD international conference on management of data. Association for Computing Machinery, Amsterdam, Netherlands, pp 1933–1936

88. Jensen SK, Pedersen TB, Thomsen C (2021) Scalable model-based management of correlated dimensional time series in modelardb+. In: 2021 IEEE 37th international conference on data engineering (ICDE). IEEE, Chania, Greece, pp 1380–1391

89. Jensen SK, Thomsen C, Pedersen TB (2023) ModelarDB: integrated model-based management of time series from edge to cloud. Springer, Berlin, pp 1–33

90. Jensen SK, Thomsen C, Pedersen TB, Muñiz-Cuza CE, Abduvakhobov A (2024) Why model-based lossy compression is great for wind turbine analytics. In: 2024 IEEE 40th international conference on data engineering (ICDE). IEEE, Utrecht, Netherlands, pp 5667–5668

91. Cuza CEM, Jensen SK, Brusokas J, Ho N, Pedersen TB (2024) Evaluating the impact of error-bounded lossy compression on time series forecasting. In: Proceedings 27th international conference on extending database technology, EDBT 2024, Paestum, Italy, March 25–March 28, OpenProceedings.org, Paestum, Italy, pp 650–663

92. Tirupathi S, Salwala D, Zizzo G, Rawat A, Purcell M, Kejser Jensen S, Thomsen C, Ho N, Muñiz-Cuza CE, Brusokas J, Bach Pedersen T, Alexiou G, Giannopoulos G, Gidarakos P, Kalimeris A, Maroulis S, Papastefanatos G, Psarros I, Stamatopoulos V, Terrovitis M (2022) Machine learning platform for extreme scale computing on compressed IoT data. In: 2022 IEEE international conference on big data (Big Data). IEEE, Osaka, Japan, pp 3179–3185

Chapter 6
MLOps Systems for Developing ML Pipelines

Antonios Kontaxakis⊙, Dimitris Sacharidis⊙, Alkis Simitsis⊙,
Alberto Abelló⊙, and Sergi Nadal⊙

Abstract Applications based on machine learning (ML) have revolutionized many domains, including economics, health care, security, etc., which includes an ever-increasing need for fast and efficient development of pipelines. This is a trial-and-error process that involves constantly creating, executing, and evaluating new pipelines until an acceptable level of model performance is achieved. Recently, industry and research have developed MLOps systems to support data scientists in these tasks. These systems assist by automating pipeline design, providing execution environments, and tracking valuable metadata to support assessment. However, despite recent advancements, most MLOps systems only support parts of the design, execution, and evaluation cycle, while recent surveys raise concerns about their sustainability and cost-efficiency. In this chapter, we first describe the main challenges MLOps systems must address at different stages of ML pipeline development. We then outline techniques used and state-of-the-art systems that implement these techniques to tackle these challenges. Lastly, we highlight the remaining challenges in the field and present future research directions.

Keywords MLOps systems · ML pipelines · Automation · Optimization · AutoML

A. Kontaxakis (✉) · D. Sacharidis
Department of Computer & Decision Engineering (CoDE) CP 165/15, Université libre de Bruxelles, Brussels, Belgium
e-mail: antonios.kontaxakis@ulb.be

D. Sacharidis
e-mail: dimitris.sacharidis@ulb.be

A. Simitsis
Athena Research Center, Athens, Greece
e-mail: alkis@athenarc.gr

A. Abelló · S. Nadal
Universitat Politècnica de Catalunya, Barcelona, Spain
e-mail: alberto.abello@upc.edu

S. Nadal
e-mail: sergi.nadal@upc.edu

© The Author(s) 2026
G. Dejaegere et al. (eds.), *Data Engineering for Data Science*,
https://doi.org/10.1007/978-3-032-18765-9_6

6.1 Introduction

In recent years, we have seen an unprecedented rise in the demand for machine learning (ML) pipelines [1] that leverage advancements in natural language models, deep learning, and traditional ML to address a wide range of problems such as classification, regression, anomaly detection, clustering, and recommendation. This growing demand has significantly increased the workload for data science teams and the cost of developing ML pipelines. Additionally, the complexity of these pipelines has grown, requiring more sophisticated tools and methodologies to efficiently manage data preprocessing, model training, and evaluation.

Various ML operations (MLOps) systems have been developed to support data scientists by reducing the manual effort required to manage growing workloads. These systems include frameworks, platforms, or toolchains–a set of interconnected tools–that automate and manage the end-to-end lifecycle of machine learning models, providing features that support the design, execution, and evaluation of ML-based workloads. These systems aim to enhance productivity, improve reproducibility, and ensure scalability in ML workflows. They generally fall into one or both of the following broad categories.

AutoML systems, including Auto-sklearn [2], TPOT [3], and H2O [4], automate the process of designing ML pipelines by performing hyperparameter tuning, feature selection, and model selection. These systems iteratively explore different models and preprocessing steps to identify the most effective solutions for a given task. Some, like Auto-sklearn [2], incorporate meta-learning to transfer knowledge from past tasks, while others, such as TPOT, use evolutionary algorithms to optimize pipelines over successive iterations.

ML Workflow management systems, such as MLflow [5] and Kubeflow [6], streamline ML development by providing automated provenance tracking, metadata collection and storage for models and datasets. These systems track the lineage of experiments, ensuring that model versions, hyperparameters, and data transformations are recorded and reproducible. Furthermore, they facilitate collaboration by integrating with cloud-based platforms and version control systems, enabling teams to share and manage ML artifacts (models and datasets).

While these systems are valuable for the automation they provide, they offer little to no features that consider the cost associated with executing the workloads that data scientists create. One could argue that, in fact, they often lead to increased costs as pipelines are developed at an ever-increasing rate and need to be evaluated, raising the computational cost. Moreover, many adopt a "store-everything" policy, continuously increasing storage requirements. There is an apparent lack of systems that provide automation while leveraging available optimization opportunities.

For many years, various optimization techniques–such as materialization, reuse, operator selection, and flow optimization–have been introduced within the fields of computer science and database research to tackle the challenge of reducing execution and storage costs. These techniques have proven effective in traditional computing environments. However, they need to be adapted to keep pace with current trends in

ML automation, which have introduced new challenges. The dynamic and iterative nature of ML workflows and the need for continuous retraining and monitoring require specific optimization strategies that can intelligently balance performance, cost, and storage efficiency. Lately, Several Systems have been developed to address this lack of optimization and have shown promising results [7, 8]. We will refer to them as **ML execution optimizers**.

To our knowledge, no commercial system effectively addresses the increasing cost of ML model development while maintaining automation. Most available solutions prioritize automation at the expense of optimization, leading to inefficiencies in resource utilization. These issues emphasize the need to incorporate effective optimization techniques into MLOps systems, making their advancement more critical than ever and presenting new opportunities for research in optimization and automation. By integrating cost-aware automation strategies, the next generation of MLOps systems can bridge the gap between efficiency and scalability, ensuring sustainable ML model development [9].

In what follows, we first describe the stages and challenges of developing ML models. Section 6.3 presents a taxonomy of available methods and techniques that address these challenges. Next, in Sect. 6.4, we review the state-of-the-art systems (AutoML, ML execution optimizers, and ML Workflow management) that leverage techniques presented in Sect. 6.3. We conclude by discussing unresolved challenges and outlining directions for future research in Sect. 6.5.

6.2 The Stages and Challenges of Developing ML Models

The process of developing an ML model is an interactive trial-and-error process with three stages: design, execution, and tracking. In what follows, we describe each stage and outline its challenges.

6.2.1 Design

The first stage involves designing a pipeline as a set of building blocks known as operators, which perform essential tasks such as data preprocessing, feature selection, and model training and are interconnected in a Directed Acyclic Graph (DAG), that ensures a logical flow of data. Recent advancements in pipeline design may also require the training of multiple models, either in a single operator (like in the case of using an ensemble predictor [10, 11]) or in multiple operators (like feature selection, labeling, or imputation). Additionally, each operator includes hyperparameters that require careful configuration to optimize performance.

The collection of all possible design options, including the reorganization of operators and their hyperparameters, is called the search space. The search space can be homogeneous, in which all operators require the same execution environment (e.g., implemented in Python), or heterogeneous. Additionally, researchers may rely on prior knowledge to construct a search space with operators likely to yield effective results [2]. As a result, the design stage involves (1) an ever-expanding pool of operators, their combinations, and the associated hyperparameters and (2) making use of the search space to extract feasible and compatible DAG of operators that can perform all the required tasks.

The primary challenges in this stage include:

D1-Designing the search space : Constructing a search space that is both comprehensive and feasible is a significant challenge. A poorly designed search space can lead to suboptimal results or excessive computational demands, especially as the number of operators and hyperparameters grows exponentially [12].

D2-Defining an effective search strategy : Choosing a search strategy that balances exploration and exploitation is critical to optimizing the design process. The strategy must ensure design diversity to avoid local optima while focusing computational resources on promising configurations [12].

D3-Incorporating prior knowledge effectively : Leveraging domain expertise or information from previous evaluations to prioritize specific operators or configurations can improve efficiency. However, integrating this knowledge should avoid introducing bias or excluding promising solutions [13].

6.2.2 Execution

Once a set of pipelines is designed and passed to an execution environment, we get a set of evaluation metrics measuring their efficiency and effectiveness. Execution is the most resource-intensive stage, and it heavily depends on selecting an evaluation strategy and the optimization techniques that can be exploited. Moreover, each pipeline may need to be executed multiple times due to the need for the model to be retrained when sufficient new data have been acquired.

Various evaluation strategies have been developed over the years, which can be broadly categorized based on their focus on quality or execution time. Depending on the task's requirements, these strategies measure the effectiveness of the pipeline in one way or another. Optimization techniques can also be employed at this stage to minimize computational costs. Examples include reordering operators [14–16], selecting efficient implementations [8, 17, 18], and reusing precomputed intermediate results from previous evaluations [7], which can significantly reduce redundant computations and enhance efficiency.

The primary challenges in this stage include:

E1-Selecting an appropriate evaluation strategy : Choosing the right evaluation strategy (e.g., cross-validation, hold-out testing, or racing algorithms) is critical to balance accuracy and computational efficiency. Misaligned strategies can cause unnecessary overhead (e.g., high crossover requiring multiple training runs on data subsets) or unreliable evaluation (e.g., racing algorithms skipping crossover, which helps in complex tasks) [19].

E2-Incorporating optimization techniques : Optimization techniques, such as operator reordering or reusing precomputed results, can reduce evaluation costs. However, these techniques must be carefully implemented to maintain the validity and reliability of results [8].

6.2.3 Tracking

After execution, the pipelines are compared with each other and, if possible, with those evaluated in past iterations of the design-execute cycle to assess progress and identify areas of improvement. This can be achieved by collecting metadata and the provenance of the pipelines. This essential information includes the datasets, operators, and configurations used, enabling reproducibility and traceability of experiments. The insights gathered during the execution guide the refinement of pipelines in future iterations, driving systematic improvements and enabling a more efficient development process.

Moreover, during this stage, materialization policies are often employed to store computed results, such as feature transformations or partially trained models, for reuse in subsequent iterations. These policies can significantly accelerate development cycles and reduce resource consumption by avoiding redundant computations.

A1-Ensuring reproducibility and traceability : Managing metadata and provenance can become complex as the number of iterations increases, especially when using large datasets and distributed workflows. Properly tracking changes in datasets, parameters, and configurations is critical [20].

A2-Efficient materialization policies : Deciding which intermediate results to materialize and how to store them without incurring excessive storage costs or slowing down workflows is a non-trivial optimization problem [7, 21].

A3-Insight extraction for iterative improvements :   Extracting   actionable insights from the results and metadata to guide subsequent iterations can be difficult, mainly when dealing with conflicting metrics or multi-objective optimization tasks [22].

6.3 The Landscape of MLOps

MLOps systems assist data scientists in developing ML pipelines across one or more of the key stages mentioned above. In the following sections, we first outline the core functionalities offered by MLOps for each stage, followed by an exploration of advanced features and techniques, which are highlighted in academia. Figure 6.1 shows a taxonomy of the stages, systems category, and techniques.

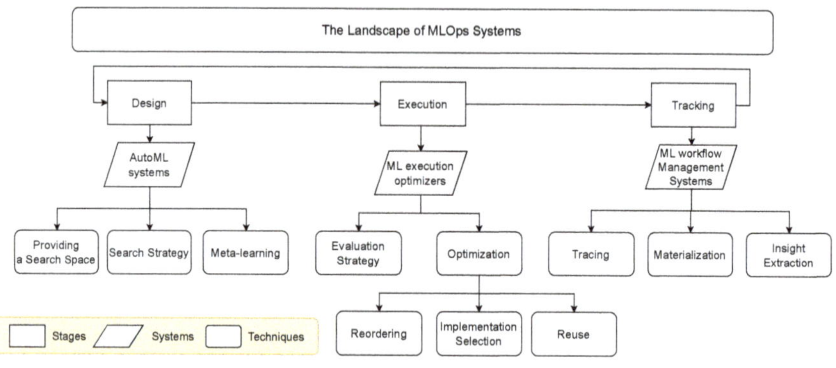

Fig. 6.1 Taxonomy of MLOps

6.3.1 Design

To support the design of pipelines, modern MLOps provide domain-specific languages (DSL) that allow users to express their workloads as a set of building blocks with defined dependencies. Some systems extend this by offering libraries of popular, pre-implemented techniques. For instance, TPOT [3] includes a predefined catalog of operators for solving classification problems, as shown in Table 6.1. More advanced systems even take this further by providing predefined search spaces tailored to specific problem types (e.g., classification, regression).

All together enables MLOps to efficiently define search spaces and implement search strategies. Additionally, prior knowledge is often incorporated through meta-learning techniques, which leverage insights from previous evaluations to optimize the design process. These techniques address design challenges described in **D1**, Sect. 6.2.1.

Table 6.1 Example list of operators for preprocessing, feature selection, and learners

Preprocessing	Feature selection	Learners
Binarizer	SelectFwe	GaussianNB
StandardScaler	SelectPercentile	XGBClassifier
OneHotEncoder	VarianceThreshold	LinearSVC
MaxAbsScaler	RFE	DecisionTreeClassifier
MinMaxScaler	SelectFromModel	ExtraTreesClassifier
ZeroCount	PCA	KNeighborsClassifier
Normalizer	PolynomialFeatures	RandomForestClassifier
Nystroem	FastICA	GradientBoostingClassifier
RBFSampler	FeatureAgglomeration	SGDClassifier

6.3.1.1 Providing a Search Space

MLOps systems construct a search space as a catalog of popular building blocks, often selected based on their widespread usage [23], while allowing users to implement custom operators incorporating domain knowledge. Common building blocks in ML include algorithms such as Random Forest, Support Vector Machines, Gradient Boosting, and preprocessing techniques like Normalization and Feature Selection.

While many MLOps leverage a single library (e.g., scikit-learn), others support multiple libraries and execution environments [24], enabling the design of heterogeneous applications. This approach allows for a broader search space, as not all popular techniques are available in a single library or execution environment. For instance, a search space might include Scikit-learn's 'RandomForestClassifier', Keras's 'LSTM-TextClassifier', and XGBoost's 'XGBClassifier', each requiring distinct backends, dependencies, or configurations.

Thus, MLOps often include additional operators that transform data from the data structure of one library to another to accommodate data compatibility between components [24]. For example, while scikit-learn primarily uses NumPy arrays and TensorFlow relies on Tensors, integrating the two requires transforming the former into the latter and vice versa. Similarly, applications that combine scikit-learn for preprocessing, TensorFlow for neural networks, and PyTorch for advanced model tuning must handle diverse data formats and ensure execution environment compatibility, such as managing dependencies. Achieving this level of flexibility relies on metadata describing operators' input/output formats and environment dependencies, ensuring that potential solutions are feasible and well-integrated across various frameworks and environments.

6.3.1.2 Search Strategy

The search strategy determines how the search space is explored to identify optimal solutions. Various techniques have been developed, each suited to different objectives, resource constraints, and problem complexities. Bayesian optimization is a widely used iterative strategy for efficiently exploring high-dimensional spaces. By modeling the objective function, it focuses on promising regions, making it particularly effective for tasks where evaluations are computationally expensive [25]. In contrast, random search is a simpler approach that explores the search space by randomly sampling configurations. It is particularly suitable for problems where the search space is less structured [26]. Additionally, random search can take advantage of maximum parallelism since it does not rely on iterative processes. Evolutionary algorithms, such as genetic algorithms, are another type of iterative strategy that simulates natural selection to explore the search space. These methods evolve solutions over generations, making them highly effective for complex and non-linear search spaces [27].

The choice of search strategy depends on factors such as whether an iterative approach is desired, the number of configurations that can be evaluated per iteration, and the nature of the search space.

6.3.1.3   Incoporating Prior Knowledge

Meta-learning leverages prior knowledge from related tasks to guide and optimize the search process in MLOps systems. To effectively utilize meta-learning, the system must have access to performance data from other pipelines on similar datasets. This historical knowledge enables the system to make informed decisions about which configurations are likely to perform well.

Meta-learning techniques are integrated into MLOps in two primary ways: (1) by building a portfolio of high-quality pipelines, which are then tested on new datasets to identify the best-performing configurations, or (2) by training a predictive model that uses features from both the pipeline (e.g., structure, hyperparameters) and the dataset (e.g., size, feature types, values distribution). These approaches allow MLOps to efficiently explore the search space by focusing on configurations with higher probabilities of success.

By identifying patterns in previously successful configurations, meta-learning not only reduces the time and computational resources required to find effective solutions but also improves the robustness of the design process. This approach is particularly valuable when dealing with large and complex search spaces, where exhaustive exploration would be impractical. Furthermore, as the system gathers more data from its evaluations, meta-learning models can be continuously refined, further enhancing their predictive power [28].

6.3.2   Execution

MLOps systems are often equipped with an execution environment. Once the user specifies a dataset, a pipeline, and an evaluation strategy, the system executes the application and retrieves the evaluation metrics. Some systems accelerate the evaluation process by leveraging specialized hardware like GPUs or frameworks like Dask.[1] Additionally, many systems support distributed pipeline evaluation to further enhance scalability and efficiency. More advanced systems also allow multiple evaluation strategies and optimization techniques, further increasing efficiency.

---

[1] https://www.dask.org/.

6.3.2.1 Evaluation Strategy

Evaluation strategies are essential for assessing the performance of configurations during the search process. Common techniques include cross-validation, which provides robust performance estimates by partitioning the data into training and testing subsets [29], and racing algorithms, which terminate poorly performing candidates early, focusing resources on the most promising ones [30].

Another approach, orthogonal to the ones mentioned above, involves sampling a subset of the data for initial evaluations, allowing quicker assessments when dealing with large datasets or resource constraints. These strategies may trade off accuracy for a substantial reduction of computational cost.

6.3.2.2 Optimization Techniques

Optimization techniques are crucial for minimizing computational overhead during the execution process. Reuse strategies aim to avoid redundant computations by leveraging intermediate results–loading them instead of recomputing from previous evaluations. This involves solving two optimization problems: identifying whether an identical or equivalent result exists in storage [31] and deciding if it is beneficial to load it rather than recomputing it [21]. A system's capability to reuse results heavily depends on its materialization policy. This is why systems that support reuse are often accompanied by a materialization policy [7, 8, 21]. However, materialization is a different optimization problem–typically guided by one or multiple objectives (not always related to reuse potential)–that determines which results should be stored. For this reason, materialization is presented as a separate technique in the next Sect. 6.3.3. Implementation selection involves choosing the most efficient implementation of a given operation, such as selecting between CPU and GPU. Equivalence-based techniques identify configurations that produce the same outcomes, allowing the system to skip redundant evaluations and focus on exploring unique solutions. Together, these approaches enhance the efficiency of the execution process while maintaining the consistency of the results.

6.3.3 Tracking

MLOps systems support the tracking stage by tracing, storing, and querying valuable information that provides insights and enhances reproducibility [20]. This is achieved by maintaining a lineage graph, represented as a directed acyclic bipartite graph with two types of nodes: artifacts (e.g., datasets and models) and the building blocks used to produce these artifacts (e.g., transformations, training steps), along with their configurations. The edges in this graph encode dependencies between artifacts and operations, enabling the system to track all processes and artifacts required to reproduce a given asset, ensuring comprehensive end-to-end traceability.

6.3.3.1　Tracing

Tracing in MLOps involves capturing detailed information about the data and processes within machine learning pipelines to enhance observability, reliability, and reproducibility [32]. Unlike traditional system tracing, MLOps tracing focuses on tracking data transformations, monitoring workflow progress, and diagnosing issues in the data itself rather than just the code. This level of observability ensures that both data and model changes are accurately recorded and monitored, providing users with actionable insights into how artifacts are produced and how different components interact.

Techniques such as data-centric tracing–introduced in Dagger [33]–focus on identifying anomalies and inconsistencies in data rather than code execution, enabling rapid debugging and correction of data errors that could compromise model performance. Additionally, fine-grained lineage tracing allows for precise tracking of transformations and dependencies within machine learning pipelines, helping to reconstruct execution history and identify failure points [34, 35]. This approach captures the lineage of artifacts, linking datasets, models, and configurations to their generating processes, thus supporting reproducibility and rollback when errors are detected.

6.3.3.2　Materialization

Materialization policies are essential for managing and preserving artifacts. These policies define which artifacts–such as preprocessed datasets, feature transformations, and trained models–should be stored to support various objectives, such as reuse potential, version control, and facilitating pipeline review and debugging.

A well-designed materialization strategy can adapt to different objectives depending on the system's goals. Store-everything approaches prioritize comprehensive coverage but may quickly exceed storage budgets and significantly increase storage costs [8]. Version-based policies ensure that multiple versions of models or datasets are kept to track changes and allow rollbacks when needed. Quality-based strategies focus on storing high-performing models or features that can be reused for solving other problems [36, 37], while intermediate-state materialization saves partial pipeline results for quick inspection and reuse. Materialization objectives can include minimizing error propagation, enabling rollbacks to earlier versions, and improving transparency in pipeline execution. Balancing these diverse objectives with storage limitations requires adaptive policies that align with the system's long-term goals and resource constraints.

6.3.3.3　Extracting Insights

MLOps enables users to derive actionable insights from previously evaluated pipelines by visualizing or querying the stored information [22]. Techniques for extracting insights include tracking and comparing quality and performance metrics, and analyzing dependencies.

Visualization techniques are essential for assessing pipeline progress and outcomes. By presenting metrics such as model accuracy, loss, or execution time in intuitive formats, users can identify trends, bottlenecks, and anomalies. On the other hand, dependency analysis provides a comprehensive view of how artifacts and operations are interconnected, allowing users to trace data lineage, evaluate the impact of specific configurations, and ensure reproducibility.

Querying metadata is another basic but powerful technique for insight extraction. Users can analyze relationships between various building blocks, compare configurations, and assess resource utilization. Additionally, tracking the performance of individual components helps identify inefficiencies and optimize pipelines. By leveraging these techniques, users can refine their pipelines, improve efficiency, and ensure the reliable deployment of results.

6.4   State of the Art

This section briefly describes the most representative MLOps systems, the challenges they address, and the techniques they employ. It is important to note that not all MLOps systems support every stage of the development process; in fact, most systems primarily focus on a specific stage. Therefore, we categorize them as follows: (1) AutoML systems that primarily focus on the design stage, automating model selection, hyperparameter tuning, and pipeline construction; (2) ML execution optimizers that provide optimization techniques to accelerate evaluation and reduce computational costs; and (3) ML workflow management systems that focus on assessing, tracing, and metadata collection, ensuring model lineage, experiment reproducibility, and insight extraction. Lastly, we present a set of commercial MLOps systems, predominantly classified as ML workflow management systems. While some of these incorporate AutoML features, most still lack optimization capabilities.

6.4.1   AutoML Systems

AutoML systems aim to discover high-performing solutions within extremely large search spaces.

6.4.1.1   Providing a Search Space

Different AutoML frameworks adopt distinct types of search spaces based on their optimization strategies. Some frameworks use hierarchical or grammar-based search spaces, while others focus on simpler hyperparameter tuning or predefined pipelines.

For example, TPOT [3] implements a grammar-based search space, using a context-free grammar (CFG) to generate valid and meaningful pipelines from scikit-learn components. This ensures that only syntactically correct combinations of preprocessing steps and models are explored. Auto-sklearn [2] adopts a hierarchical search space with conditional dependencies, where specific hyperparameters become relevant only for selected algorithms. In contrast, H2O [4] AutoML focuses on a flat search space that emphasizes hyperparameter tuning for individual models. These different search space designs reflect the unique goals and trade-offs of each AutoML system.

6.4.1.2 Search Strategy

Various search strategies have been employed, including genetic algorithms (e.g., TPOT [3]), random search (e.g., H2O [4]), and Bayesian optimization (e.g., Hyper-Opt [38]). Genetic algorithms and random search are easily parallelizable and require minimal adjustments to handle multiple objectives. In contrast, Bayesian optimization is often faster and more cost-efficient, but it introduces added complexity and is typically applied to single-objective optimization tasks. The main challenge with parallelizing Bayesian optimization lies in its sequential nature: it builds a surrogate model (e.g., Gaussian Process) that relies on previously evaluated points to predict the next most promising point. Since new evaluations depend on updated information from the surrogate model, parallelization requires careful coordination and approximation techniques, which complicate its implementation.

Early AutoML systems faced significant limitations due to the computational intensity required to train and evaluate thousands of models before identifying a good one. For example, the 2019 AutoML benchmark [39] demonstrated that simply training a Random Forest model could outperform these systems in many scenarios. To address these inefficiencies, recent advancements have focused on techniques such as meta-learning to improve them.

Recent trends in AutoML aim to reduce total execution time while maintaining high performance [40]. These advancements achieve efficiency by minimizing the number of pipelines or models proposed for evaluation. Notable works include Alpine Meadow [41], Smart-ML [42], and Bazaar [24]. Systems like SmartML [42] mitigate the evaluation cost by aggressively narrowing the search space, first selecting the building blocks of a pipeline and then searching for the optimal hyperparameters. For designing advanced pipelines based on model ensembles, approaches like those in [36, 37] leverage the model's performance or domain knowledge extracted from raw data [36] and essential model features [37]. Some AutoML systems allow multiple objectives, including cost, diversity and performance. A critical cost factor in AutoML is the need to evaluate many pipelines, driven by the large search space of possible solutions and the computational expense of training on large datasets.

While many efforts have focused on reducing the overall cost of AutoML, few approaches effectively balance the dual objectives of evaluating both inexpensive and promising candidates. For instance, Snoek et al. [25] introduced a Bayesian optimization approach that modifies the objective function to optimize expected improvement per second, targeting hyperparameter optimization problems. FlaML [43] extends this idea to include algorithm selection alongside hyperparameter optimization. However, these methods are limited in scope, as they do not accommodate all pipelines, including preprocessing, feature selection, and learners. Additionally, FlaML relies on fixed cost estimations that fail to account for variations in learner configurations, while the approach by Snoek et al. tends to over-prioritize cheap candidates at the expense of promising ones.

Another emerging trend in AutoML is the focus on predicting diversity [44, 45], which is increasingly recognized as a crucial metric for successful systems, particularly for constructing robust model ensembles, as it helps mitigate overfitting and improves generalization.

6.4.1.3 Meta-Learning

Meta-learning is an alternative approach to improve the search order in AutoML systems [2, 41, 46–48]. This technique is built on the assumption that it is possible to collect a large number of datasets and corresponding experiments to train a meta-learner. The core idea is that the performance of a model on one dataset can be a reliable predictor of its performance on another similar dataset.

Meta-learning approaches can be broadly categorized into two types: portfolio-based methods and model-training-based methods. Portfolio-based methods rely on pre-existing knowledge, maintaining a library of high-performing configurations, which can be directly applied to new tasks. Examples include Auto-sklearn [2] and Alpine Meadow [41], which prioritize pre-evaluated applications from similar tasks. In contrast, model-training-based approaches train a meta-learner that predicts the performance of pipelines or configurations based on the features of datasets and applications [47, 48]. These methods require extensive meta-training with numerous datasets and experiments.

Despite their potential, meta-learning approaches often lack adaptability. Retraining the meta-learner for new objectives or changes in the search space requires collecting additional datasets and experiments, making it computationally expensive and resource-intensive.

6.4.2 ML Execution Optimizers

ML Execution Optimizers aim to manage the available resource budget and find optimal execution plans by using several optimization techniques.

6.4.2.1 Evaluation Strategy

Several approaches have been proposed to optimize the evaluation strategy for ML applications. Sampling techniques reduce the size of training data, thereby lowering computational costs [49–51]. Additionally, racing algorithms, such as Hyperband [19] and successive halving [52], enable early stopping for candidates that are dominated by others, significantly improving evaluation efficiency. While these strategies are practical and effective, they offer limited guarantees that better solutions do not exist elsewhere in the search space.

6.4.2.2 Reuse

Lima [35] provides artifact reuse within single applications. However, it is unable to reuse artifacts across applications. Feature selection systems such as Columbus [53] and model databases like ModelDB [54] have explored artifact reuse, focusing on retrieval to optimize application execution. However, these systems are specialized in specific types of artifacts, with Columbus focusing on features [53] and ModelDB on models [54].

Systems such as Helix [21], and Collab [7] extend artifact reuse across multiple pipelines and enable the reuse of any intermediate artifacts. Helix models the reuse problem as a project selection problem solvable in polynomial time [55], while Collab uses a linear-time algorithm. Although Collab's algorithm is computationally efficient, it may not always yield optimal results. Additionally, Helix restricts reuse to artifacts from the immediately preceding application, whereas Collab builds an experimental graph that incorporates reuse of artifacts from all prior applications, thus expanding reuse opportunities.

These optimization systems are limited to simply supporting reuse, making them inefficient in scenarios where multiple implementations of the same logical operator exist or when operator reordering could be applied.

6.4.2.3 Equivalence and Operator Reordering

The concept of multiple physical implementations for equivalent logical operators, rooted in relational query optimization (e.g., the Volcano optimizer [56] and query rewriting [57]), has been adapted to optimize ML pipelines. KeystoneML [17] maps logical sequences of operators to physical implementations by employing a cost model for each operator to determine the most efficient one. This process involves evaluating various physical implementations on a subset of the data before executing the pipeline on the full dataset.

Operator reordering is another critical optimization technique designed to enhance computational efficiency by rearranging operations to minimize redundant computations. The system described in [58] demonstrates the advantages of performing feature selection before costly join operations, reducing intermediate data size and improving overall performance. Similarly, the work in [59] highlights how propagating a compact data structure across data sources enables efficient feature selection without requiring joins upfront. Extending this concept to inference pipelines, [60] introduces techniques to optimize inference by skipping unnecessary computations while maintaining accuracy. Furthermore, [61] presents *Raven*, a system that integrates data processing and ML operations into a unified framework, employing operator reordering and execution environment optimizations to streamline complex pipelines. Additionally, [31] focuses on detecting equivalences between sequential executions of an application and reusing prior results, reducing redundant computations and enhancing the efficiency of iterative analytics. Collectively, these works illustrate the diverse pipelines and significant performance gains achieved through operator reordering. Recent research has extended equivalence principles to other domains, including video analysis [62], user-defined function (UDF) optimization [63–66], analytical workflows [67, 68], and streaming big data applications [69].

6.4.3 ML Workflow Management Systems

In 2015, Google's report on the technical debt of ML [70] identified two critical challenges for the Tracking stage.

6.4.3.1 Tracing

The first challenge, described as "changing something may change everything, and changing nothing may change everything," highlights the impact of evolving data behavior over time. This observation emphasized the need for systems that carefully track the lineage of data. Lineage information answers questions such as "Which data was used to train this model?" and "What data was derived from these inputs?" Several systems address this issue by automatically tracking data provenance and maintaining execution traces.

For example, noWorkflow [71] and Vasma [72] automatically capture execution traces and store lineage data, enabling users to track how inputs influence model outputs. Other platforms, such as Google's ML Metadata (MLMD)[2] and Databricks' MLflow [5], extend lineage tracking beyond datasets to include changes in pipeline operators, hyperparameters, and model performance over time. These systems help trace how pipeline components evolve and how modifications impact downstream results.

More advanced systems enhance observability and debugging capabilities in ML workflows. MLtrace [32] provides a comprehensive framework for real-time tracing across different pipeline stages, detecting anomalies and surfacing operational issues in production ML workflows. MISTIQUE [73] focuses on storing and querying model intermediates, capturing snapshots of training states to facilitate debugging and fine-grained analysis. MLDebugger [34] goes further by diagnosing failures using provenance tracking, automatically inferring root causes and presenting succinct explanations of errors.

Additionally, tools such as Data Debugging [33] enhance transparency by tracing inconsistencies in data transformations and flagging potential sources of bias or mislabeled training data. Together, these tracing systems provide the foundation for robust, reproducible, and explainable ML workflows.

6.4.3.2 Extracting Insights

The second challenge, referred to as "Pipeline Jungles," highlights the difficulty and cost of managing complex workflows, detecting errors, and recovering from failures. While tracing systems capture execution details, extracting insights from these traces is essential for optimizing pipelines, diagnosing errors, and making informed decisions to improve ML models.

Several ML artifact management systems (AMSs) address this need by offering tools for analyzing stored execution data. A recent survey [74] lists over 60 AMSs that manage raw data, intermediate results, and trained models, collectively called artifacts or assets. These systems help streamline workflow management and improve reproducibility by providing performance monitoring, error detection, and pipeline optimization capabilities.

For example, MLflow [5] not only tracks pipeline changes but also allows users to compare model performance across experiments, identifying configurations that yield the best results. Kubeflow[3] and Airflow[4] integrate visual dashboards that high-light workflow inefficiencies and performance bottlenecks. MISTIQUE [73], in addition to storing intermediates, enables users to query model states, extracting useful insights about training dynamics and performance trends. Similarly, MLDebugger [34] helps users extract error explanations from provenance data, allowing them to refine their pipelines by identifying critical failure points.

Beyond industry-developed solutions, research-driven platforms like REANA[5] and Kepler (developed by NCEAS) focus on enabling reproducible research by allowing scientists to analyze past workflows and optimize computational experiments. These systems provide structured execution histories, helping users make evidence-based adjustments to their pipelines.

By integrating tracing data with performance analysis, error detection, and visualization tools, these systems help users diagnose inefficiencies and refine hyperparameter configurations, leading to better-performing ML pipelines.

---

[2] https://www.tensorflow.org/tfx/guide/mlmd.

[3] https://www.kubeflow.org/.

[4] https://reana.cern.ch/signin.

[5] https://airflow.apache.org/.

6.4.3.3 Materialization

Workflow management systems emphasize reproducibility and lineage tracking, often leading to a "store everything" policy that can quickly become unsustainable and increase costs, as demonstrated in HYPPO [8]. This issue arises because the intermediate results generated during the evaluation of an application often have a size equal to or greater than the original raw data. While comprehensive storage ensures that past results can be readily accessed, it creates significant storage overhead and fails to account for the trade-off between storage constraints and execution efficiency.

To address these challenges, various systems have explored artifact materialization strategies to optimize storage. Feature selection systems such as Columbus [53] and model databases like ModelDB [54] focus on efficient storage of specific types of artifacts. Columbus prioritizes the materialization of extracted features, ensuring that useful transformations are retained, while ModelDB specializes in storing trained models for reuse in different evaluation settings. In contrast, MISTIQUE takes a more dynamic approach by selectively storing critical model intermediates, such as weights, gradients, and activations, which are particularly relevant for diagnosing errors, monitoring convergence, and analyzing performance bottlenecks. By prioritizing artifacts that contribute to debugging and optimization, MISTIQUE reduces redundant storage while maintaining critical insights for improving machine learning workflows.

Research on materialization [7, 8, 21] has shown that storing everything imposes serious storage requirements and is usually suboptimal. These works emphasize that artifacts can be recreated if needed, as systems track sufficient metadata to regenerate previous results. By solving an optimization problem, these systems determine which artifacts are the most beneficial to keep materialized across executions, balancing storage constraints and execution efficiency.

6.4.4 Commercial MLOps Systems

Commercial MLOps platforms play a crucial role in automating various stages of the ML lifecycle. These systems are widely adopted in industry to streamline model development, deployment, and monitoring. As shown in Table 6.2, commercial MLOps platforms primarily focus on workflow management and AutoML capabilities, with varying levels of support for execution optimization and tracking.

Among these platforms, services such as SageMaker,[7] Azure ML,[8] and Vertex AI[9] provide strong AutoML capabilities, including model selection and hyperparameter tuning. However, their execution optimization features remain limited, often requiring manual intervention for the reuse of intermediate results. Pipeline reordering in these systems is also not fully automated, requiring users to make manual optimizations to improve efficiency. In contrast, Kubeflow[3] and TFX[6] stand out by incorporating both AutoML functionalities and extensive execution optimization

Table 6.2 Design, optimization, and tracking capabilities in commercial MLOps systems

	Design	Execution optimization			Tracking	
System	AutoML	Reuse	Reordering	Equivalence	Tracing	Querying
Kubeflow[2]	o	✓	✓	✗	✓	✓
TFX[6]	✓	✓	✓	✓	✓	✓
SageMaker[7]	✓	o	o	✗	o	✓
Azure ML[8]	✓	o	o	✗	o	✓
Vertex AI[9]	✓	o	o	✗	o	✓
Weights & Biases[10]	✗	✗	✗	✗	o	o
Comet ML[11]	✗	✗	✗	✗	o	o
Neptune.ai[12]	✗	✗	✗	✗	o	o
Airflow[5]	o	o	o	✗	✗	✗
Prefect[13]	o	o	o	✗	✗	✗
Metaflow[14]	✗	✓	o	✗	o	✗

Legend: ✓ = Supported, ✗ = Not supported, o = Partial or manually supported

techniques. While Kubeflow supports automated reuse and reordering, TFX provides the most comprehensive support, including equivalence detection, which is the least supported execution optimization feature among commercial MLOps platforms.

Beyond pipeline design and execution, MLOps platforms also differ in their tracking capabilities. Kubeflow and TFX offer the most comprehensive execution tracing and querying support, enabling structured metadata storage and retrieval through ML Metadata (MLMD). While SageMaker, Azure ML, and Vertex AI provide partial tracking and querying support, they lack the detailed lineage tracing found in TFX. Experiment tracking platforms such as Weights & Biases,[10] Comet ML,[11] and Neptune.ai[12] offer some tracing functionality, but they focus primarily on logging and visualization, rather than pipeline lineage tracking. Similarly, workflow orchestration frameworks such as Airflow,[5] Prefect,[13] and Metaflow[14] provide minimal support for execution tracking and querying, as their primary function is workflow automation rather than metadata management.

Overall, commercial MLOps platforms provide automation for model design and tracking but vary significantly in execution optimization and tracking depth. While some platforms, such as TFX and Kubeflow, integrate extensive lineage tracking and execution optimization, many industry solutions prioritize ease of use and experiment tracking over execution optimization. The low adoption of optimization techniques highlights an opportunity for future advancements in optimization-driven MLOps systems that enhance efficiency and cost-aware automation.

6.5 Open Problems and Future Directions

In this section, we identify the unresolved challenges in the field and propose future directions for advancement.

To the best of our knowledge, current AutoML systems do not adapt their search space and strategy to the complexity or nuances of the problem at hand. One of the latest surveys on AutoML systems [39], focused on the classification problem, highlights that, in many cases, they can be outperformed by simply training a random forest classifier. Additionally, the survey reveals that no single strategy consistently outperforms others. We believe the next generation of AutoML systems should be more adaptive to the task at hand, tailoring their search space and strategy dynamically. Similarly, in terms of evaluation strategies, no existing system optimizes its search strategy based on the specific requirements of the problem.

Despite significant advancements, optimizing ML pipelines remains challenging. Future work should focus on integrating reuse, materialization, equivalence, and operator reordering techniques under unified frameworks. Incorporating adaptive cost models, exploring hybrid strategies that combine racing algorithms with artifact reuse, and extending equivalence principles to address real-time constraints are promising directions for advancing pipeline execution efficiency. A unified optimizer for the optimization techniques outlined in Sect. 6.3 is still needed. Such an optimizer would draw inspiration from related work and integrate the techniques discussed into a single cohesive system. However, this is non-trivial, as the optimization problems grow increasingly complex as more techniques are added. Furthermore, there is limited research on when and how different optimization techniques can be applied to ensure robust results.

---

[2] https://www.tensorflow.org/tfx/guide/mlmd.

[3] https://www.kubeflow.org/.

[4] https://reana.cern.ch/signin.

[5] https://airflow.apache.org/.

[6] https://www.tensorflow.org/tfx.

[7] https://aws.amazon.com/sagemaker/.

[8] https://azure.microsoft.com/en-us/products/machine-learning.

[9] https://cloud.google.com/vertex-ai.

[10] https://wandb.ai/site.

[11] https://www.comet.com/site.

[12] https://neptune.ai/.

[13] https://www.prefect.io/.

[14] https://metaflow.org/.

Finally, as discussed throughout this chapter, developing ML applications involves multiple stages. To this day, no system fully supports the entire pipeline. Most systems focus on specific stages, leaving developers with no option but to learn and integrate multiple tools.

6.6　Conclusions

The field of automation and optimization in the development of ML pipelines is rich with fascinating challenges and innovative techniques. It is of significant importance as its advancements increasingly influence various industries and applications. Over the past decade, numerous techniques have been introduced, pushing the boundaries of what is possible. However, it is now imperative to shift our focus toward building comprehensive systems that support developers throughout every stage of the lifecycle.

Moreover, there has been a disproportionate emphasis on accelerating the development process, often at the expense of addressing its associated costs. While speed is undoubtedly valuable, it should not come with an unsustainable price tag. Striking a balance between efficiency and cost-effectiveness will drive meaningful progress in this domain.

References

1. Schwartz R, Dodge J, Smith NA, Etzioni O (2020) Green AI. Commun ACM 63(12):54–63
2. Feurer M, Eggensperger K, Falkner S, Lindauer M, Hutter F (2022) Auto-sklearn 2.0: hands-free automl via meta-learning. J Mach Learn Res 23(1):1–61
3. Olson RS, Bartley N, Urbanowicz RJ, Moore JH (2016) Evaluation of a tree-based pipeline optimization tool for automating data science. In: Proceedings of the 2016 genetic and evolutionary computation conference (GECCO '16). Association for Computing Machinery, USA, pp 485–492
4. LeDell E, Poirier S (2020) H2o automl: scalable automatic machine learning. In: Hutter F, Vanschoren J, Lindauer M, Weill C, Eggensperger K, Feurer M (eds) Proceedings of the 7th ICML workshop on automated machine learning (AutoML 2020), ICML, Online, pp 1–5
5. MLflow (2023) An open source platform for the machine learning lifecycle. https://mlflow.org/
6. Kubeflow (2023) The machine learning toolkit for kubernetes. https://www.kubeflow.org/
7. Derakhshan B, Rezaei Mahdiraji A, Abedjan Z, Rabl T, Markl V (2020) Optimizing machine learning workloads in collaborative environments. In: Maier D, Pottinger R, Doan A, Tan WC, Alawini A, Ngo HQ (eds) Proceedings of the 2020 ACM SIGMOD international conference on management of data. Association for Computing Machinery, USA, pp 1701–1716
8. Kontaxakis AI, Sacharidis D, Simitsis A, Abelló A, Nadal S (2024) Hyppo: using equivalences to optimize pipelines in exploratory machine learning. In: Das G, Sellis T, He B (eds) Proceedings of the 40th IEEE international conference on data engineering (ICDE 2024). IEEE, Netherlands, pp 221–234
9. Chadli K, Botterweck G, Saber T (2024) The environmental cost of engineering machine learning-enabled systems: a mapping study. In: Yoneki E, Payberah AH (eds) Proceedings

of the 4th workshop on machine learning and systems (EuroMLSys '24). Association for Computing Machinery, Rome, Italy, pp 200–207

10. Kuncheva LI, Whitaker CJ (2015) The role of diversity in machine learning ensembles: a survey. Int J Mach Learn Cybern 6(3):439–455

11. Rokach L (2017) Ensemble methods in machine learning. In: Webb GI, Sammut C (eds) Encyclopedia of machine learning and data mining. Springer, pp 393–402

12. Elsken T, Metzen JH, Hutter F (2019) Neural architecture search: a survey. J Mach Learn Res 20(55):1–21

13. von Rueden L, Mayer S, Beckh K, Georgiev B, Giesselbach S, Heese R, Kirsch B, Pfrommer J, Pick A, Ramamurthy R, Walczak M, Garcke J, Bauckhage C, Schuecker J (2023) Informed machine learning: a taxonomy and survey of integrating prior knowledge into learning systems. IEEE Trans Knowl Data Eng 35(6):4914–4932

14. Kougka G, Gounaris A, Simitsis A (2018) The many faces of data-centric workflow optimization: a survey. Int J Data Sci Anal 6(2):81–107

15. Simitsis A, Vassiliadis P, Sellis TK (2005a) Optimizing ETL processes in data warehouses. In: Proceedings of the 21st international conference on data engineering, ICDE. IEEE Computer Society, pp 564–575

16. Simitsis A, Vassiliadis P, Terrovitis M, Skiadopoulos S (2005b) Graph-based modeling of ETL activities with multi-level transformations and updates. In: Data warehousing and knowledge discovery, 7th international conference DaWaK. Lecture notes in computer science, vol 3589. Springer, pp 43–52

17. Sparks ER, Venkataraman S, Kaftan T, Franklin MJ, Recht B (2017) KeystoneML: optimizing pipelines for large-scale advanced analytics. In: Baru C, Thuraisingham B, Papakonstantinou Y, Diao Y (eds) 2017 IEEE 33rd international conference on data engineering (ICDE). IEEE, USA, pp 535–546

18. Simitsis A, Wilkinson K, Castellanos M, Dayal U (2012) Optimizing analytic data flows for multiple execution engines. In: Proceedings of the ACM SIGMOD international conference on management of data, SIGMOD. ACM, pp 829–840

19. Li L, Jamieson K, DeSalvo G, Rostamizadeh A, Talwalkar A (2018) Hyperband: a novel bandit-based approach to hyperparameter optimization. J Mach Learn Res 18(1):6765–6816

20. Isdahl R, Gundersen OE (2019) Out-of-the-box reproducibility: a survey of machine learning platforms. In: Altintas I, Deelman E, Jha S, Parashar M (eds) Proceedings of the 2019 IEEE 15th international conference on eScience (eScience). IEEE, San Diego, CA, USA, pp 86–95

21. Xin D, Macke S, Ma L, Liu J, Song S, Parameswaran A (2018) HELIX: holistic optimization for accelerating iterative machine learning. Proc VLDB Endow 12(4):446–460

22. Idowu SO, Osman O, Strüber D, Berger T (2024) Machine learning experiment management tools: a mixed-methods empirical study. Empir Softw Eng 29(4):1–34

23. Psallidas F, Zhu Y, Karlaš B, Henkel J, Interlandi M, Krishnan S, Kroth B, Emani V, Wu W, Zhang C, Weimer M, Floratou A, Curino C, Karanasos K (2022) Data science through the looking glass: analysis of millions of github notebooks and ML.NET pipelines. SIGMOD Rec 51(2):30–37

24. Smith MJ, Sala C, Kanter JM, Veeramachaneni K (2020) The machine learning bazaar: harnessing the ml ecosystem for effective system development. In: Maier D, Pottinger R, Doan A, Tan WC (eds) Proceedings of the 2020 ACM SIGMOD international conference on management of data. Association for Computing Machinery, USA, pp 785–800

25. Snoek J, Larochelle H, Adams RP (2012) Practical bayesian optimization of machine learning algorithms. In: Bartlett PL, Pereira FCN, Burges CJC, Bottou L, Weinberger KQ (eds) Advances in neural information processing systems, vol 25. Curran Associates Inc., USA, pp 2951–2959

26. Bergstra J, Bengio Y (2012) Random search for hyper-parameter optimization. J Mach Learn Res 13(2):281–305

27. Fortin FA, De Rainville FM, Gardner MA, Parizeau M, Gagné C (2012) Deap: evolutionary algorithms made easy. J Mach Learn Res 13(Jul):2171–2175

28. Vanschoren J (2018) Meta-learning: a survey. arXiv:1810.03548

29. Stone M (1974) Cross-validatory choice and assessment of statistical predictions. J Roy Stat Soc Ser B (Methodol) 36(2):111–133
30. Maron O, Moore AW (1997) The racing algorithm: model selection for lazy learners. In: Jensen DD, Langley P (eds) Proceedings of the 6th international workshop on artificial intelligence and statistics. Morgan Kaufmann, USA, pp 101–108
31. Negi P, Wu C, Vasireddy S, Kumar A (2023) Raven: accelerating execution of iterative data analytics by reusing results of previous equivalent versions. In: Candan KS, Amer-Yahia S (eds) Proceedings of the 2023 ACM SIGMOD international conference on management of data. ACM, USA, pp 512–526
32. Sambasivan R, Zhang H, Crankshaw D, Joseph A (2022) Towards observability for production machine learning pipelines. Proc VLDB Endow 15(11):3123–3136
33. Rezig EK, Cao L, Simonini G, Schoemans M, Madden S, Tang N, Ouzzani M, Stonebraker M (2020) Dagger: a data (not code) debugger. In: CIDR 2020: 10th conference on innovative data systems research, www.cidrdb.org, Netherlands
34. Lourenço R, Freire J, Shasha D (2019) Debugging machine learning pipelines. In: Binnig C, Kumar A (eds) Proceedings of the 3rd international workshop on data management for end-to-end machine learning (DEEM). Association for Computing Machinery, The Netherlands, pp 1–10
35. Phani A, Rath B, Boehm M (2021) Lima: Fine-grained lineage tracing and reuse in machine learning systems. In: Milo T, Tan WC, Olteanu D (eds) Proceedings of the 2021 international conference on management of data. Association for Computing Machinery, USA, pp 1426–1439
36. Cui JW, Lu W, Zhao X, Du XY (2021) Efficient model store and reuse in an OLML database system. J Comput Sci Technol 36:792–805
37. Zhao M, Chen L, Yang K, Du Y, Gao Y (2023) Finding materialized models for model reuse. IEEE Trans Knowl Data Eng 35(12):1–16
38. Bergstra J, Yamins D, Cox DD (2013) Making a science of model search: hyperparameter optimization in hundreds of dimensions for vision architectures. In: Dasgupta S, McAllester D (eds) Proceedings of the 30th international conference on machine learning. PMLR, USA, pp 115–123
39. Gijsbers P, LeDell E, Thomas J, Poirier S, Bischl B, Vanschoren J (2019) An open source automl benchmark. In: 6th ICML workshop on automated machine learning (AutoML)
40. He X, Zhao K, Chu X (2021) Automl: a survey of the state-of-the-art. Knowl Based Syst 212:106622
41. Shang Z, Zgraggen E, Buratti B, Kossmann F, Eichmann P, Chung Y, Binnig C, Upfal E, Kraska T (2019) Democratizing data science through interactive curation of ml pipelines. In: Boncz PA, Manegold S, Ailamaki A, Deshpande A, Kraska T (eds) Proceedings of the 2019 international conference on machine learning, USA, pp 1171–1188
42. Patel D, Shrivastava S, Gifford W, Siegel S, Kalagnanam J, Reddy C (2020) Smart-ml: a system for machine learning model exploration using pipeline graph. In: Wu X, Jermaine C, Xiong L, Hu X, Kotevska O, Lu S, Xu W, Aluru S, Zhai C, Al-Masri E, Chen Z, Saltz J (eds) 2020 IEEE international conference on big data (big data). IEEE, Atlanta, GA, USA, pp 1604–1613
43. Wang C, Wu Q, Weimer M, Zhu E (2021) Flaml: A fast and lightweight automl library. In: Smola A, Dimakis A, Stoica I (eds) Proceedings of machine learning and systems 3 (MLSys). MLSys, USA, pp 772–788
44. Purucker LO, Schneider L, Anastacio M, Beel J, Bischl B, Hoos HH (2023) Q(d)o-es: Population-based quality (diversity) optimisation for post hoc ensemble selection in automl. In: Hutter F, Lindauer M, Vanschoren J (eds) Proceedings of the 2nd international conference on machine learning (AutoML 2023). PMLR, Germany, pp 10/1–10/34
45. Shen Y, Lu Y, Li Y, Tu Y, Zhang W, Cui B (2022) Divbo: diversity-aware cash for ensemble learning. In: Oh AH, Agarwal A, Belgrave D, Cho K (eds) Advances in neural information processing systems 35 (NeurIPS 2022). Curran Associates Inc, USA, pp 2958–2971
46. Redyuk S, Kaoudi Z, Schelter S, Markl V (2024) Assisted design of data science pipelines. VLDB J 33(5):1129–1153

47. Zimmer L, Lindauer M, Hutter F (2020) Auto-pytorch tabular: multi-fidelity metalearning for efficient and robust autodl. arXiv:2006.13799
48. Helali M, Mansour E, Abdelaziz I, Dolby J, Srinivas K (2022) A scalable automl approach based on graph neural networks. Proc VLDB Endow 15(11):2428–2436
49. Zogaj F, Cambronero JP, Rinard MC, Cito J (2021) Doing more with less: characterizing dataset downsampling for automl. Proc VLDB Endow 14(11):2059–2072
50. Lazebnik T, Somech A, Weinberg AI (2022) Substrat: a subset-based optimization strategy for faster automl. Proc VLDB Endow 16(4):772–780
51. Pérez D, García S, Herrera F (2022) Auto-cve: an automl framework with dynamic sampling holdout. SN Comput Sci 3(5):1–14
52. Jamieson K, Talwalkar A (2016) Non-stochastic best arm identification and hyperparameter optimization. In: Gretton A, Robert CC (eds) Proceedings of the 19th international conference on artificial intelligence and statistics, PMLR, Spain, pp 240–248
53. Zhang C, Kumar A, Ré C (2016) Materialization optimizations for feature selection workloads. ACM Trans Database Syst 41(1):2:1–2:32
54. Vartak M, Subramanyam H, Lee WE, Viswanathan S, Husnoo S, Madden S, Zaharia M (2016) ModelDB: a system for machine learning model management. In: Binnig C, Fekete AD, Nandi A (eds) Proceedings of the workshop on Human-In-the-Loop Data Analytics (HILDA). Association for Computing Machinery, USA, p 14
55. Kleinberg JM, Tardos É (2006) Algorithm design. Addison-Wesley
56. Graefe G, McKenna W (1993) The volcano optimizer generator: extensibility and efficient search. In: Golshani F, Tjoa AM, Elmagarmid AK, Neuhold EJ (eds) Proceedings of the 9th international conference on data engineering (ICDE). IEEE, Vienna, Austria, pp 209–218
57. Cohen S, Nutt W, Serebrenik A (1999) Rewriting aggregate queries using views. In: Beeri C, Buneman P (eds) Proceedings of the 18th ACM SIGMOD-SIGACT-SIGART symposium on principles of database systems. Association for Computing Machinery, USA, pp 155–166
58. Kumar A, Naughton J, Patel JM, Zhu X (2016) To join or not to join? Thinking twice about joins before feature selection. In: Ozcan F, Koutrika G, Madden S (eds) Proceedings of the 2016 ACM SIGMOD international conference on management of data. ACM, USA, pp 19–34
59. Chaudhuri S, Ganti V, Motwani R (2005) Pushing feature selection ahead of join. In: Kargupta H, Srivastava J, Kamath C, Goodman A (eds) Proceedings of the 2005 SIAM international conference on data mining, SIAM, USA, pp 536–540
60. Kang D, Raghavan D, Bailis P, Zaharia M (2020) Willump: a statistically-aware end-to-end optimizer for machine learning inference. In: Proceedings of the 2020 MLSys conference
61. Shi Y, Kumar A (2022) End-to-end optimization of machine learning prediction queries. In: Bonifati A, Abbadi AE (eds) Proceedings of the 2022 ACM SIGMOD international conference on management of data. ACM, USA, pp 33–47
62. Xu Z, Kakkar GT, Arulraj J, Ramachandran U (2022) EVA: a symbolic approach to accelerating exploratory video analytics with materialized views. In: Bonifati A, Abbadi AE (eds) Proceedings of the 2022 ACM SIGMOD international conference on management of data. Association for Computing Machinery, USA, pp 602–616
63. Ramjit L, Interlandi M, Wu E, Netravali R (2019) Acorn: aggressive result caching in distributed data processing frameworks. In: Taft R (ed) Proceedings of the ACM symposium on cloud computing. Association for Computing Machinery, Santa Cruz, CA, USA, pp 206–219
64. LeFevre J, Sankaranarayanan J, Hacigumus H, Tatemura J, Polyzotis N, Carey MJ (2014) Opportunistic physical design for big data analytics. In: Dyreson CE, Li F, Özsu MT (eds) Proceedings of the 2014 ACM SIGMOD international conference on management of data. Association for Computing Machinery, Snowbird, UT, USA, pp 851–862
65. Foufoulas Y, Simitsis A (2023) Efficient execution of user-defined functions in SQL queries. Proc VLDB Endow 16(12):3874–3877
66. Foufoulas YE, Simitsis A, Stamatogiannakis E, Ioannidis YE (2022) YeSQL: "you extend SQL" with rich and highly performant user-defined functions in relational databases. Proc VLDB Endow 15(10):2270–2283

67. Simitsis A, Wilkinson K, Dayal U, Hsu M (2013) HFMS: managing the lifecycle and complexity of hybrid analytic data flows. In: 29th IEEE international conference on data engineering. IEEE Computer Society, ICDE, pp 1174–1185
68. Jovanovic P, Simitsis A, Wilkinson K (2014) Engine independence for logical analytic flows. In: Cruz I, Markl V, Christophides V, Mokbel M, Sellis T (eds) Proceedings of the 2014 IEEE 30th international conference on data engineering (ICDE). IEEE, USA, pp 1060–1071
69. Giatrakos N, Arnu D, Bitsakis T, Deligiannakis A, Garofalakis M, Klinkenberg R, Konidaris A, Kontaxakis A, Kotidis Y, Samoladas V, Simitsis A, Stamatakis G, Temme F, Torok M, Yaqub E, Montagud A, Ponce de León M, Arndt H, Burkard S (2020) INforE: interactive cross-platform analytics for everyone. In: d'Aquin M, Dietze S (eds) Proceedings of the 29th ACM international conference on information and knowledge management. Association for Computing Machinery, USA, pp 3389–3392
70. Sculley D, Holt G, Golovin D, Davydov E, Phillips T, Ebner D, Chaudhary V, Young M, Crespo JF, Dennison D (2015) Hidden technical debt in machine learning systems. In: Advances in neural information processing systems 28 (NeurIPS 2015). Curran Associates Inc., Canada, pp 2503–2511
71. Pimentel JF, Murta L, Braganholo V, Freire J (2017) noWorkflow: a tool for collecting, analyzing, and managing provenance from python scripts. Proc VLDB Endow 10:1841–1844
72. Namaki MH, Floratou A, Psallidas F, Krishnan S, Agrawal A, Wu Y, Zhu Y, Weimer M (2020) Vamsa: automated provenance tracking in data science scripts. In: Gupta R, Liu Y, Tang J, Prakash BA (eds) Proceedings of the 26th ACM SIGKDD international conference on knowledge discovery and data mining. Association for Computing Machinery, USA, pp 1542–1551
73. Vartak M, da Trindade JMF, Madden S, Zaharia M (2018) MISTIQUE: a system to store and query model intermediates for model diagnosis. In: Das G, Jermaine CM, Bernstein PA (eds) Proceedings of the 2018 international conference on management of data. Association for Computing Machinery, USA, pp 1285–1300
74. Schlegel M, Sattler KU (2023) Management of machine learning lifecycle artifacts: a survey. SIGMOD Rec 51(4):18–35

Chapter 7
Workload Placement and Scheduling on Heterogeneous CPU-GPU Architectures

Marcos N. L. Carvalho, **Anna Queralt**, **Oscar Romero**, **and Alkis Simitsis**

Abstract The need for heterogeneous CPU-GPU processing has significantly grown in recent years. The efficient utilization of such heterogeneous resources requires data processing systems to employ efficient workload placement strategies to assign the appropriate amount of compute to the right processor. However, identifying an optimal placement strategy is challenging due to several complex and conflicting trade-offs involving the characteristics of processors, the nature of the workload, and data locality. Additionally, placement decisions affect workload runtime and performance costs, while also relying on the availability of potentially different implementations for CPUs and GPUs, adding more complexity in such heterogeneous environments. In this chapter, we review and compare state-of-the-art strategies for workload placement and scheduling on heterogeneous CPU-GPU architectures, techniques for run-time prediction, and methods to support multi-device code.

Keywords Heterogeneous computing · CPU-GPU architectures · Workload placement · Task scheduling

M. N. L. Carvalho (✉)
NKUA & Athena RC, Athens, Greece
e-mail: marcos.nogueira@upc.edu

A. Simitsis
Athena Research Center, Athens, Greece
e-mail: alkis@athenarc.gr

A. Queralt · O. Romero
Universitat Politécnica de Catalunya, Barcelona, Spain
e-mail: anna.queralt@upc.edu

O. Romero
e-mail: oscar.romero@upc.edu

© The Author(s) 2026
G. Dejaegere et al. (eds.), *Data Engineering for Data Science*,
https://doi.org/10.1007/978-3-032-18765-9_7

7.1 Overview

Over the past decade, graphics processing units (GPUs) have become increasingly popular in a wide range of fields, including Data Science, Artificial Intelligence, and High-Performance Computing (HPC). Due to core architectural design aspects (e.g., *Single Instruction, Multiple Thread (SIMT)* execution model, high-bandwidth memory), GPUs provide a high level of parallel processing power to accelerate applications. However, they also present three core limitations: (*i*) *Data transfer bottleneck:* data must be moved between CPUs and GPUs, usually via a low-bandwidth system bus, which adds communication overhead. (*ii*) *Limited memory size:* Despite the significantly greater bandwidth of GPU memory, its capacity is much smaller than the main memory available for CPUs, limiting the possibility of processing large volumes of data on GPUs; and (*iii*) *CPUs and GPUs are designed for different purposes:* CPUs excel at processing low-latency operations with low degree of thread parallelism, whereas GPUs are most efficient at processing regular (i.e., without branch divergence) operations with a high degree of thread parallelism (Table 7.1).

Therefore, heterogeneous CPU-GPU architectures have emerged as a strategy to mitigate the limitations of GPUs, while leveraging the strengths of both CPUs and GPUs. Optimizing the utilization of heterogeneous CPU-GPU systems is a challenging research problem that requires (a) finding the optimal workload placement strategy (i.e., deciding on which kind of processor, CPU or GPU, to run each part of an application), (b) estimating runtime costs over different hardware architectures, and (c) managing multi-device code [1].

Identifying the most efficient placement strategy is not trivial and requires balancing the trade-off between optimal execution (i.e., assign a workload to a processor) and mitigated data transfers (i.e., employ a single processor to avoid moving data) [2]. Additionally, the large search space of non-linear execution parameters (a.k.a. factors) involved in the design process and the absence of concrete guidelines often result in inefficient placement decisions that cause load imbalance, waste of resources, and amplify the data transfer bottleneck. For estimating runtime costs, existing works consider a blend of techniques that include heuristics, cost models, and learned models. Regarding the management of heterogeneous CPU-GPU code, kernel templates, compilers, and raw kernels are the common options.

Related Work. Previous surveys [3, 4] provided an excellent overview of the existing work about heterogeneous CPU-GPU computing techniques. Mittal and Vetter [3] proposed an extensive taxonomy including the categorization of (*i*) *CPU-GPU partitioning techniques,* to introduce two key dimensions related to *scheduling time* (i.e., when scheduling decisions are made) and *scheduling strategy* (i.e., what are the criteria used for scheduling); (*ii*) *CPU-GPU programming languages, frameworks, and compiler,* to classify multi-device implementation aspects; (*iii*) *Techniques for energy efficiency,* to analyze the solutions for saving energy and classify them in terms of the core approach used; (*iv*) *GPU architectural differences,* to compare the features and limitations of works using dedicated and integrated GPU architectures; and (*v*)

Application areas and benchmarks, to classify works according to their application field and discuss benchmark suites to compare CPUs and GPUs [3].

A more recent survey [4] focused on heterogeneous CPU-GPU query processing and extended the taxonomy introduced in [3] on scheduling by including the dimensions (*i*) *Processor usage,* to classify works using GPUs only for specific tasks, generic tasks, or hybrid by combining the two approaches; (*ii*) *Workload distribution,* to categorize the strategies used to divide the workload according to the processor usage; (*iii*) *Task granularity,* to present different strategies used to schedule fine-grained or coarse-grained tasks; and (*iv*) *Data partitioning,* to classify the partitioning techniques used into horizontal or vertical.

Our contributions. Workload placement and scheduling are similar concepts, but they have different design purposes. While placement strategies decide whether to run a workload in a CPU or in a GPU processor, scheduling strategies decide where (the specific CPU core or GPU device), and when (the execution order considering resource utilization) to run a workload. In general, placement strategies work on an optimization layer providing hints to schedulers about the ideal processor type to run a workload. Although past surveys covered a broad range of the CPU-GPU scheduling approaches, none focused on the specifics of the CPU-GPU workload placement techniques. Also there is a gap in current surveys about task scheduling on generic task-based applications.

An initial effort in this direction is [5], which provides an extensive analysis of the related work and proposes a taxonomy with three dimensions: (*i*) *Workload placement strategies*; (*ii*) *Techniques used to estimate costs for placement decisions*; and (*iii*) *Strategies to manage heterogeneous CPU-GPU code*. We present an extended overview of these dimensions in Sect. 7.3. Additionally, this chapter presents in depth the more recent advances on CPU-GPU workload placement, also including an analysis of the state-of-the-art on task scheduling, as a complementary material to the existing surveys.

Scope. The body of research on heterogeneous CPU-GPU processing is extensive and several approaches have explored different alternatives to *GPU usage*, i.e., as primary processor, as accelerator, or heterogeneous CPU-GPU processing, and *GPU integration* architectures, i.e., integrated GPU (iGPU) or dedicated GPU (dGPU, simply referred as GPU hereafter). Given the practical challenges of reviewing such wide range of research works about heterogeneous CPU-GPU processing, we focus on heterogeneous CPU-dGPU processing solution approaches as this is the most popular setting in both server and high performance computing infrastructures [6]. Here, the term 'workload' covers a variety of data processing functions, including database queries, data flows/pipelines, or general apps and programs.

Structure. We organized this chapter as follows. Section 7.2 discusses the different alternatives to use and integrate GPUs, as well as the data transfer bottleneck. Section 7.3 introduces and describes each dimension of our proposed taxonomy. Section 7.4 presents an overview of the state-of-the-art on task scheduling. Section 7.5 discusses open issues and research directions and, finally, Sect. 7.6 concludes the paper.

7.2 The Landscape of CPU-GPU

GPU usage. There are three main ways of using GPUs. The first one is using a GPU as the primary processor, meaning that the GPU processes the entire workload. An alternative is to use a GPU as an accelerator, where the GPU processes specific portions of the workload. Finally, a third option is to use heterogeneous CPU-GPU processing, where both CPU and GPU can process the entire or specific portions of the workload [7, 8]. Applications using GPUs, either as a primary processor or as an accelerator, typically assume that the workload placement decision is done beforehand (static placement) [e.g., 9–11]. However, most practical applications use heterogeneous CPU-GPU processing, which requires dynamic (and often adaptive) workload placement, which is a challenging problem [4].

GPU integration and data transfer bottleneck. iGPUs have limited memory bandwidth (they share the main memory with CPUs) and reduced processing capacity compared to dGPUs. However, iGPUs are more energy efficient and do not face a data transfer bottleneck. Hence, iGPUs are popular in fine-grained workloads (e.g., stream processing or processing in edge devices) [4]. On the other hand, dGPUs have limited memory size and suffer from a data transfer bottleneck, as their memory is decoupled from the main memory, hence, requiring data transfer over a low-bandwidth interconnect. However, their high memory bandwidth and processing capacity make dGPUs ideal for processing coarse-grained and compute-intensive workloads (e.g., machine learning, numeric algorithms) [4]. Although the limited memory size of dGPUs is gradually increasing (e.g., NVIDIA H200 offers 141GB memory [12]), it still remains much smaller than the system main memory, which can be in the order of terabytes in modern servers. Price-wise, the current \$/GB ratio of GPU memory, which is around 100\$/GB, resembles the price per GB of DRAM a decade ago (60\$/GB), and it is steadily improving [13].

Data transfer techniques and workload placement. Several software/hardware solutions aim at mitigating the CPU-GPU data transfer bottleneck [4]. For example, data-locality-based solutions avoid data transfer by placing workloads on processors where the input data is already located [7, 14–17]. Such solutions use off-chip caching policies to avoid slow disk accesses and are typically used in placement strategies that consider data locality [4]. Other placement strategies ignore data locality and focus on placing workloads on the fastest processor, potentially at the cost of a more expensive CPU-GPU data transfer.

7.3 State of the Art

We analyzed 77 papers published between 2009 and 2024 on workload placement on heterogeneous CPU-GPU systems, identifying several recurring factors across the studies. Considering these factors, we classify the state of the art according to the following dimensions: (a) placement strategy, (b) off-chip cache policy, (c) placement granularity, (d) placement time, (e) placement decision, (f) placement prediction model, (g) monitored metrics, and (h) GPU programming method. Table 7.1 shows the classification of the related work into our taxonomy and also how many papers consider the dimension (the *#papers* row—e.g., 9 papers employ function shipping). We will present them when detailing each dimension in the next sections.

7.3.1 Workload Placement Strategies

In this subsection we categorize the workload placement strategies used in the related work into five dimensions: placement strategy, off-chip cache policy, placement granularity, placement time, and placement decision.

7.3.1.1 Placement Strategy

This dimension defines the placement order of a function and its input data. There are two placement strategies: *(i) data shipping* and *(ii) function shipping* [82, 83].

Data shipping places functions first and then ships data to where functions are located. Functions are placed on processors according to their relative performance (e.g., compute-bound functions are preferably placed on GPUs, while I/O-bound functions are placed on CPUs). Although this strategy prioritizes the best processor to execute a function, data locality is not a priority, resulting in an increased amount of data transfers.

In contrast, function shipping places data first and then ships functions to where data is located. In this case, data is placed and kept on off-chip memory of the processor based on a cache eviction policy (cache policies are detailed in Sect. 7.3.1.2). Data locality is the priority and hence, the amount of data transfers is reduced. However, this strategy might not always use the most efficient processor to process a function (e.g., compute-bound functions might be placed on CPU instead of GPU). Next we detail the techniques used in each strategy.

Data shipping. This approach ships all required data from CPU to GPU so that a function is fully processed on the GPU [e.g., 1, 41, 49]. Although this technique leads to an intuitive placement decision and implementation, it has three main limitations as it moves the entire data to the GPU. First, as the function will be processed only on the GPU, the CPU will be idle and vice-versa. Second, the amount of data per data transfer is increased, potentially saturating the bus interconnect. Third, the limited size of the GPU memory might not be sufficient, increasing the risk of out-of-memory (OOM) errors. An alternative technique involves partitioning the data so that independent functions can be processed in parallel on both CPU and GPU enabling device parallelism [e.g., 15, 56, 60, 78]. As a consequence of the device parallelism and the control of data granularity to ship to each processor, this technique allows load balance between CPUs and GPUs. However, the amount of data transfers is

Table 7.1 Taxonomy of approaches for workload placement on CPU-GPU systems

| Dimension Group | Workload Placement Strategies | | | | | | | | | | | | | | | | | | | Cost Estimation | | | | | | | Het. Code Mgmt. | | |
| Dimensions | Plmt. Strtg. | | Off-chip Cache Policy | | | | Plmt. Granularity | | | | | | | Plmt. Time | | | Plmt. Decision | | | Plmt. Pred. Model | | | Monitored Metrics | | | | GPU Prog. Model | | |
Features	Data shipping	Function shipping	Semantic-aware	Frequency-based	Greedy	No caching	Job	Application	Pipeline	Task	Function	Segment	Bit	Runtime	Compilation time	Hybrid	Automatic	Semi-automatic	Manual	Cost model	Heuristic model	Learn model	Latency	Throughput	Energy consumption	Monetary price	Kernel template	Compiler	Raw kernel
#papers	68	9	5	10	13	49	3	9	7	49	2	5	2	26	35	20	25	42	19	11	46	26	75	21	7	1	30	14	41
Qilin [18]	X					X				X				X				X				X	X		X				X
StarPU [19]	X			X						X				X				X			X		X	X			X		X
GDB [11]	X					X				X					X		X			X			X						X
Kerr et al. [20]	X					X		X							X			X			X	X	X				X	X	
Ravi et al. [21]	X					X				X				X			X				X		X	X				X	
Heterogeneous Linpack [22]	X					X				X				X			X			X			X	X			X		
MEGHA [23]	X					X				X					X			X			X		X					X	
Rectangle method [24]	X					X						X			X			X			X		X	X					X
Grewe and O'Boyle [6]	X					X				X					X			X			X		X	X					X
Ravi et al. [25]	X					X	X								X			X			X		X	X			X		
Pirk [26]	X					X							X		X			X			X		X						X
SnuCL [27]	X				X					X				X				X			X		X	X					X
SCCG [28]	X					X			X							X		X			X		X	X					X
GreenGPU [29]	X					X				X				X				X			X		X		X				X
DAGuE [30]	X					X				X				X				X			X		X	X			X		
AHP [31]	X					X			X							X		X				X	X					X	
Clarke et al. [32]	X					X				X				X			X			X			X				X		
MC-MOC [33]	X					X				X				X			X			X			X						X
Legion [34]	X					X				X					X		X		X		X		X						X
Extended TOTEM [35]	X					X				X					X				X		X		X		X				X
Kofler et al. [36]	X					X				X						X		X				X	X					X	
SKMD [37]	X					X				X				X				X				X	X					X	
A&R [38]	X			X									X		X			X			X		X						X
Cumming et al. [39]	X				X			X							X				X		X		X		X				X
Ocelot+HyPE [1]	X			X						X				X				X			X		X						X
Cheng et al. [40]	X					X				X					X				X		X		X		X				X
Karnagel et al. [41]	X					X				X						X		X				X	X				X		X
Dask [42]	X			X						X					X				X		X		X				X		
TensorFlow [43]	X				X					X					X		X		X		X		X	X			X		
HyGraph [44]	X				X							X		X	X		X		X	X	X		X						X
SÃ®rbu and Babaoglu [45]	X					X	X								X			X				X	X		X				
Spark-GPU [46]	X			X						X					X			X		X		X	X			X			
GFlink [55]	X			X						X					X				X			X	X					X	
Gowanlock et al. [56]	X					X				X				X				X		X			X				X		X
Troodon [57]	X					X	X								X			X				X	X	X					X
PLB-HAC [58]	X					X				X					X			X				X	X				X		
PyTorch [59]	X				X					X					X		X	X				X	X	X			X		
HetExchange [60]	X					X			X								X	X				X	X	X				X	
Symphony [61]	X					X				X				X				X					X						
Crystal [9]	X					X		X							X				X	X			X			X			X
MCL Schedulers [62]		X			X					X				X	X		X	X			X		X					X	X
Lutz et al. [63]	X				X					X						X	X				X		X	X					X
LDWP-IAWP [64]	X					X					X					X	X					X	X	X			X		
ZeRO-Offload [65]	X				X					X					X		X				X		X	X			X		
Sigmoid [66]	X					X				X				X			X			X			X		X				X
Li et al. [67]	X					X				X				X				X			X		X				X		
Lee and Park [68]	X					X				X					X				X		X		X				X		
RateupDB [13]	X			X				X							X		X				X		X	X					X
READYS [69]	X					X				X				X				X			X		X				X		X
Compressed Crystal [70]	X					X		X							X			X		X	X		X					X	
GAP [71]	X					X		X							X			X			X		X						X
GHive [72]	X					X			X						X			X			X		X				X		X
Xekalaki et al. [73]	X					X				X					X		X				X		X					X	
Mordred [7]		X	X									X				X	X			X	X		X				X		
Parla [16]	X				X					X				X	X		X	X			X		X				X	X	
GaccO [74]		X	X					X								X		X			X				X				X
Extended MCL [75]		X			X					X				X	X		X	X			X		X					X	
DBD [76]	X					X				X							X	X			X		X						X
CoTrain [15]	X				X					X								X		X	X		X	X			X		
HetCache [14]		X	X						X					X				X			X		X	X				X	
CGgraph [77]		X	X							X							X	X			X		X						X
Kroviakov et al. [78]	X				X							X		X			X		X		X	X						X	
gSWORD [79]	X					X						X				X	X				X		X						X
FusionFlow [80]	X					X				X				X				X			X		X	X			X		
NeutronOrch [81]		X	X							X				X				X			X		X				X		
Carvalho et al. [8]	X					X				X					X				X		X		X				X		

increased and extra implementation complexity is necessary to merge output data processed by each processor in a final result.

Function shipping. This technique prioritizes the processors where the entire data to process a function is local [e.g., 2, 14, 17, 62, 74, 75, 77, 81]. By keeping data locally in the off-memory cache, the amount of data transfers is reduced, but it requires a portion of the limited GPU memory. This technique can also result in load imbalance, since one processor might be used more than the other, depending on where the data is located. An optimized technique prioritizes the processor where only fine-grained data is local [7]. That is, only the specific portion of data that is effectively used to process a function is kept in memory. Although this technique provides an efficient use of the GPU memory, it increases the amount of data transfers to move the partitioned data. The implementation complexity is also increased to manage intermediate materialization and merge the outputs of partitioned data processing.

7.3.1.2 Off-Chip Cache Policy

Function shipping approaches typically employ caches to keep a subset of data on off-chip memory (i.e., host main memory for CPU or device global memory for GPU) and reduce data movement [7, 17]. The GPU cache has higher bandwidth but smaller size than the CPU cache. Thus, besides finding a cache policy that selects a subset of data to fit into the limited cache size, choosing a cache type is also a challenge. In our study, we identify four strategies to manage off-chip caching: *(i) no caching*; *(ii) greedy caching*; *(iii) frequency-based caching*; and *(iv) semantic-aware caching*.

Most related work does not report any cache strategy [e.g., 48, 68, 69, 73], which is mainly due to the following reasons. First, caching demands multiple complex challenges, such as defining the ideal cache size and implementing an efficient cache policy. Second, with the emergence of high-bandwidth storage alternatives (such as NVMe), caching performance improvements are diminished. Nevertheless, out-of-memory data fetches can severely hinder the performance of certain workloads (e.g., analytical queries) and therefore high-performance analytics still requires caching data [14]. Works using a greedy caching strategy [e.g., 15, 16, 44, 78] try to keep as much data as possible in memory, assuming that cache benefits are linear to the cache size. Although greedy caching is intuitive to implement, it might cache more data than necessary. By occupying the limited cache size with non-relevant data, this strategy potentially results in an inefficient use of cache and OOM errors.

On the other hand, frequency-based and semantic-aware caching implement more sophisticated algorithms to control the cache eviction policy. Frequency-based caching keeps data in memory based on its access frequency [17, 51] or recency [e.g., 2, 13, 17, 55] of use. This strategy relies on popular algorithms (LRU, LFU, and FIFO) and is suitable when storage is a significant bottleneck. However, this strategy is not appropriate in high-bandwidth storage interconnects, or in compute-bound workloads. For instance, caching data involved in a compute-bound workload offers less benefit than caching data for an IO-bound workload, even with equal access frequency. Therefore, the benefit of caching data depends on how the data is used in the workload [14]. This is the main motivation of semantic-aware cache approaches [7, 14, 74, 77, 81]. Besides considering workload aspects (e.g., compute-bound, IO-bound), this strategy also takes into account CPU-GPU heterogeneity aspects, such as different memory capacity and bandwidth. These features result in an efficient cache utilization at the cost of requiring complex and oftentimes ad hoc implementations. Our subsequent discussion will delve into the techniques used in frequency-based and semantic-aware caching.

Frequency-based caching. The primary objective of existing frequency-based caching policies is to minimize storage accesses by maximizing the cache hit rate. Least Recently Used (LRU) is the most used algorithm in the related work [1, 2, 13, 17, 38, 42, 46, 54]. By evicting least recently used data based on a timestamp, this algorithm is suitable for data frequently used during a period and then usage drops. However, LRU needs to update timestamps and maintain access order of data, resulting in extra memory overhead. Least Frequently Used (LFU) is another caching algorithm that evicts least frequently used data using a frequency counter [17, 51]. Although LFU reduces the memory overhead by replacing costly timestamp updates by lightweight frequency counter updates, it is inefficient for sudden drops on frequent-accessed data. A last example of a frequency-based algorithm is the First In First Out (FIFO) [55], which evicts the oldest data in the cache queue. Compared to LRU and LFU, the FIFO algorithm is simpler to implement and does not require extra memory to monitor data access statistics. Nevertheless, FIFO is not suitable when data access patterns do not match with the order of insertion of data.

Semantic-aware caching. Different from the previous strategies, semantic-aware caching keeps in memory the data with higher impact on the workload execution performance. However, approaches using such cache strategy are dependent on the application. For OLAP (online analytical processing) query processing we have identified two works: HetCache [14] and Mordred [7]. HetCache [14] caches in GPU memory only table columns that accelerate analytical queries with respect to the query processing throughput. Conversely, Mordred [7] implements an adapted version of LFU with weighted frequency counters representing the potential benefits of caching a segment (i.e., a subset of a table column). In the context of OLTP query processing applications, GaccO [74] caches in GPU memory only the tables used in dominant OLTP transactions, which are manually defined at compilation time by the DBMS administrator. In machine learning training applications, NeutronOrch [81] uses a frequency-based cache in CPU memory. However, if the GPU is idle, Neu-tronOrch caches hot vertices (i.e., frequently accessed vertices) in GPU memory and ships embedding computations from the bottom layer of Graph Neural Network (GNN) training to GPU. NeutronOrch stops assigning hot vertices to a GPU when its memory is exhausted or the GPU idle time reaches zero. Finally, CGgraph [77] implements a semantic-aware cache in graph processing applications. In particular, CGgraph [77] keeps only active edges and vertices of graphs in GPU memory.

7.3.1.3 Placement Granularity

This dimension defines the level of abstraction and size of the functions and/or data to place. The related work tackles seven different placement granularities: *(i) job*; *(ii) application*; *(iii) pipeline*; *(iv) task*; *(v) function*; *(vi) segment*; and *(vii) bit*.

Entire applications can be placed on either CPU or GPU [e.g., 9, 13, 20, 39, 47, 49, 70, 71, 74]. Generally at this granularity all input data is moved to the GPU memory at the beginning of the execution and all outputdata is moved back to the CPU at the end, hence, reducing the amount of data transfers. However, this coarse-grained granularity has three main drawbacks. First, the memory required to store all data is increased, which oftentimes cannot fit into the limited GPU memory. Second, not all parts of an application might be suitable for GPU acceleration, causing limited parallelism opportunities. Third, a processor is busy while the other remains idle during the entire application processing, leading to severe load imbalance. There are also works [25, 45, 57] on scheduling jobs in secenarios with multiple concurrent applications, but the placement is still at the application level.

Applications are typically represented by a Directed Acyclic Graph (DAG) where vertices are tasks (i.e., sequence of operations to be executed in an application) and edges are task dependencies [84, 85]. Works operating at a task granularity [e.g., 8, 41, 62, 68, 78, 81] enable many parallelism opportunities, as specific tasks can be placed on GPUs. Additionally, working at a task level improves load balance as both CPUs and GPUs can be used to process different tasks. A potential drawback of this approach is the increased amount of data transfers required. To reduce data transfers, tasks can also be grouped or pipelined and placed together at a coarser granularity (i.e., pipeline placement granularity) [10, 14, 28, 31, 52, 60, 72].

Tasks can further be logically partitioned into functions (i.e., sequence of operations to be executed within a task) [2, 64]. Function placement granularity provides a high degree of code specialization, which improves even more the parallelism opportunities on GPUs. This granularity also reduces the complexity of the code, leading to less runtime cost estimation variation (more details about cost estimation techniques in Sect. 7.3.1.5). At the same time, this granularity increases the amount of data transfers (to move data for each GPU-accelerated function inside each task) and also causes potential load imbalance due to the highly specialized code implementing functions, being available either for both CPU and GPU.

Tasks can also be physically partitioned and placed at segment granularity [7, 24, 44, 78, 79]. In this case, the input data is partitioned into multiple segments and each segment is processed by a processor in different sub-tasks. Besides the efficient memory use by storing data on both the limited GPU memory and the large system main memory, segment placement granularity also improves load balance by processing segments on both CPUs and GPUs. However, merge and synchronization overheads are included in the runtime to merge outputs and synchronize each sub-task processing different segments. Additionally, the amount of data movement increases, as segments of data must be transferred inside each task. To reduce the volume of data per transfer, tasks can be further partitioned into bits, at the cost of increasing the implementation complexity [26, 38].

Next we detail common techniques used in application, task, function, segment, and bit-level placement granularities.

Application. A popular technique consists in implementing specific GPU kernels per application [e.g., 20, 39, 74]. A key advantage of this approach is fine-tuned optimization, as kernels can be developed targeting specific performance issues of an application (e.g., memory limitation, low streaming processing (SM) utilization). However, this technique oftentimes requires complex and ad hoc implementation that limits the kernel reuse in other applications. An alternative is implementing applications via block-wide function kernels [e.g., 9, 13, 47, 49, 70, 71]. This technique provides opportunities for kernel reuse and also simplifies the code implementation. However, the main drawback of this approach is the potential increase of kernel warm-up overheads (to compile the GPU code, create CUDA contexts, etc.).

Task. When tasks are placed without data partitioning [e.g., 41, 48, 61, 62], the amount of data transfers is expected to be reduced. However, tasks can require amounts of data that exceed the GPU memory size, and load imbalance is caused due to the processing of the entire task on either CPU or GPU. An optimization technique to reduce the task memory footprint consists on partitioning the tasks data and applying horizontal device parallelism [e.g., 15, 16, 18, 60, 77, 78, 80, 81]. In this technique, the task input data is partitioned in such a way that both CPUs and GPUs are used at the same time to process the same task, but over different data partitions. Although load balance is improved, data locality is not a priority and hence the amount of data transfers is increased. Moreover, additional merge and synchronization overheads are necessary when combining the outputs of tasks executed by each processor. An alternative approach is to prioritize data locality in favour of load imbalance [e.g., 8, 42, 46, 51, 68]. In this technique, the task input data is partitioned in such a way that either CPU or GPU is used to process tasks over different partitions.

Function. Function-based approaches are highly dependent on the application. In stream processing applications, LDWP-IAWP [64] places functions with smallest predicted GPU over CPU speedups on the CPU, while the remaining tasks are placed on the GPU. In OLAP query processing, HERO [2] functions are defined as sub-operators that are recurrently executed in OLAP query operators. Placing such functions improves the execution time predictions of their placement algorithms.

Segment. Segment-based approaches are also highly dependent on the application. In graph processing gSWORD [79] places sampling tasks on the GPU and enumeration tasks on the CPU to balance execution time and accuracy in a subgraph counting algorithm. HyGraph [44] divides the vertices of the graph into consecutive fixed-size segments called blocks. Each task, depending on the graph processing algorithm, is performed on each vertex block. In stream processing, Verner et al. [24] places streams partitions ensuring both deadline constraint and throughput requirements. In OLAP query processing, Kroviakov et al. [78] defines segments as fragments (i.e., horizontal partitions of a table with a fixed number of tuples) and enables the execution of aggregations on different fragments. Mordred [7] allows different query plans to execute different segments of a column based on the locality of segments.

Bit. Bitwise decomposition and partition of data has been proposed as a compression technique for query processing [26, 38]. The key idea is to place higher/lower bits on GPU/CPU reducing the memory footprint on GPUs. More specifically, the Approximate & Refine (A&R) query plan model [38] uses GPU to compute approximate results and uses CPUs to refine the GPU approximations into exact results. This technique enables query acceleration on data greater than the GPU memory and minimizes CPU-GPU communication.

7.3.1.4 Placement Time

This dimension defines when the placement decisions are made: *(i)* runtime; *(ii)* compilation time; and *(iii)* hybrid.

In runtime approaches (a.k.a. local or dynamic) [e.g., 14, 18, 80, 81]placement decisions are made right before workload execution. In this case, only one function is placed at a time, resulting in a small search space of the amount of available compute units. Additionally, since the decision is taken at runtime, its overhead must be minimal. Therefore, placement algorithms are generally simple and rely on precise function input data. In this way, data transfer overheads can be precisely known and eventual unforeseen events (e.g., OOM errors) can be treated in advance. However, because placement decisions are not fully informed (e.g., the DAG structure and task dependencies are not known in advance), a placement decision can be worse than keeping the entire execution on a single processor if the cost of moving data to another processor type exceeds the potential gains of placing a function on the most fit processor.

On the other hand, in compilation time approaches [e.g., 43, 46, 53], the placement decision is taken during the compilation phase, and the decision is fully informed. Consequently, the worst-case placement is better than single processor execution. However, compilation time placement deals with a huge search space of possible placement options because the placement should be done for multiple functions considering the entire DAG. It also relies on complex placement algorithms that result in high computational overhead. Additionally, data transfer overheads are not precisely known because function input data must be estimated, which makes this approach unflexible to react to unforeseen events.

Finally, hybrid approaches [e.g., 15, 60, 77] are a combination of runtime and compilation time approaches.

Next we will delve into the techniques used in each strategy.

Runtime. A first runtime approach is to perform placement based on load balance [e.g., 16, 44, 47, 56, 62, 80, 81]. As this technique requires runtime resource statistics, such as processor and memory utilization, it is commonly implemented in schedulers. As an advantage, it maximizes resource utilization and processing throughput by using both CPUs and GPUs in parallel. However, this technique not always uses the best processor and often results in an increased amount of data transfers to share data among all available processors. Another strategy performs placement based on precise input data size information [e.g., 14, 16, 17, 62, 67,

81]. A key feature of this technique is the precise knowledge of data transfer costs, resulting in a reduced amount of data transfers. Nevertheless, the most appropriate processor might not always be used, potentially resulting in load imbalance issues. A last technique performs placement based on the GPU memory size. Although it avoids GPU OOM errors, it may not reuse GPU in future executions to avoid extra data transfers, based on the assumption that data transfer times exceed the gains of computing functions in the GPU.

Compilation time. The only key compilation time technique identified in the related work performs placement based on precise data locality information [e.g., 43, 46, 53]. Since data dependencies are known in advance, the amount of data transfers is reduced. However, the search space of possible placement plans is huge and requires complex algorithms, resulting in non-negligible overheads.

Hybrid. In hybrid approaches an initial placement is defined at compilation time and updated at runtime. As in runtime approaches, there are hybrid placement techniques based on load balance [e.g., 15, 28, 31, 36, 58, 60, 63, 77] and on GPU memory utilization [e.g., 7, 13, 15, 60, 63, 74]. Alternatively, there are hybrid placement techniques that are based on data locality, similar to compilation time approaches [e.g., 2, 7, 15, 31, 60, 77].

7.3.1.5 Placement Decision

This dimension defines the degree of automation of placement decisions, resulting in three categories: *(i) automatic*; *(ii) semi-automatic*; and *(iii) manual*. It is possible to combine multiple options in this dimension.

Automatic placement decisions [e.g., 13, 73, 78, 80] do not require any human interaction and achieve near-optimal placement decisions. However, this approach requires complex placement algorithms, generally designed ad hoc.

Semi-automatic placement decisions [e.g., 2, 7, 15, 57, 74] require some human intervention for tuning the decisions taken automatically. The main caveat of this approach is the need of partial human in the loop for optimization, which is oftentimes a limitation for scalability.

Manual placement decisions [e.g., 43, 44, 59, 78] have no automation and are fully dependent on human input. Although no implementation effort for placement algorithm is required, this approach requires extensive manual profiling. Additionally, placement is statically defined (i.e., not adaptive) and are not scalable to complex setups.

Next we detail the key techniques used in each strategy.

Automatic. A common technique performs data partition between CPUs and GPUs automatically [e.g., 32, 44, 63, 66, 78, 80]. Since both CPUs and GPUs are used at the same time, this technique ensures load balance and high utilization of resources. The main challenge is related to the implementation effort necessary to estimate the most efficient data partition. An alternative technique performs automatic processor affinity classification [e.g., 13, 76]. This technique exploits the fact that certain functions and data fit better either on CPU or GPU, and ensures that each processor will process workloads accordingly. However, a workload can be mostly or entirely suitable to a specific processor, causing load imbalance between CPUs and GPUs. Another technique automatically identifies GPU kernel implementation availability [73] and only places functions on GPU that have been implemented for GPU, at the cost of not always using the most fit processor to run a function.

Semi-automatic. Semi-automatic techniques are very diverse and tailored to specific problems. For instance, placement can be done based on the specification of function parameters (e.g., placing in GPUs compute-intensive tasks in deep learning training [15] or dominant OLTP transactions in query processing [74]) or based on data parameters (e.g., CPU-GPU placement based on data type specification [43, 59]). Other semi-automatic techniques rely on workload profiling to characterize the workload on each processor and optimize placement decisions [e.g., 18, 46, 49, 56, 57]. Other semi-automatic approaches require a training phase to learn the best placement decision [e.g., 1, 2, 10, 36, 37, 57, 72].

Manual. Manual approaches are divided into two distinct techniques. Users can explicitly specify data partition [e.g., 35, 40, 44, 78] and distribution, or they can explicitly specify the tasks to be placed on each processor [e.g., 9, 16, 34, 42, 43, 51, 54, 55, 59, 62]. The latter is typically done with the support of specific API features (e.g., task annotations [51], or configuration parameters [46]).

7.3.2 Cost Estimation

In this subsection we categorize the strategies used to estimate costs for placement decisions into two dimensions: prediction model and monitored metrics.

7.3.2.1 Placement Prediction Model

This dimension defines the nature of the model used to estimate the performance metrics to support placement decisions. There are three options: *(i) cost model*; *(ii) heuristic model*; and *(iii) learn model*. It is possible to combine multiple options in this dimension.

Cost-based models (a.k.a. white box) [e.g., 7, 9, 15, 32, 44, 56, 66] build a quantitative performance formulation considering hardware and/or algorithm internals. Such formulation provides near-optimal placement decisions with minimal latency. However, its ad hoc design makes it complex to generalize to other scenarios, such as different hardware or algorithms. Moreover, it also requires complex implementation and deep knowledge about the hardware and the algorithm to find potential tuning spots.

Heuristic-based models [e.g., 15, 16, 27, 43, 81] devise a set of predefined rules, which are typically implemented as optimization algorithms. Although heuristics provide a flexible and generalizable design, it might result in sub-optimal placement decisions due to their imprecision in the estimation of performance metrics.

Learn-based models (a.k.a. black box) [e.g., 1, 2, 24, 39, 69, 72, 78] perform data-driven predictions to estimate performance metrics based on historical data. This approach is adaptive, and is able to identify complex patterns with no prior knowledge about the hardware or the algorithm. Nevertheless, expensive training (e.g., big training data size, training time, energy footprint) is required to provide useful placement decisions.

Next we detail the key techniques used in each strategy.

Cost Model. Several works in the database community devise cost-based models to estimate task (here understood as query operator) execution time to support placement decisions [9, 11, 70] or caching decisions [7]. Another line of research proposes cost-based models to estimate data partitioning in a wide range of cases, including functions (i.e., database primitives) [56], deep learning training tasks [15], distributed scientific application tasks [22, 33, 66], matrix multiplication tasks [32], and graph processing application tasks [44].

Heuristic Model. Heuristic-based approaches are very diverse and include several techniques. Heuristics have been proposed to improve data locality and load balance via locality-aware data partitioning [15, 60] or processor ranks based on load balance and data locality scores [16, 62]. There are heuristics focused on the improvement of only data locality via semantic-aware caching [7, 14, 81] or task placement based on processor distance to the target data [15, 27, 43]. There are also heuristics focused on the improvement of load balance via data partitioning optimizations [29] or on-demand task placement based on processor status (i.e., idle or busy) [19, 28, 34, 49, 77, 80]. Some heuristics aim to overcome the GPU memory capacity constraints by performing smart placement decisions capable to avoid GPU OOM errors [13, 26, 38, 65]. Alternative heuristics have been proposed based on processor affinity classification in terms of data [21, 26, 38, 63, 76] or function [13, 23, 28, 46, 47, 49, 50, 59, 65, 74, 79]. Finally, heuristics have also been proposed to reduce energy consumption [40].

Learn Model. A popular learn-based model technique consists on performing numerical analysis over past data for placement decisions, being simple to maintain and providing fast estimations. However, data size and execution time are assumed to scale linearly and therefore are only applicable to scenarios that fit in such assumption. Examples of numerical analysis include estimating task execution time with interpolation [1, 2, 17, 31, 41], task execution time for data partitioning with hyperplane fitting [78], data partitioning [18, 24] and application energy consumption estimations [39] with curve fitting. Alternatively, works using supervised learning [6, 20, 36, 37, 48, 57, 58, 67, 71, 72] provide a wider model applicability, keeping intuitive learn models implemented in popular algorithms. However, such approach is not appropriate to identify very complex patterns, such as task scheduling (more details about task scheduling in Sect. 7.4). Targeting scenarios involving complex patterns, learn-based models using reinforcement learning (RL) have been proposed by previous works [10, 53, 61, 64, 69]. Although reinforcement learning models are self-adaptive, they require expensive training time, complex models and algorithms with difficult interpretability and explainability.

7.3.2.2   Monitored Metrics

This dimension defines the target metrics used to guide and/or compare CPU-GPU placement decisions: *(i) latency*; *(ii) throughput*; *(iii) energy consumption*; and *(iv) monetary price*. It is possible to combine multiple options in this dimension.

Latency [e.g., 15, 18, 37, 46, 57] monitors the workload execution time (total or specific portions) on each processor. It provides several placement performance insights and it is easy to measure and interpret. Hence, this is the most popular performance metric. However, it has a limited scope, especially if the target optimization goal of the placement algorithm is not focused on performance.

Throughput [e.g., 14, 22, 47, 63, 64] monitors the volume of data processed per time unit (e.g., GB/s) on each processor. Different than latency, throughput provides more detailed insights about the utilization of resources, such as processor, memory, storage, and bus interconnects. However, since throughput considers multiple resources, it requires a more complex analysis.

Energy consumption [e.g., 29, 35, 39, 40, 45, 66] monitors the energy required to run a workload on each processor. Being physically determined as the product of the processor power dissipation and execution time, energy consumption provides insights about the energy footprint, revealing energy-saving opportunities. It might be complex to measure though, depending on the measurement strategy used (e.g., external power meters or specialized software).

Monetary price [9, 86] monitors the cost-efficiency of each processor. Although it is the least popular metric, it is gaining traction in the last years with the popularization of heterogeneous cloud services (e.g., Google Cloud Platform [87] and Amazon EC2 [88]), which charge clients according to their resource consumption [86]. Although monetary prices provide a notion of the economical impacts of opting for a processor, it is also complex to measure due to dynamic pricing policies and the different hardware models and settings commercially available.

Next we detail each metric.

Latency. Latency can be directly used as the placement criterion by opting for the processor that results in the fastest execution time [11, 23, 25, 31, 50, 57, 64]. Alternatively, latency can also be used to define data partition in two techniques. First, the partition ratio that makes concurrent CPU-GPU tasks finish at the same time [15, 18, 29, 32, 33, 44, 56, 58]. Second, the partition that results in minimal penalty (i.e., data transfer and merge times) for concurrent CPU-GPU tasks [37, 67].

Throughput. Throughput is applied in three different scenarios. First, it can be used directly as a placement criterion for function-level granularity [28, 47, 64]. Second, it can be used to define data partitioning. For instance, the technique proposed in [63] places grouped or individual chunks of data on CPU and GPU to minimize the estimated throughput execution skew between CPU and GPU tasks, whereas Yang et al. [22] partitions CPU-GPU task data proportionally to the computed throughput from the previous execution. Third, it is used to support caching decisions, where the per-device task processing throughput is inferred to determine if a query is interconnect-bandwidth or storage-bandwidth to support the semantic-aware cache [14].

Energy Consumption. Approaches based on energy consumption must define a measurement technique that balances precision and that is practical to measure at the same time. There are three techniques in this group. First measurements can be done using external power meters, which provide high precision, but they are laborious to obtain. This technique has been employed in latency-driven data partitioning to minimize the energy spent by an idle processor [18, 29], as well as in empirical energy consumption analysis studies [35, 40]. Second, measurements can be taken using software tools based on internal sensors, which also provide high precision and are more practical to measure as they do not require any plug of external power meters [66]. Third, data-driven approaches estimate energy [45] or power consumption [39] based on past execution time measurements, being less precise, but considerably more practical to measure since they rely only on latency metrics.

Monetary Price. Although not commonly used for placement decisions, monetary price has been considered in terms of purchase cost [9] and renting costs in cloud providers [9, 86]. According to a study done in [9], GPUs have superior performance (25x) than CPUs for data analytics. However, CPU/GPU cost ratio is around 6x (i.e., GPUs are 6x more expensive than CPUs). Hence, a factor of 4x improvement in cost-effectiveness of GPU over CPU is achievable. The authors argue that similar cost-effectiveness ratios should be obtained across different CPU and GPU models.

7.3.3 Heterogeneous Code Management

In this subsection we categorize the strategies used to manage heterogeneous CPU-GPU code. More specifically, we focus the discussion on the dimension regarding GPU programming methods.

7.3.3.1 GPU Programming Method

This dimension defines the programming abstractions and coding paradigms to enable workload execution on CPU and GPU. There are three options: *(i) kernel template*; *(ii) compiler*; and *(iii) raw kernel*. It is possible to combine multiple options in this dimension. As shown in Fig. 7.1, each GPU programming method provides a different balance between performance, productivity, and portability. Next we detail each method.

Kernel Template. Kernel template offers pre-defined kernels via public libraries or frameworks. It is the easiest method to use because programmers simply call the desired kernel template implementation as a function. Hence, it provides a high level of code productivity and portability. Because of its flexible design, it supports multiple platforms (e.g., GPU vendors and models) with minimal code change. Although kernel templates are typically highly optimized, they lack the capability of performance tuning. Table 7.2 shows that kernel templates used in the related work have

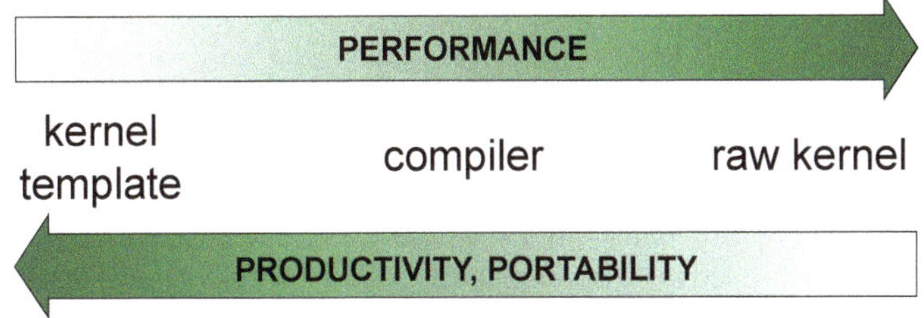

Fig. 7.1 Overview of performance, productivity, and portability of GPU programming methods

Table 7.2 Overview of kernel templates used in the related work

Kernel template	Field	Examples
CUDA-X (e.g., cuBLAS [89], cuDNN [90])	Data processing, AI, HPC	[19, 30, 32, 43, 58, 59, 67]
CuPy [91] (atop CUDA-X)	Data processing, AI, HPC	[5, 16, 42, 51]
RAPIDS [92] (atop CUDA-X)	Data processing, AI, HPC	[42, 68]
Crystal [9]	Query processing	[7, 9, 70]
Ocelot [1]	Query processing	[1, 20]
Panda [72]	Query processing	[72]
ACML-GPU [93]	Math (mainly linear algebra)	[22]
MATE-CG [94]	Map-Reduce applications	[25]
OpenCV [95]	Image processing, computer vision	[64]
Thrust [96]	Data structures	[56]
MAGMA [97]	Linear algebra	[69]
DGL framework [98]	Graph neural network	[81]
NVIDIA DALI [99]	Data pre-processing for Deep learning	[80]

enabled programmers to use GPU kernel templates in a wide range of application fields.

Compiler. Compilers enable the implementation of user-defined function (UDF) kernels using high-level sequential code. This GPU programming method provides the most balanced level of performance, productivity, and portability compared to kernel templates and raw kernels. LLVM [100] is a state-of-the-art compiler that converts high-level code into a hardware-agnostic intermediate representation. Some approaches use LLVM to generate GPU code [20, 21, 37, 48, 60], while other works use LLVM-based compilers, such as Numba (JIT compiler for Python code) [16, 51], NVIDIA NVCC (compiler for C, C++, and Fortran code) [31], and Intel HDK (JIT compiler for OLAP queries expressed as relational algebra or SQL) [78]. Compilers not using LLVM were also observed in some works such as JIT compiler for Java code (TornadoVM) [73], JIT compiler for MATLAB code [23], and domain-specific language compilers (e.g., TACO [101]) [62, 75].

Raw kernel. Raw kernels allow programmers to implement their own kernels. This GPU programming method provides the highest level of performance optimization through low-level kernel programming fine tuning. However, implementing GPU kernels from scratch is a time-consuming and labor-intensive process, especially if several complex kernels need to be implemented. It also typically results in low portability because of its ad hoc design. The related work opting for this programming method is divided into implementing kernels using NVIDIA CUDA [102] or OpenCL [103]. Works using NVIDIA CUDA develop kernels for NVIDIA GPUs [e.g., 9, 17, 18, 29, 39, 63]. On the other hand, works using OpenCL [e.g., 6, 36, 40, 41, 48, 62] aim to write kernels once and run on any device (i.e., portable on multiple processor vendors).

7.4 DAG Task Scheduling Strategies

A number of cost-based models have been proposed to solve the task scheduling problem in heterogeneous platforms. Two of the main classical cost-based algorithms for static scheduling are the Heterogeneous Earliest Finish Time (HEFT) and the Critical-Path-on-a-Processor (CPOP) and both select the processor based on tasks' ranks [104]. Popular algorithms proposed to address the dynamic task scheduling are the Balanced Minimum Completion Time (BMCT) [105] and the Dynamic Level Scheduling (DLS) [106]. However, these solutions are restricted to specific application scenarios and cannot be generalized to more complex heterogeneous hardware and constantly changing applications [53]. Moreover, such cost-based approaches require mastering the internals of hardware and algorithms, which is time consuming and demands a significant effort from experts.

The DAG scheduling problem is NP-complete and for the dynamic approach it has a stochastic nature, as the exact execution time of tasks and data communication is unknown beforehand. Nonetheless, estimations of task execution times can be made based on previous executions. When several constraints (e.g., task dependency, resource availability, communication costs) are considered, the task scheduling problem becomes a combinatorial problem, which is NP-hard [69]. RL is ideal for this kind of problem, because it can dynamically adapt to different execution conditions, dealing with unplanned situations. Furthermore, RL properly handles the stochastic nature of dynamic schedulers as a series of scheduling decisions are taken considering the current state of tasks and resources [69]. In this sense, RL algorithms have been successfully used in approaches such as Adaptive DAG Tasks Scheduling (ADTS) [53]—for static scheduling—and READYS [69]—for dynamic scheduling—to design adaptive DAG schedulers from past executions. Although these solutions show promising results, they do not consider the integration with real-world applications, such as Database Management Systems (DBMS) or other data processing engines. Table 7.3 summarizes typical works based on the schedul-

Table 7.3 State-of-the-art DAG scheduling approaches for dynamic and static scheduling

	Static scheduling	Dynamic scheduling
Cost-based models	HEFT/CPOP [104]	BMCT [105]; DLS [106]
Learn-based models (RL)	ADTS [53]	READYS [69]

ing category (static or dynamic) and types of models (cost-based heuristics and RL-based).

Reinforcement learning for task scheduling. Recently, RL has been successfully used to optimize static (e.g., ADTS) and dynamic (e.g., READYS) scheduling on heterogeneous CPU-GPU environments. ADTS uses Deep Reinforcement Learning (DRL) with a policy gradient modeled as Monte-Carlo trials. Although empirical experiments showed a high overhead, the ADTS algorithm can learn very fast and, once trained, it can outperform HEFT and CPOP in terms of average speedup. READYS models the adaptive dynamic scheduling problem as a Markov Decision Problem (MDP) and combines Graph Convolutional Networks (GCN) with an Actor-Critic Algorithm (A2C) to represent the scheduling problem and learn a scheduling policy at runtime. Experiments showed that READYS is able to process DAG tasks faster than state-of-art heuristic-based algorithms HEFT and BMCT.

7.5   Open Issues and Future Directions

Open issues. We identified five key open problems in this work. *(i) Non-linear, conflicting correlation of factors*: performance depends on the combination of multiple factors related to the application/algorithm, dataset, hardware resources, and processing system. A poor balance between such factors leads to CPU-GPU load imbalance and waste of resources. To further complicate the situation, there is a huge design space of tuning knobs that require considerable effort from developers to identify efficient execution settings; *(ii) Different software and hardware offerings requiring specialized treatment*: a plethora of programming models and processors are available nowadays, being often unclear to programmers the link between GPU platforms and supported programming methods. An initial study [107] for hardware/software compatibility spots has been recently published, opening the path towards this research direction; *(iii) Expensive and energy-intensive infrastructure*: GPUs provide astonishing speedups, but are more expensive and consume more energy than CPUs. However, because of their considerable speedups, GPUs compensate their high energy and monetary costs if placed on the right workloads. Finding a balance between energy efficiency and compute performance between CPUs and GPUs is a tricky task; *(iv) High complexity to predict performance metrics*: making accurate predictions is a challenge due to dynamic execution environment and processing of skewed data. Learn-based approaches also suffer the challenge of transfer learning for new data and hardware; and *(v) Shortage on GPU developers*: GPU kernels are harder to code, to optimize, to debug and less predictable than sequential code. Such programming characteristics require in-depth knowledge of parallel programming. Therefore, programmers not familiar with parallel programming need methods and abstractions to facilitate GPU kernel development.

Research directions. The insights of this work lead to four key future research directions. *(i) Automated, multi-objective, multi-dimensional placement to optimize the trade-off between processor choice and data transfer:* to avoid the costly host-device round trips for memory allocation, the most fit processor to run a task might be sacrificed. Additionally, current approaches focus on optimizing a single or dependent performance metrics. For instance, how to automatically balance performance, energy consumption, and monetary costs is an open question; *(ii) Automated fine-tuning of execution parameters:* empirically determining and tuning parameters, such as partition size, degree of parallelism, cache size, overlap of data transfer and compute, etc. is labor-intensive, requiring automated solutions to find optimal solutions. Additionally, being aware of kernel-level optimizations (e.g., Matmul-free Language Models kernels [108]) is another desirable feature for future automatic solutions to exploit further performance improvements by choosing the best GPU kernel to solve a problem; *(iii) Adaptive cache-aware workload placement:* cache is fast, but very limited in size, especially in GPUs. Allocating GPU memory for cache can significantly reduce the amount of data available to be processed by GPUs, limiting processing throughput. Knowing when CPU-GPU cache is worth on generic workloads is another open question; and *(iv) Suitable heterogeneous CPU-GPU benchmarks and simulators to compare different strategies*: Build upon initial efforts for benchmarks (e.g., NPBench [109], RAFT [110], and MLPerf [111] for Python; SHOC [112], Parboil [113], Rodinia [114], and Valar [115] for C++ (OpenCL and CUDA); NDS [116] and NDS-H [117] are benchmarks tailored for distributed query processing) and simulators (e.g., gem5-gpu [118], Multi2Sim [119]) is another promising research direction.

7.6   Conclusions

In this chapter, we provide a comprehensive analysis of existing strategies for workload placement and scheduling on heterogeneous CPU-GPU architectures. We identified key trends in efficient techniques and devised a taxonomy to categorize the state of the art in terms of workload placement strategies, cost estimation, and heterogeneous code management. An overview of DAG task scheduling solutions is also provided. Additionally, we highlighted key open issues and potential research directions to improve workload placement and scheduling on heterogeneous CPU-GPU systems.

References

1. Breß S, Köcher B, Heimel M, Markl V, Saecker M, Saake G (2014) Ocelot/hype: optimized data processing on heterogeneous hardware. Proc VLDB Endow 7(13):1609–1612
2. Karnagel T, Habich D, Lehner W (2017) Adaptive work placement for query processing on heterogeneous computing resources. Proc VLDB Endow 10(7):733–744
3. Mittal S, Vetter JS (2015) A survey of cpu-gpu heterogeneous computing techniques. ACM Comput Surv (CSUR) 47(4):1–35
4. Rosenfeld V, Breß S, Markl V (2022) Query processing on heterogeneous cpu/gpu systems. ACM Comput Surv (CSUR) 55(1):1–38
5. Carvalho MNL, Simitsis A, Queralt A, Romero O (2024) Workload placement on heterogeneous cpu-gpu systems. Proc VLDB Endow 17(12):4241–4244
6. Grewe D, O'Boyle MF (2011) A static task partitioning approach for heterogeneous systems using opencl. In: Compiler construction: 20th international conference, CC 2011, Held as part of the joint European conferences on theory and practice of software, ETAPS 2011, Saarbrücken, Germany, March 26–April 3, 2011. Proceedings 20, Springer, Cham, Switzerland, pp 286–305
7. Yogatama BW, Gong W, Yu X (2022) Orchestrating data placement and query execution in heterogeneous cpu-gpu dbms. Proc VLDB Endow 15(11):2491–2503
8. Carvalho M, Queralt Calafat A, Romero Moral Ó, Simitsis A, Tatu C, Badia Sala RM (2024) Performance analysis of distributed gpu-accelerated task-based workflows. In: Proceedings 27th international conference on extending database technology (EDBT 2024): Paestum, Italy, March 25–March 28, OpenProceedings.org, Paestum, Italy, pp 690–703
9. Shanbhag A, Madden S, Yu X (2020) A study of the fundamental performance characteristics of gpus and cpus for database analytics. In: Proceedings of the 2020 ACM SIGMOD international conference on management of data. Association for Computing Machinery, New York, NY, USA, pp 1617–1632
10. Addanki R, Venkatakrishnan SB, Gupta S, Mao H, Alizadeh M (2018) Placeto: efficient progressive device placement optimization. In: NIPS machine learning for systems workshop. NeurIPS/workshop series. Montreal, Canada, pp 1–8
11. He B, Lu M, Yang K, Fang R, Govindaraju NK, Luo Q, Sander PV (2009) Relational query coprocessing on graphics processors. ACM Trans Database Syst (TODS) 34(4):1–39
12. NVIDIA (2024) Nvidia h200. https://www.nvidia.com/en-eu/data-center/h200/
13. Lee R, Zhou M, Li C, Hu S, Teng J, Li D, Zhang X (2021) The art of balance: a rateupdb^{TM} experience of building a cpu/gpu hybrid database product. Proc VLDB Endow 14(12):2999–3013
14. Nicholson H, Raza A, Chrysogelos P, Ailamaki A (2023) Hetcache: synergising nvme storage and gpu acceleration for memory-efficient analytics. In: Proceedings of the 13th conference on innovative data systems research (CIDR 2023). CIDR, Amsterdam, The Netherlands, p 84
15. Li Z, Cao Q, Chen Y, Yan W (2023) Cotrain: Efficient scheduling for large-model training upon gpu and cpu in parallel. In: Proceedings of the 52nd international conference on parallel processing. ACM, Salt Lake City, Utah, USA, pp 92–101
16. Lee H, Ruys W, Henriksen I, Peters A, Yan Y, Stephens S, You B, Fingler H, Burtscher M, Gligoric M et al (2022) Parla: a python orchestration system for heterogeneous architectures. SC22: international conference for high performance computing. Networking, storage and analysis. IEEE, Dallas, TX, USA, pp 1–15
17. Breß S, Funke H, Teubner J (2016) Robust query processing in co-processor-accelerated databases. In: Proceedings of the 2016 international conference on management of data. Association for Computing Machinery, New York, NY, USA, pp 1891–1906
18. Luk CK, Hong S, Kim H (2009) Qilin: exploiting parallelism on heterogeneous multiprocessors with adaptive mapping. In: Proceedings of the 42nd Annual IEEE/ACM international symposium on microarchitecture. IEEE, Piscataway, NJ, USA, pp 45–55

19. Augonnet C, Thibault S, Namyst R, Wacrenier PA (2009) Starpu: a unified platform for task scheduling on heterogeneous multicore architectures. In: Euro-Par 2009 parallel processing: 15th international Euro-Par conference, Delft, The Netherlands, August 25-28, 2009. Proceedings 15, Springer, Cham, Switzerland, pp 863–874

20. Kerr A, Diamos G, Yalamanchili S (2010) Modeling gpu-cpu workloads and systems. In: Proceedings of the 3rd workshop on general-purpose computation on graphics processing units. Association for Computing Machinery, New York, NY, USA, pp 31–42

21. Ravi VT, Ma W, Chiu D, Agrawal G (2010) Compiler and runtime support for enabling generalized reduction computations on heterogeneous parallel configurations. In: Proceedings of the 24th ACM international conference on supercomputing. Association for Computing Machinery, New York, NY, USA, pp 137–146

22. Yang C, Wang F, Du Y, Chen J, Liu J, Yi H, Lu K (2010) Adaptive optimization for petascale heterogeneous cpu/gpu computing. In: 2010 IEEE international conference on cluster computing. IEEE, Piscataway, NJ, USA, pp 19–28

23. Prasad A, Anantpur J, Govindarajan R (2011) Automatic compilation of matlab programs for synergistic execution on heterogeneous processors. In: Proceedings of the 32nd ACM SIGPLAN conference on programming language design and implementation. Association for Computing Machinery, New York, NY, USA, pp 152–163

24. Verner U, Schuster A, Silberstein M (2011) Processing data streams with hard real-time constraints on heterogeneous systems. In: Proceedings of the international conference on supercomputing. Association for Computing Machinery, New York, NY, USA, pp 120–129

25. Ravi VT, Becchi M, Jiang W, Agrawal G, Chakradhar S (2012) Scheduling concurrent applications on a cluster of cpu-gpu nodes. In: 2012 12th IEEE/ACM international symposium on cluster, cloud and grid computing (ccgrid 2012). IEEE, Piscataway, NJ, USA, pp 140–147

26. Pirk H (2012) Efficient cross-device query processing. In: Proceedings of the VLDB Ph.D. workshop 2012. VLDB Endowment, San Francisco, CA, USA

27. Kim J, Seo S, Lee J, Nah J, Jo G, Lee J (2012) Snucl: an opencl framework for heterogeneous cpu/gpu clusters. In: Proceedings of the 26th ACM international conference on supercomputing. Association for Computing Machinery, New York, NY, USA, pp 341–352

28. Wang K, Huai Y, Lee R, Wang F, Zhang X, Saltz JH (2012) Accelerating pathology image data cross-comparison on cpu-gpu hybrid systems. In: Proceedings of the VLDB endowment international conference on very large data bases. NIH Public Access, Bethesda, MD, USA, p 1543

29. Ma K, Li X, Chen W, Zhang C, Wang X (2012) Greengpu: A holistic approach to energy efficiency in gpu-cpu heterogeneous architectures. In: 2012 41st international conference on parallel processing. IEEE, Piscataway, NJ, USA, pp 48–57

30. Bosilca G, Bouteiller A, Danalis A, Herault T, Lemarinier P, Dongarra J (2012) Dague: a generic distributed dag engine for high performance computing. Parallel Comput 38(1–2):37–51

31. Pienaar JA, Chakradhar S, Raghunathan A (2012) Automatic generation of software pipelines for heterogeneous parallel systems. In: SC'12: proceedings of the international conference on high performance computing, networking, storage and analysis. IEEE, Piscataway, NJ, USA, pp 1–12

32. Clarke D, Ilic A, Lastovetsky A, Sousa L (2012) Hierarchical partitioning algorithm for scientific computing on highly heterogeneous cpu+ gpu clusters. In: Euro-Par 2012 parallel processing: 18th international conference, Euro-Par 2012. Rhodes Island, Greece, August 27–31, 2012. Proceedings 18. Springer, Cham, Switzerland, pp 489–501

33. Lu F, Song J, Yin F, Zhu X (2012) Performance evaluation of hybrid programming patterns for large cpu/gpu heterogeneous clusters. Comput Phys Commun 183(6):1172–1181

34. Bauer M, Treichler S, Slaughter E, Aiken A (2012) Legion: expressing locality and independence with logical regions. In: SC'12: Proceedings of the international conference on high performance computing, networking, storage and analysis. IEEE, Piscataway, NJ, USA, pp 1–11

35. Gharaibeh A, Santos-Neto E, Costa LBa, Ripeanu M (2013) The energy case for graph processing on hybrid cpu and gpu systems. In: Proceedings of the 3rd workshop on irregular applications: architectures and algorithms. Association for Computing Machinery, New York, NY, USA

36. Kofler K, Grasso I, Cosenza B, Fahringer T (2013) An automatic input-sensitive approach for heterogeneous task partitioning. In: Proceedings of the 27th international ACM conference on international conference on supercomputing. Association for Computing Machinery, New York, NY, USA, pp 149–160

37. Lee J, Samadi M, Park Y, Mahlke S (2013) Transparent cpu-gpu collaboration for data-parallel kernels on heterogeneous systems. In: Proceedings of the 22nd international conference on Parallel architectures and compilation techniques. IEEE, Piscataway, NJ, USA, pp 245–255

38. Pirk H, Manegold S, Kersten M (2014) Waste not... efficient co-processing of relational data. In: 2014 IEEE 30th international conference on data engineering. IEEE, Piscataway, NJ, USA, pp 508–519

39. Cumming B, Fourestey G, Fuhrer O, Gysi T, Fatica M, Schulthess TC (2014) Application centric energy-efficiency study of distributed multi-core and hybrid cpu-gpu systems. In: SC'14: Proceedings of the international conference for high performance computing, networking, storage and analysis. IEEE, Piscataway, NJ, USA, pp 819–829

40. Cheng X, He B, Lau CT (2015) Energy-efficient query processing on embedded cpu-gpu architectures. In: Proceedings of the 11th international workshop on data management on new hardware. Association for Computing Machinery, New York, NY, USA

41. Karnagel T, Habich D, Lehner W (2015) Local versus global optimization: operator placement strategies in heterogeneous environments. In: Proceedings of the EDBT/ICDT 2015 joint conference workshops, CEUR workshop proceedings. Brussels, Belgium, pp 48–55

42. Rocklin M (2015) Dask: parallel computation with blocked algorithms and task scheduling. In: Proceedings of the 14th python in science conference (SciPy 2015). SciPy.org, Austin, TX, USA, pp 126–132

43. Abadi M, Barham P, Chen J, Chen Z, Davis A, Dean J, Devin M, Ghemawat S, Irving G, Isard M, et al. (2016) Tensorflow: a system for large-scale machine learning. In: Proceedings of the 12th USENIX symposium on operating systems design and implementation (OSDI 16). USENIX Association, CA, USA, pp 265–283

44. Heldens S, Varbanescu AL, Iosup A (2016) Dynamic load balancing for high-performance graph processing on hybrid cpu-gpu platforms. In: 2016 6th workshop on irregular applications: architecture and algorithms (IA3). IEEE, Piscataway, NJ, USA, pp 62–65

45. Sîrbu A, Babaoglu O (2016) Power consumption modeling and prediction in a hybrid cpu-gpu-mic supercomputer. In: Euro-Par 2016: parallel processing: 22nd international conference on parallel and distributed computing, Grenoble, France, August 24–26, 2016, Proceedings, vol. 22. Springer, Cham, Switzerland, pp 117–130

46. Yuan Y, Salmi MF, Huai Y, Wang K, Lee R, Zhang X (2016) Spark-gpu: An accelerated in-memory data processing engine on clusters. In: 2016 IEEE international conference on big data (Big Data). IEEE, Piscataway, NJ, USA, pp 273–283

47. Koliousis A, Weidlich M, Castro Fernandez R, Wolf AL, Costa P, Pietzuch P (2016) Saber: window-based hybrid stream processing for heterogeneous architectures. In: Proceedings of the 2016 international conference on management of data. Association for Computing Machinery, New York, NY, USA, pp 555–569

48. Wen Y, O'Boyle MF (2017) Merge or separate? multi-job scheduling for opencl kernels on cpu/gpu platforms. In: Proceedings of the general purpose GPUs. Association for Computing Machinery, New York, NY, USA, pp 22–31

49. Appuswamy R, Karpathiotakis M, Porobic D, Ailamaki A (2017) The case for heterogeneous htap. In: Proceedings of the 8th biennial conference on innovative data systems research (CIDR 2017), CIDR Conference. Chaminade, CA, USA

50. Mayer R, Mayer C, Laich L (2017) The tensorflow partitioning and scheduling problem: it's the critical path! In: Proceedings of the 1st workshop on distributed infrastructures for deep learning. Association for Computing Machinery, New York, NY, USA, pp 1–6

51. Amela R, Ramon-Cortes C, Ejarque J, Conejero J, Badia RM (2018) Executing linear algebra kernels in heterogeneous distributed infrastructures with pycompss. Oil Gas Sci. Technol.- Revue d'IFP Energies Nouvelles 73:47
52. Mirhoseini A, Goldie A, Pham H, Steiner B, Le QV, Dean J (2018) A hierarchical model for device placement. In: International conference on learning representations. Curran Associates, Inc., NY, USA
53. Wu Q, Wu Z, Zhuang Y, Cheng Y (2018) Adaptive dag tasks scheduling with deep reinforcement learning. In: Algorithms and architectures for parallel processing: 18th international conference, ICA3PP 2018, Guangzhou, China, November 15–17, 2018, Proceedings, Part II 18. Springer, Cham, Switzerland, pp 477–490
54. Moritz P, Nishihara R, Wang S, Tumanov A, Liaw R, Liang E, Elibol M, Yang Z, Paul W, Jordan MI, Stoica I (2018) Ray: a distributed framework for emerging AI applications. In: 13th USENIX symposium on operating systems design and implementation (OSDI 18). USENIX Association, Carlsbad, CA, pp 561–577
55. Chen C, Li K, Ouyang A, Zeng Z, Li K (2018) Gflink: an in-memory computing architecture on heterogeneous cpu-gpu clusters for big data. IEEE Trans Parallel Distrib Syst 29(6):1275–1288
56. Gowanlock M, Karsin B, Fink Z, Wright J (2019) Accelerating the unacceleratable: Hybrid cpu/gpu algorithms for memory-bound database primitives. In: Proceedings of the 15th international workshop on data management on new hardware. Association for Computing Machinery, New York, NY, USA, pp 1–11
57. Khalid YN, Aleem M, Ahmed U, Islam MA, Iqbal MA (2019) Troodon: a machine-learning based load-balancing application scheduler for cpu-gpu system. J Parallel Distrib Comput 132:79–94
58. Sant'Ana L, Cordeiro D, de Camargo RY (2019) Plb-hac: dynamic load-balancing for heterogeneous accelerator clusters. In: Euro-Par 2019: parallel processing: 25th international conference on parallel and distributed computing, Göttingen, Germany, August 26–30, 2019, Proceedings 25. Springer, Cham, Switzerland, pp 197–209
59. Paszke A, Gross S, Massa F, Lerer A, Bradbury J, Chanan G, Killeen T, Lin Z, Gimelshein N, Antiga L, et al. (2019) Pytorch: an imperative style, high-performance deep learning library. Adv Neural Inf Process Syst 32
60. Chrysogelos P, Karpathiotakis M, Appuswamy R, Ailamaki A (2019) Hetexchange: encapsulating heterogeneous cpu-gpu parallelism in jit compiled engines. Proc VLDB Endow 12(5):544–556
61. Banerjee S, Jha S, Kalbarczyk Z, Iyer R (2020) Inductive-bias-driven reinforcement learning for efficient schedules in heterogeneous clusters. In: III HD, Singh A (eds) International conference on machine learning. PMLR, Online/Virtual, pp 629–641
62. Kamatar AV, Friese RD, Gioiosa R (2020) Locality-aware scheduling for scalable heterogeneous environments. In: 2020 IEEE/ACM International workshop on runtime and operating systems for supercomputers (ROSS). IEEE, Piscataway, NJ, USA, pp 50–58
63. Lutz C, Breß S, Zeuch S, Rabl T, Markl V (2020) Pump up the volume: processing large data on gpus with fast interconnects. In: Proceedings of the 2020 ACM SIGMOD international conference on management of data. Association for Computing Machinery, New York, NY, USA, pp 1633–1649
64. Zhang H, Geng X, Ma H (2020) Learning-driven interference-aware workload parallelization for streaming applications in heterogeneous cluster. IEEE Trans Parallel Distrib Syst 32(1):1–15
65. Ren J, Rajbhandari S, Aminabadi RY, Ruwase O, Yang S, Zhang M, Li D, He Y (2021) {Zero-offload}: Democratizing {billion-scale} model training. In: 2021 USENIX annual technical conference (USENIX ATC 21). USENIX Association, Berkeley, CA, USA, pp 551–564
66. Pérez B, Stafford E, Bosque JL, Beivide R (2021) Sigmoid: an auto-tuned load balancing algorithm for heterogeneous systems. J Parallel Distrib Comput 157:30–42
67. Li Z, Zhang Y, Ding A, Zhou H, Liu C (2021) Efficient algorithms for task mapping on heterogeneous cpu/gpu platforms for fast completion time. J Syst Archit 114:101936

68. Lee S, Park S (2021) Performance analysis of big data etl process over cpu-gpu heterogeneous architectures. In: 2021 IEEE 37th international conference on data engineering workshops (ICDEW). IEEE, Piscataway, NJ, USA, pp 42–47

69. Grinsztajn N, Beaumont O, Jeannot E, Preux P (2021) Readys: A reinforcement learning based strategy for heterogeneous dynamic scheduling. In: 2021 IEEE international conference on cluster computing (CLUSTER). IEEE, Piscataway, NJ, USA, pp 70–81

70. Shanbhag A, Yogatama BW, Yu X, Madden S (2022) Tile-based lightweight integer compression in gpu. In: Proceedings of the 2022 international conference on management of data. Association for Computing Machinery, New York, NY, USA, pp 1390–1403

71. Xing H, Agrawal G, Ramnath R (2023) Gpu adaptive in-situ parallel analytics (gap). In: Proceedings of the international conference on parallel architectures and compilation techniques. Association for Computing Machinery, New York, NY, USA, pp 467–480

72. Liu H, Tang B, Zhang J, Deng Y, Yan X, Zheng X, Shen Q, Zeng D, Mao Z, Zhang C, You Z, Wang Z, Jiang R, Wang F, Yiu ML, Li H, Han M, Li Q, Luo Z (2022) Ghive: accelerating analytical query processing in apache hive via cpu-gpu heterogeneous computing. In: Proceedings of the 13th symposium on cloud computing. Association for Computing Machinery, New York, NY, USA, pp 158–172

73. Xekalaki M, Fumero J, Stratikopoulos A, Doka K, Katsakioris C, Bitsakos C, Koziris N, Kotselidis C (2022) Enabling transparent acceleration of big data frameworks using heterogeneous hardware. Proc VLDB Endow 15(13):3869–3882

74. Boeschen N, Binnig C (2022) Gacco - a gpu-accelerated oltp dbms. In: Proceedings of the 2022 international conference on management of data. Association for Computing Machinery, New York, NY, USA, pp 1003–1016

75. Kamatar A, Friese R, Gioiosa R (2023) A task based approach for co-scheduling ensemble workloads on heterogeneous nodes. In: 2023 IEEE international parallel and distributed processing symposium workshops (IPDPSW). IEEE, St. Petersburg, Florida, USA, pp 6–15

76. Park T, Kang S, Jang MH, Kim SW, Park Y (2023) Orchestrating large-scale spgemms using dynamic block distribution and data transfer minimization on heterogeneous systems. In: 2023 IEEE 39th international conference on data engineering (ICDE). IEEE, Dallas, USA, pp 2456–2459

77. Cui P, Liu H, Tang B, Yuan Y (2024) Cggraph: an ultra-fast graph processing system on modern commodity cpu-gpu co-processor. Proc VLDB Endow 17(6):1405–1417

78. Kroviakov A, Kurapov P, Anneser C, Giceva J (2024) Heterogeneous intra-pipeline device-parallel aggregations. In: Proceedings of the 20th international workshop on data management on new hardware. Association for Computing Machinery, New York, NY, USA, pp 1–10

79. Ye C, Li Y, Sun S, Guo W (2024) gsword: Gpu-accelerated sampling for subgraph counting. Proc ACM Manag Data 2(1):1–26

80. Kim T, Park C, Mukimbckov M, Hong H, Kim M, Jin Z, Kim C, Shin JY, Jeon M (2023) Fusionflow: accelerating data preprocessing for machine learning with cpu-gpu cooperation. Proc VLDB Endow 17(4):863–876

81. Ai X, Wang Q, Cao C, Zhang Y, Chen C, Yuan H, Gu Y, Yu G (2024) Neutronorch: rethinking sample-based gnn training under cpu-gpu heterogeneous environments. Proc VLDB Endow 17(8):1995–2008

82. Simitsis A, Wilkinson K, Castellanos M, Dayal U (2012) Optimizing analytic data flows for multiple execution engines. In: Proceedings of the ACM SIGMOD international conference on management of data. SIGMOD 2012, Scottsdale, AZ, USA, May 20-24, 2012, ACM, pp 829–840

83. Simitsis A, Wilkinson K, Dayal U, Hsu M (2013) HFMS: managing the lifecycle and complexity of hybrid analytic data flows. In: 29th IEEE international conference on data engineering. ICDE 2013, Brisbane, Australia, April 8-12, 2013, IEEE Computer Society, pp 1174–1185

84. Simitsis A, Vassiliadis P, Sellis TK (2005a) Optimizing ETL processes in data warehouses. In: Proceedings of the 21st international conference on data engineering. ICDE 2005, 5-8 April 2005, Tokyo, Japan, IEEE Computer Society, pp 564–575

85. Simitsis A, Vassiliadis P, Terrovitis M, Skiadopoulos S (2005b) Graph-based modeling of ETL activities with multi-level transformations and updates. In: Data warehousing and knowledge discovery, 7th international conference, DaWaK 2005, Copenhagen, Denmark, August 22-26, 2005, Proceedings, Lecture Notes in Computer Science, vol 3589. Springer, pp 43–52
86. Romero F, Hauswald J, Partap A, Kang D, Zaharia M, Kozyrakis C (2022) Optimizing video analytics with declarative model relationships. Proc VLDB Endow 16(3):447–460
87. Google (2024) Google cloud platform. https://cloud.google.com/why-google-cloud/
88. Amazon (2024) Amazon ec2. https://aws.amazon.com/ec2/
89. NVIDIA (2024a) cublas. https://developer.nvidia.com/cublas
90. NVIDIA (2024b) cudnn. https://developer.nvidia.com/cudnn
91. Networks P (2024) Cupy. https://cupy.dev/
92. NVIDIA (2024) Rapids. https://rapids.ai/
93. AMD (2024) Amd core math library for graphic processors—acml-gpu. https://archive.org/details/manuallib-id-2071936
94. Jiang W, Agrawal G (2012) Mate-cg: a map reduce-like framework for accelerating data-intensive computations on heterogeneous clusters. In: 2012 IEEE 26th international parallel and distributed processing symposium. IEEE, Shanghai, China, pp 644–655
95. OpenCV (2024) Open source computer vision library. https://opencv.org/
96. NVIDIA (2024) Thrust. https://developer.nvidia.com/thrust
97. Tomov S, Dongarra J, Baboulin M (2010) Towards dense linear algebra for hybrid gpu accelerated manycore systems. Parallel Comput 36(5–6):232–240
98. DGL (2024) Deep graph library. https://www.dgl.ai/
99. NVIDIA (2024) Nvidia data loading library—dali. https://developer.nvidia.com/dali
100. LLVM (2024) The llvm project. https://llvm.org/
101. Kjolstad F, Kamil S, Chou S, Lugato D, Amarasinghe S (2017) The tensor algebra compiler. Proc ACM Program Lang 1(OOPSLA):1–29
102. NVIDIA (2024) Cuda c++ programming guide. https://docs.nvidia.com/cuda/cuda-c-programming-guide/
103. Inc TKG (2024) Open computing language opencl. https://www.khronos.org/opencl/
104. Topcuoglu H, Hariri S, Wu MY (2002) Performance-effective and low-complexity task scheduling for heterogeneous computing. IEEE Trans Parallel Distrib Syst 13(3):260–274
105. Sakellariou R, Zhao H (2004) A hybrid heuristic for dag scheduling on heterogeneous systems. In: 18th international parallel and distributed processing symposium, 2004. Proceedings, IEEE, Santa Fe, NM, USA, p 111
106. Sih GC, Lee EA (1993) A compile-time scheduling heuristic for interconnection-constrained heterogeneous processor architectures. IEEE Trans Parallel Distrib Syst 4(2):175–187
107. Herten A (2023) Many cores, many models: Gpu programming model versus vendor compatibility overview. In: Proceedings of the SC '23 workshops of the international conference on high performance computing, network, storage, and analysis. Association for Computing Machinery, New York, NY, USA, pp 1019–1026
108. Zhu RJ, Zhang Y, Sifferman E, Sheaves T, Wang Y, Richmond D, Zhou P, Eshraghian JK (2024) Scalable matmul-free language modeling. CoRR. arxiv:abs/2406.02528
109. Ziogas AN, Ben-Nun T, Schneider T, Hoefler T (2021) Npbench: a benchmarking suite for high-performance numpy. In: Proceedings of the 35th ACM international conference on supercomputing. Association for Computing Machinery, New York, NY, USA, pp 63–74
110. NVIDIA (2024) Rapids ai: Raft ann benchmarks. https://docs.rapids.ai/api/raft/stable/raft_ann_benchmarks/
111. Reddi VJ, Cheng C, Kanter D, Mattson P, Schmuelling G, Wu CJ, Anderson B, Breughe M, Charlebois M, Chou W et al (2020) Mlperf inference benchmark. In: 2020 ACM/IEEE 47th annual international symposium on computer architecture (ISCA). IEEE, Piscataway, NJ, USA, pp 446–459
112. Danalis A, Marin G, McCurdy C, Meredith JS, Roth PC, Spafford K, Tipparaju V, Vetter JS (2010) The scalable heterogeneous computing (shoc) benchmark suite. In: Proceedings of the 3rd workshop on general-purpose computation on graphics processing units. Association for Computing Machinery, New York, NY, USA, pp 63–74

113. Stratton JA, Rodrigues C, Sung IJ, Obeid N, Chang LW, Anssari N, Liu GD, Hwu WmW (2012) Parboil: A revised benchmark suite for scientific and commercial throughput computing. Technical Report IMPACT-12-01, University of Illinois at Urbana-Champaign, Center for Reliable and High-Performance Computing
114. Che S, Boyer M, Meng J, Tarjan D, Sheaffer JW, Lee SH, Skadron K (2009) Rodinia: a benchmark suite for heterogeneous computing. In: 2009 IEEE international symposium on workload characterization (IISWC). IEEE, Piscataway, NJ, USA, pp 44–54
115. Mistry P, Ukidave Y, Schaa D, Kaeli D (2013) Valar: a benchmark suite to study the dynamic behavior of heterogeneous systems. In: Proceedings of the 6th workshop on general purpose processor using graphics processing units. Association for Computing Machinery, New York, NY, USA, pp 54–65
116. NVIDIA (2024a) Nvidia decision support (nds) benchmark. https://github.com/NVIDIA/spark-rapids-benchmarks/tree/dev/nds
117. NVIDIA (2024b) Nvidia decision support-h (nds-h) benchmark. https://github.com/NVIDIA/spark-rapids-benchmarks/tree/dev/nds-h
118. Power J, Hestness J, Orr MS, Hill MD, Wood DA (2014) gem5-gpu: a heterogeneous cpu-gpu simulator. IEEE Comput Archit Lett 14(1):34–36
119. Ubal R, Jang B, Mistry P, Schaa D, Kaeli D (2012) Multi2sim: a simulation framework for cpu-gpu computing. In: Proceedings of the 21st international conference on parallel architectures and compilation techniques. Association for Computing Machinery, New York, NY, USA, pp 335–344

Part III
Preparation

Chapter 8
Privacy-Preserving Blockchain-Based Federated Learning

Eros Fabrici⊙, Besim Bilalli⊙, Josep Lluís Berral García⊙,
and Minos Garofalakis⊙

Abstract The growing need to analyze distributed sensitive data while preserving privacy has led to the emergence of Federated Learning (FL), a paradigm enabling collaborative model training without centralizing data. However, traditional FL architectures face challenges including single points of failure, potential privacy leaks through model updates, and vulnerabilities to malicious participants. This chapter explores how blockchain technology can address these challenges, presenting a comprehensive analysis of privacy-preserving blockchain-based federated learning systems. We examine various architectural approaches, from decoupled systems where blockchain and training nodes are separate, to coupled systems with dynamic role assignment. We provide a taxonomy of privacy-preserving mechanisms, including differential privacy and cryptographic techniques, analyzing their integration with blockchain-based FL. We discuss both theoretical foundations and practical implementations, highlighting how different approaches balance privacy guarantees with system performance.

Keywords Privacy-preserving data · Federated learning · Blockchain · Secure data sharing

E. Fabrici (✉) · M. Garofalakis
Athena Research and Innovation Center, Marousi, Greece
e-mail: eros.fabrici@athenarc.gr

M. Garofalakis
e-mail: minos@athenarc.gr

B. Bilalli · J. L. Berral García
Universitat Politècnica de Catalunya, Barcelona, Spain
e-mail: besim.bilalli@upc.edu

J. L. Berral García
e-mail: josep.ll.berral@upc.edu

G. Dejaegere et al. (eds.), *Data Engineering for Data Science*,
https://doi.org/10.1007/978-3-032-18765-9_8

8.1 Introduction

The term Big Data refers to datasets that are too large or complex to be handled with traditional data processing software. In particular, Big Data captures 4 dimensions: Velocity, Variety, Veracity, and Volume. This large quantity of data available nowadays has a lot of statistical and business value, therefore, its analysis is of core importance for business decision-making. Nevertheless, the data involved in those processes may contain a lot of sensitive information regarding individuals (e.g., the health sector). Therefore, the need to guarantee the privacy of individuals while being able to extract insights and patterns from the data has become an important requirement.

Sensitive data holders who want to collaborate for a specific learning objective may also not be allowed to move their data. This led to abundant research in the context of private data mining and Machine Learning (ML), including the birth of a new ML Engineering paradigm called Federated Learning (FL), which consists of training a global model across different data owners without moving the data to a central server. A traditional FL architecture is composed of *clients* or *data holders*, who train their local models over their own datasets, and a coordinating *server*, which orchestrates the whole process and takes care of merging the local models into a global one. Consider a practical healthcare use case: multiple hospitals across different regions wish to collaborate on developing a machine learning model for early disease detection. Each hospital has thousands of patient records containing vital signs, lab results, diagnostic images, and treatment outcomes. Traditional approaches would require centralizing this data, but this raises several critical issues:

1. **Privacy Regulations**: healthcare providers must comply with strict regulations (e.g., the GDPR in Europe [1]) that restrict the sharing of patient data.
2. **Data Security**: centralizing sensitive medical data creates a single point of vulnerability for potential breaches.
3. **Data Ownership**: hospitals need to maintain control over their patient data while still contributing to the collective learning process.
4. **Trust**: without transparency in the model training process, hospitals cannot verify that their data are being used appropriately.

Although FL is a good fit for the above scenario, if the orchestrator fails, the whole system is compromised. In particular:

1. The need for a trusted central server in traditional FL architectures creates a single point of failure.
2. Model updates exchanged during training might still leak sensitive information.
3. Ensuring the integrity and transparency of the training process across multiple participants is challenging.
4. Protecting against malicious participants who might attempt to poison the model requires adopting specific algorithms that may affect the model's performance.

Therefore, the research community has demonstrated a growing interest in using blockchain and privacy-enhancing technologies to tackle the above-mentioned challenges.

This chapter explores the intersection of federated learning, blockchain technology, and privacy-preserving mechanisms. We begin by examining the fundamental concepts of federated learning and blockchain technology. We then delve into various approaches for preserving privacy in blockchain-based FL systems, including Differential Privacy and cryptographic techniques. Throughout the chapter, we analyze different system architectures, privacy guarantees, and security mechanisms. We provide a comprehensive overview of the current state-of-the-art while highlighting open challenges and future directions.

8.2 Background

The topic covered in this book chapter comprises three main different research topics. This section will cover the background of federated learning, blockchain, privacy and security.

8.2.1 Federated Learning

Federated Learning (FL) is a distributed machine learning technique, where a group of clients collectively trains a model under the orchestration of a central server. Google introduced it in 2016, where the authors presented Federated Averaging [2]. The algorithm consists of 4 steps (shown also in Fig. 8.1):

1. The server distributes the global model to the clients.
2. The clients proceed by training the model with their local data.
3. The clients send the trained models to the server.
4. The server aggregates the collected local models by averaging them; if the number of maximum iterations is reached, the FL process is terminated, otherwise the process iterates by going back to step 1.

The main advantage of federated learning is privacy, as the data stays with the local clients and only the models are exchanged. In the specific case of Federated Averaging, the privacy guarantee is what we call an ad-hoc guarantee, which will be described in Sect. 8.3.

There are many open problems in federated learning [3]:

1. *Improving the effectiveness of the models*: learning models in a federated context will lead to a decreased accuracy for various reasons, e.g., handling non-ID data, using optimization algorithms that are for the centralized setting, etc.
2. *Preserving the privacy of the user data*: reasoning about the federated model f:

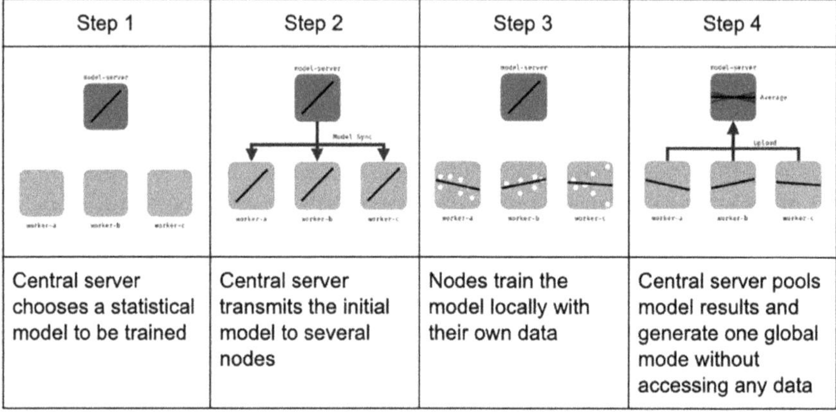

Step 1	Step 2	Step 3	Step 4
Central server chooses a statistical model to be trained	Central server transmits the initial model to several nodes	Nodes train the model locally with their own data	Central server pools model results and generate one global mode without accessing any data

Fig. 8.1 The classic federated learning process [4]

- How f is computed? Using techniques like **Secure Multiparty Computation** SMC and homomorphic encryption is the state-of-the-art to mask the data.
- What f could reveal about the users? Differential privacy represents the state of the art to limit the information leaked by f and prevent inference attacks.

3. *Security*: How to design a FL protocol to be robust against attacks. In particular, how to prevent attacks from malicious adversaries?
4. *Software systems for federated learning.*
5. *Guaranteeing fairness in federated learning.*

8.2.2 Blockchain

Blockchain technology is a distributed ledger that records transactions across multiple users in a way that ensures transparency, security, and immutability. It was introduced in the context of cryptocurrencies, which are digital currencies that make use of blockchain and cryptography for security, making it computationally infeasible to counterfeit. The first cryptocurrency is Bitcoin [5]. Bitcoin uses a clever system of decentralized trust-less verification, based on digital signatures and hash functions, and consensus algorithms to keep a consistent state across the network peers.

Digital Signatures

A digital signature is a public key infrastructure composed of a triple of efficient algorithms (KeyGen, Sign, Verify) defined as follows:

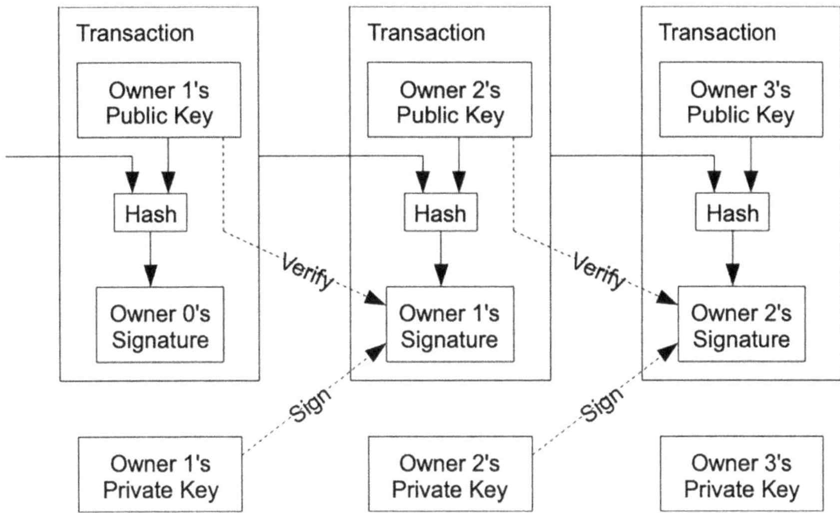

Fig. 8.2 Bitcoin transaction signing and verification [5]

- keygen(λ) $\rightarrow$ (pk, sk): is a key generation algorithm which, given a security parameter, outputs a public and a private key (pk, sk).
- Sign(sk,m) $\rightarrow \sigma$ is a signing algorithm that, given a secret key sk and a message m it outputs a signature σ.
- Verify(pk, m, σ) $\rightarrow \{0, 1\}$ is a deterministic verification algorithm that, given the public key pk and a signature σ, outputs 1 if the signature verifies or 0 if it does not verify.

This scheme guarantees the authenticity of a message, namely that the message has been produced by a certain user, by signing it with its private key. Other users will be able to verify the authenticity with the public key of the signer. Figure 8.2 shows how transactions are signed and verified in the Bitcoin system.

Proof-of-Work and Proof-of-Stake

Consensus algorithms are a class of algorithms used in distributed systems to reach consistency about a state change. A classic algorithm is Raft [6], used in many distributed database systems for the replication of data. Raft is designed for small clusters as its complexity grows substantially with the number of parties in the system. In the context of cryptocurrencies, the number of parties is incredibly high. Therefore specific consensus algorithms were designed: Proof-of-Work and Proof-of-Stake.

Proof-of-Work (PoW) was proposed in [5] in order to provide an efficient and secure consensus mechanism for approving new transactions and as a mechanism to avoid fraudulent behaviors. The key idea is that, when a party proposes a new transaction, e.g., Alice sends Bob 5 Bitcoins, the transaction is encapsulated in a block, which is broadcast to the network. Then, the network members compete in finding a number, the *nonce*, such that the `hash(concat(transaction, nonce))` starts with a predefined number of sequential zeros. The party that manages to find the nonce, wins the PoW, e.g., the hash signature, then it pretends the proof-of-work of the last block registered and broadcasts the new blockchain to the network. The block creator receives a reward in Bitcoin for winning the competition. Suppose there are conflicting blockchains, for instance, a party receives two updated blockchains with conflicting transaction histories. In that case, the party chooses the longest one, reaching therefore a decentralized consensus.

The main disadvantage of PoW is the amount of energy consumed to find the nonce. An alternative is Proof-of-Stake (PoS), which is a cryptographic consensus mechanism that uses a validator-based transaction verification process. In PoS, network participants are selected to create new blocks with a probability proportional to the quantity of cryptocurrency they stake as collateral. Unlike PoW, which requires substantial computational resources, PoS achieves network security through an economic mechanism where validators have a financial incentive to act honestly, as malicious behavior results in the forfeiture of their staked coins. While PoS is a less computationally intensive solution, its disadvantage is that peers that have more stake have a higher probability of being selected, therefore, wealthy participants gain more control over the network.

Smartcontracts

An important feature of blockchains is Smartcontracts. Generally, a self-contained piece of code that is stored in every peer of the system. Any member can request its execution on the network, which leads to a modification of the state of the ledger. In the Bitcoin case, there is only one Smartcontract and it's the asset transfer. Other blockchains, like Ethereum [7], are generalized versions of transaction ledgers, therefore, it is possible to build applications on top of them.

8.2.3 Privacy and Security

Different technologies guarantee different types of guarantees. Generally, if we want to compute an output o of a specific function (e.g., the federated learning model), we should consider:

- *How o is computed?* In a distributed scenario, designing a proper protocol and using techniques like SMC and homomorphic encryption are the best techniques to hide the information involved in the protocol used for computing o. An SMC protocol permits a set of parties to compute a function on their inputs while keeping those inputs private. HE is an encryption scheme where it is possible to apply mathematical operations directly to the encrypted data, with no need to decrypt it.

- *What o could reveal about the input data?* Especially in the case of FL, o is the global model, which could reveal information about the data used to train it. In this case, **differential privacy** [8] is the de facto standard.

The main goal of federated learning is to avoid integrating the distributed sensitive datasets into a central server. Only the model weights are exchanged to protect privacy. Nevertheless, this is not enough to guarantee the privacy of the dataset records, as *membership inference* or *model inversion* attacks are possible [9, 10]. The first refers to the possibility of an attacker inferring the presence or absence of an individual's record by querying the model. The second is a class of attacks that, from the model predictions and confidence values, it is possible to reconstruct the original input used to train the model. The main technique used to protect privacy is *differential privacy*.

Differential Privacy (DP) [8] is an information-theoretic property of a randomized algorithm. More formally, a randomized algorithm $A : \mathbb{D} \to \mathbb{U}$, where $\mathbb{D}$ is the domain of all possible datasets and $\mathbb{U}$ is the co-domain of the algorithm, is ϵ-differentially private if for all datasets $D \in \mathbb{D}$, $D' \in \mathbb{D}$ where D' is the neighbor dataset of D (i.e., removing one record), and all $S \subseteq \mathbb{U}$

$$Pr[A(D) \in S] \le e^\epsilon Pr[A(D') \in S]$$

In other words, it guarantees that any sequence of outputs (response to a query) is equally likely to occur, independent of the presence or absence of any individual in the dataset. The main advantage of DP is that it is robust against membership and information inference attacks.

The term *security* refers to the resilience of a system to failures and attacks that aim to disrupt the normal execution of the system. Some attacks are:

- *Data Poisoning*: a malicious party modifies the data in order to manipulate the model performance, e.g., to prevent a model from converging.
- *Byzantine Attacks* [11]: one or more malicious parties collaborate to disrupt the functioning of a system.

Validation techniques are the best approaches for building poisoning resistance, like discarding model updates that deviate excessively from the average of the other updates, while consensus algorithms guarantee the consistency of the state in a distributed system, which guarantees Byzantine Fault Tolerance.

8.3 Privacy-Preserving Blockchain-Based Federated Learning

Typically, the whole federated learning process is supervised by a server, which initializes the model parameters, distributes the initial model to the clients, and takes care of the aggregation of the local model updates provided by them. As the server

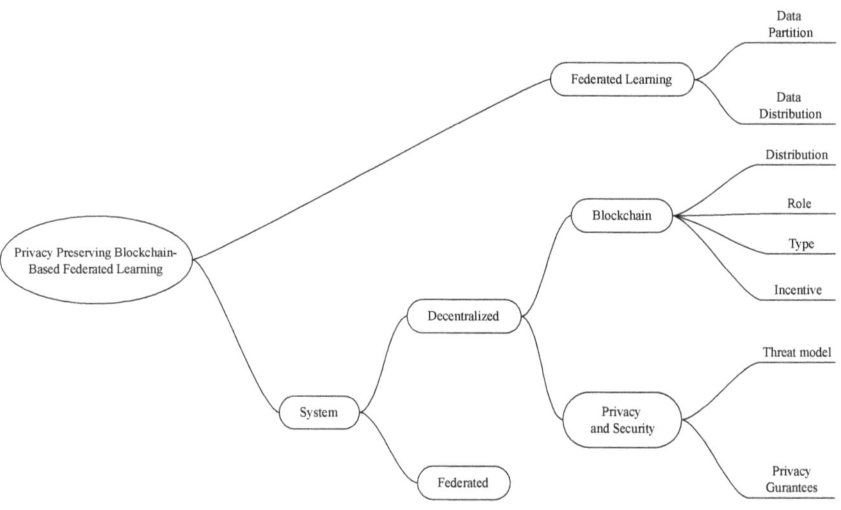

Fig. 8.3 Taxonomy of the topic

may represent a point of failure, a Blockchain can be used, which can be deployed on the same computing peers of the clients or separate ones. The framework cannot guarantee privacy and security by itself. The exchanged local models can leak information, e.g., [10] presents a model inversion attack where images are recovered in a facial recognition system. For this reason, cryptographic techniques and information-theoretic techniques can provide such privacy and security guarantees. In this section, we will present how we categorized the state-of-the-art methods that combine Blockchain and privacy. Figure 8.3 presents the taxonomy, where we show the key dimension taken into consideration in this section.

8.3.1 Federated Learning Dimensions

We identified two categories in Blockchain-Based Federated Learning: the *Data Partition* and the *Data Distribution*.

8.3.1.1 Data Partition

A typical and crucial requirement in FL is how the data is partitioned. When data features are shared between clients while data samples are not, we refer to *horizontal* partitioning. This partitioning is the predominant case as it covers most of the state-of-the-art [12–20]. The other case happens when the features are partitioned across the clients, which is called *vertical* partitioning. This is an improbable case in typical

FL scenarios, as it is more likely to happen in datasets that are part of the same organizations. Only one article covers a vertical FL scenario with blockchain [21].

8.3.1.2 Data Distribution

A crucial requirement of FL is the data distribution. Most of the works in the literature consider the case of Identically and Independently Distributed (IID) datasets, which means that the predictors and the targets are sampled from the same probability distribution. A more realistic and more challenging scenario is the non-ID one, where there are different data distribution skews:

- *Quantity skew*, when different clients have different amounts of data.
- *Label distribution skew*, where different clients have different class distributions.
- *Feature distribution skew*, when the data generation process is different across the clients.

8.3.2 Blockchain Dimensions

Regarding the Blockchain, we considered the following dimensions: the *Distribution*, the *Type*, the *Consensus Algorithm* and its *Role*.

8.3.2.1 Distribution

Federated learning is traditionally implemented as a client-server architecture. Blockchain is introduced to completely distribute the system by eliminating the figure of the server. This solves the problem of a single point of failure [3], as in the classic FL scenario, if the server fails or behaves maliciously, the whole learning process is compromised. The Blockchain-Based Federated Learning Systems are then divided into two categories:

1. *Decoupled*: a peer in the network is either a training peer or a blockchain peer [14, 16, 20–22].
2. *Coupled*: a role may change at each round [19].

8.3.2.2 Type

An important feature of Blockchain is transparency, which means that everyone with access to it can verify transactions, ensuring trust among users. The mode of accessing the blockchain defines its type, which can be: *public*, *private*, or *permissioned*.

In a *public* blockchain, anyone can access the network, with a simple public key generation algorithm producing the digital identity. After this process, the new user can interact with the blockchain by querying it or by proposing new transactions. In the context of FL, a public blockchain means that even a malicious entity can access the network. If this type of blockchain is used, it would be better to protect the privacy of any sensitive data that is published on it, like in [13, 14, 17, 19, 21, 23]. If multiple malicious parties join to disrupt the system, it is also important to implement robust consensus algorithms to avoid Sybil or Byzantine attacks [12, 14, 15, 17, 19]

If FL participants are fixed, a *private* blockchain is the ideal option. Nevertheless, this scenario is unrealistic, as it is typical for FL nodes to join and leave the system during the training phase.

A solution that meets in the middle is the *permissioned* blockchain. In this scenario, the number of initial peers is specified in the network configuration and one or more network administrators can admit new members after the system is running, making the system management more flexible [12, 22].

8.3.2.3   Consensus Algorithms

In blockchain-based federated learning, consensus is a crucial part of the federated learning process, especially if the blockchain is employed to improve the federated learning algorithms.

All of the works considered in this chapter use classic consensus algorithms like PoW, PoS, and Raft, except for Biscotti [19].

8.3.2.4   Role

In federated learning, the blockchain can play different roles. We identified three types of roles:

1. **Integrated in the FL process**. This means the blockchain is used as a substitute for the central server, which takes care of aggregation and redistribution of the aggregated model by publishing it on the blockchain [13, 14, 16, 19].
2. **Storage**. Clients may share part of their data on the blockchain to pre-train the initial model [18, 21, 24].
3. **External tasks**. In this case, the blockchain does not carry any typical FL operation, but it is used for support, like for logging or external computations [22].

8.3.3 Privacy and Security Dimensions

We captured the following dimensions regarding the privacy and security of blockchain-based federated learning: the *privacy guarantees* and the *adversarial models*.

8.3.3.1 Privacy Guarantees

There are different privacy guarantees. We define an *ad-hoc privacy* guarantee where the sensitive data or models are "embedded" in a non-human understandable format (e.g., the model weights of a deep learning model). *Formal privacy* refers to when data is obfuscated such that the privacy guarantee is mathematically quantifiable, e.g., statistical and computational indistinguishability.

Ad-Hoc Privacy

All FL solutions that make use of algorithms without any formal and rigorous privacy guarantee are labeled ad-hoc solutions. Therefore, even a deep learning model falls in this category, as it is practically an encoding of the original data. The encoding, i.e., the weight vector, does not expose any information at first glance, but it is possible to use model inversion or membership inference attacks to reconstruct the original data or to assess the presence or absence of a specific individual in the dataset.

Formal Privacy

When privacy is guaranteed by the construction of a protocol or by proof of statistical or computational indistinguishability, we label this as *formal privacy*.

Differential privacy is the most used technique. In particular, in the context of FL, Local Differential Privacy (LDP) [8] is the most used technique, as the trust boundary is between the clients and the aggregator (see Fig. 8.4). In this way, client updates are already protected at the aggregation level, which happens at the blockchain level [14, 16, 19, 21, 23].

8.3.3.2 Adversarial Models

A typical assumption that is defined when designing a secure and private system is the threat model. A threat model states how the parties of a system are going to behave:

- *Honest-but-Curious adversary*: an adversary that behaves honestly according to the protocol, but tries to learn as much as possible from it.
- *Byzantine adversary*: the adversary can act arbitrarily.

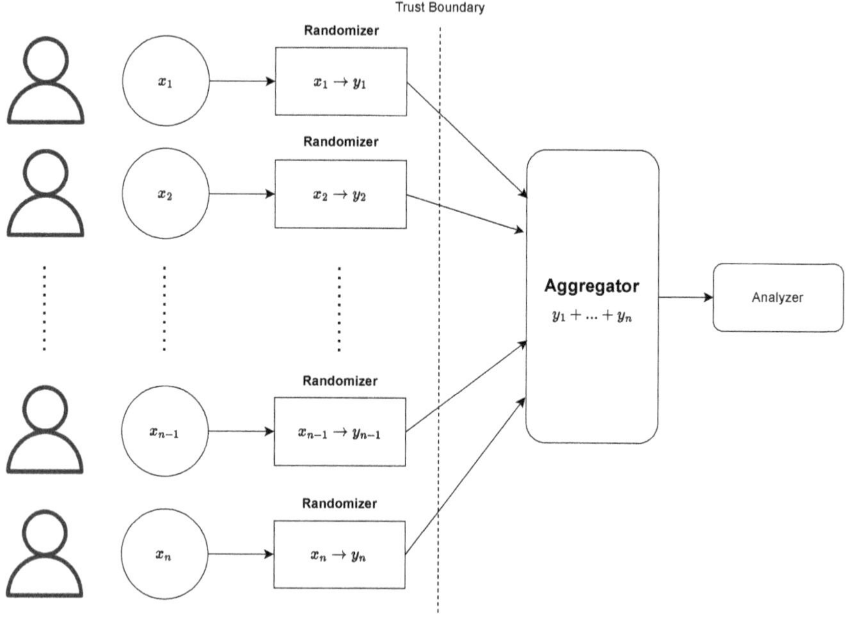

Fig. 8.4 A visual representation of the aggregation model

From a system perspective, FL presents various weak points that parties can exploit to compromise its functionality. In particular, a group of malicious clients can poison their data or models so that the global model does not converge.

8.4 State of the Art

In this section, we will focus on the most recent state-of-the-art works and we analyze them according to the taxonomy presented in the previous section.

8.4.1 Fedchain

Fedchain [22] is a clustered blockchain-based federated learning system that uses a knowledge distillation technique to mitigate the non-ID data [25]. Its architecture is represented in Fig. 8.5. In each round, a client may, with a specified probability, train its local model or request their models to its teachers to carry out the distillation process. The teachers are different in each round as they are selected by computing the similarity across clients using their feature sparsity vectors [26]. Therefore, from the *data distribution* dimension, this falls in the non-ID and, more particularly, is the

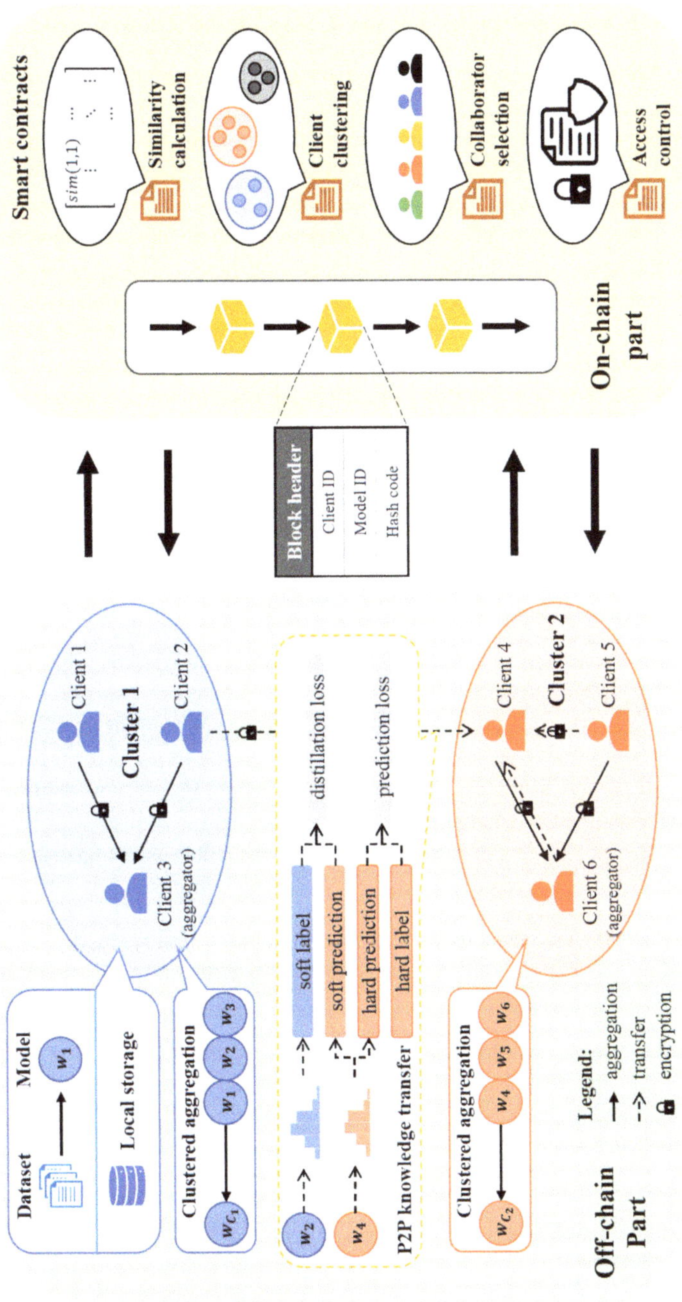

Fig. 8.5 Fedchain architecture [22]

only work that provides a solution for this type of data distribution. Regarding the data partition, the dataset is *horizontally* partitioned.

The blockchain component in Fedchain is used for the following tasks:

- *Logging*. Metadata about the local models and cluster models is stored on the blockchain. This is done because FL clients will use it to retrieve the new cluster models from the cluster leader, and also for the knowledge transfer phase.
- *Similarity Computation and Clustering*. Each client uploads a signature vector to the blockchain, which represents the sparsity of the kernel matrices extracted during the feed-forward pass of the stochastic gradient descent algorithm [26]. Whenever a new cluster update is produced, the blockchain computes the new similarity matrix using the signature vectors. Then, the agglomerative clustering Smartcontract computes the new clusters and records them on the blockchain.
- *Knowledge Transfer*. Clients have a certain probability of performing local training on their dataset or transfer learning in each local round. In the latter case, the client will query the blockchain to ask which are its teachers, but using the similarity matrix mentioned before.
- *Access Control*: when a client needs to access a teacher model, firstly, he needs permission from it, by using a simple protocol that uses a public key infrastructure, as shown in Fig. 8.6.

Fedchain's blockchain falls into the following dimensions:

- Distribution: *decoupled*, because a computing node in the system is either a training node or a blockchain node.
- Role: *External tasks*, the blockchain acts as a logging and support for the clustering process.
- Type: *permissioned*, as the system used is Hyperledger Fabric, a well-known open source permissioned blockchain.
- Incentive: no incentive mechanism is used as the consensus algorithm used is RAFT, which does not require a token for reaching consensus.

From the privacy and security perspective, Fedchain provides an *ad-hoc privacy* solution and no security guarantee. The system starts with all the clients in the same cluster:

1. Clients load the cluster model and start the local training or ask their teachers to provide their models for distillation. Clients upload their local model metadata (model info and signature vectors) when they are finished with the local round.
2. For each cluster:

    a. As soon as there are enough model updates, the cluster aggregator requests the models, aggregates the weights, and then uploads the model metadata to the blockchain.

3. The blockchain will recluster the clients using their signature vectors uploaded during the local training.

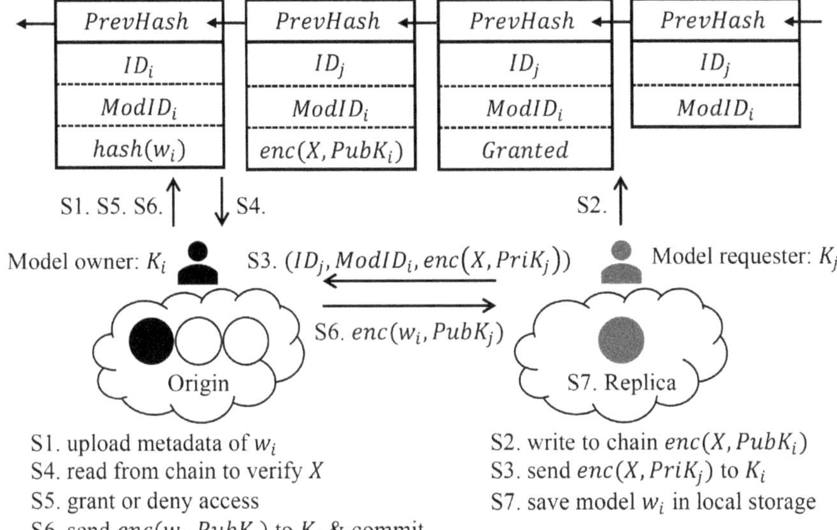

Fig. 8.6 Access control protocol in fedchain [22]

4. If clusters are stable, meaning they are not changing over the last rounds (a parameter defines how many previous rounds the system needs to check), and the system
   is converging, terminate the execution, otherwise go to step 1.

   By analyzing the protocol, it is possible to observe that privacy issues occur when:

- Clients send their local model to the cluster aggregator.
- Teachers send their models to their students.
- Clients upload the signature vector to the blockchain, even when we are unaware
  that these signature vectors, which represent the sparsity of the extracted kernels, can leak private information. The authors state that it is intuitively privacy-
  preserving as it would be difficult to invert, but there is no empirical evidence
  or formal proof of privacy [26]. It remains unknown if they can be exploited for
  membership inference attacks.

8.4.2 Biscotti

The Biscotti paper [19] describes a blockchain-based federated learning system
where privacy and security are the focus. Regarding the FL part, the *data distribution* is IID and it is an *horizontal* FL scenario.

Biscotti is a *coupled* system as it introduces four peer roles, *trainer, noisier,
verifier*, and *aggregator*, where at each round the roles are selected, using a consistent

hashing weighted in the stake of the peers. Peers can be assigned multiple roles in a given iteration, but cannot be a verifier and an aggregator in the same round.

Moreover, the authors implemented a variant of PoS, called Proof of Federation. The consensus is delegated to a subset of peers based on their stake, which is a measure of contribution to the system. A peer acquires stake through beneficial model updates or by facilitating the consensus process. The aggregators will compete to aggregate the local models and add a new block to the blockchain. Given m, the number of aggregators, the steps are the following:

1. The trainers will send the updates to the aggregators.
2. Once every aggregator has enough shares to merge them into a global update, they broadcast the sum of their accepted shares to each other.
3. Given the aggregator a has received $\frac{m}{2}$ sum of the aggregated shares, including itself, it can compute the global model.
4. Once computed, the aggregator a creates a new block and disseminates it in the network.

Regarding privacy, *formal privacy* algorithms are applied. In particular, classical LPD algorithms have been used, such as Gaussian Differential Privacy applied to Stochastic Gradient Descent [27]. Biscotti [19] uses this mechanism, which, given a gradient norm bound C, $\sigma = $ the noise scale, and L the batch size, for each training iteration t:

1. Takes a random sample (batch) S_t
2. For each $x \in S_t$ computes $g_t(x) \leftarrow \nabla_{\theta_t} \mathcal{L}(\theta_t, x)$
3. Clips the gradient: $\bar{g} \leftarrow \dfrac{g_t(x)}{max(1, \frac{\|g_t(x)\|_2}{C})}$
4. Adds Gaussian Noise: $\tilde{g}_t \leftarrow \frac{1}{L}(\sum \bar{g}(x) + \mathcal{N}(0, \sigma^2 C^2 I))$
5. Applies gradient descent.

Biscotti [19] considers a *Byzantine adversary* scenario, in which peers can act maliciously to disrupt the system. In order to guarantee security and robustness, a Multi-Krum algorithm has been used to guarantee system security, particularly for poisoning attacks. For every local update, the verifier calculates a score, which is the sum of the Euclidean distances of the local model update in question and a subset of the other closest updates. Then, a specified number of local updates with the lowest scores are selected, while the others are rejected. This verification process is run and digitally signed by multiple verifiers. Before these updates are disseminated to the aggregators, a peer needs to receive signatures from most verifiers.

8.4.3 Other Works

Voltran [12] uses a Trusted Execution Environment (TEE), specifically Intel SGX, for the aggregation phase to avoid the expensive cost of aggregating local models on the blockchain. TEE provides a safe area within the central processor. In particular, the protocol ensures *authenticity* and *confidentiality*. This first property ensures that an adversary cannot forge a participant's identity and convince another participant to accept a message that is not the expected content. The second and last, states that no one except the clients and the TEE enclave can access the local models.

Reference [14] uses an l-nearest aggregation scheme, an efficient algorithm that is resilient to Byzantine attacks. The aggregator sums the normalized local gradients and then it calculates the cosine distance between each data holder's normalized gradient and the sum vector. Then, the aggregator picks the top l-nearest data holders' gradients to update the global model.

Réynyi Differential Privacy is applied in [21], by using the Poisson-Binomial mechanism. At its core, the Poisson-Binomial Mechanism injects controlled Poisson noise into the weight vector before aggregation. Simultaneously, it quantizes the noisy weights with the Binomial distribution, converting the real values to integers.

Another discrete transformation used to guarantee DP is the Discrete M-Band Wavelet Transform [23]. It decomposes the data into a wavelet domain, separating the data into high-frequency and low-frequency sections. This is achieved by multiplying the transform matrix by the dataset to decompose it into M wavelets. Then, DP Laplace noise is added to the wavelet coefficients.

Differential Privacy protects the output of a random algorithm, while cryptography aims to protect the input or restrict its access to it. Distributed ElGamal is an extension of the classic ElGamal public key infrastructure [28]. It is used to encrypt local models, which are then sent to the blockchain, where the aggregation will be carried out on the encrypted models, as this cipher is homomorphic [20].

8.5 Open Problems and Future Directions

Privacy-preserving blockchain-based federated learning represents a promising intersection of FL, privacy-enhancing technologies, and blockchain ledgers. While significant progress has been made in this domain, several critical challenges and opportunities exist for future research. This section explores key open problems and potential future directions in the field.

8.5.1 *Space-Efficient Model Storage on Blockchain*

A fundamental challenge in blockchain-based federated learning is the efficient storage of model updates on the blockchain. Current approaches typically require storing complete model weights [19], which can be prohibitively expensive given the storage constraints of the blockchain. Future research should explore more space-efficient data structures and compression techniques to make blockchain-based federated learning more scalable and practical.

Fedchain [22] clients upload to the blockchain their signature vectors representing the data distribution of the clients, which are more space-efficient than uploading the entire model. It is unclear if these vectors are vulnerable to membership inference attacks, which is a clear open problem.

One promising direction is the application of sketching techniques to compress model updates while preserving essential information. Techniques such as Count-Sketch [29] could be adapted to create compact representations of model updates and use them for the aggregation carried out by a blockchain network.

8.5.2 *Integration of Anti-poisoning and Differential Privacy*

Combining Byzantine Fault-Tolerance algorithm mechanisms with differentially private algorithms presents a complex challenge in blockchain-based federated learning. While both mechanisms aim to protect the system, they operate under different threat models and can potentially interfere with each other's effectiveness. For example, an algorithm like Multi-Krum that is used in Biscotti [19] can interpret a DP model update as a poisoned update, by then discarding it. A solution would be to avoid using Local DP, and apply the central model (i.e., adding noise after the aggregation).

8.6 Conclusions

Privacy-preserving blockchain-based federated learning represents a significant advancement in distributed machine learning systems, prioritizing data privacy and security. Throughout this chapter, we have examined the intersection of federated learning, blockchain technology, and privacy-preserving mechanisms, highlighting both current achievements and remaining challenges.

The integration of blockchain with federated learning addresses several critical issues in traditional federated learning systems, particularly the single point of failure problem inherent in centralized architectures. Various architectural approaches have been presented, from decoupled systems where blockchain and training nodes are separate, to coupled systems where roles can dynamically change. The choice of blockchain type–public, private, or permissioned–significantly impacts the system's security properties and operational characteristics.

Privacy preservation in these systems has evolved beyond the basic privacy afforded by federated learning's data locality principle. We have seen implementations ranging from ad-hoc privacy guarantees to formal privacy mechanisms such as differential privacy and cryptographic techniques. The chapter highlights how different approaches such as Local Differential Privacy, RÃƒÂ©nyi Differential Privacy, and public key encryption can be effectively combined with blockchain-based systems to enhance privacy guarantees.

However, significant challenges remain, particularly in three key areas:

1. The efficient storage of model updates on blockchain platforms.
2. The integration of Byzantine fault tolerance with differential privacy mechanisms.
3. The development of robust privacy budget tracking systems in decentralized environments.

These challenges represent important directions for future research. Additionally, as these systems move towards practical deployment, questions of scalability, performance optimization, and real-world privacy guarantees will need continued attention.

As organizations increasingly seek to collaborate on machine learning tasks while maintaining data privacy and security, the approaches discussed in this chapter will likely play an increasingly important role in shaping the future of distributed learning systems.

References

1. European Parliament CotEU (2016) Regulation (EU) 2016/679 of the European parliament and of the council of 27 April 2016 on the protection of natural persons with regard to the processing of personal data and on the free movement of such data, and repealing directive 95/46/ec (general data protection regulation)
2. McMahan HB, Moore E, Ramage D, Hampson S, Agüera y Arcas B (2016) Communication-efficient learning of deep networks from decentralized data. In: International conference on artificial intelligence and statistics
3. Kairouz P, McMahan HB, Avent B et al (2021) Advances and open problems in federated learning. Found Trends Mach Learn 14(1–2):1–210
4. Jeromemetronome (2019) Federated learning process in central orchestrator case. https://commons.wikimedia.org/wiki/File:Federated_learning_process_central_case.png, licensed under CC BY-SA 4.0
5. Nakamoto S (2009) Bitcoin: A peer-to-peer electronic cash system
6. Ongaro D, Ousterhout J (2014) In search of an understandable consensus algorithm. In: 2014 USENIX annual technical conference (USENIX ATC 14)
7. Wood G (2014) Ethereum: A secure decentralised generalised transaction ledger
8. Dwork C, Roth A (2014) The algorithmic foundations of differential privacy. Found Trends Theor Comput Sci 9(3–4):211–407
9. Shokri R, Stronati M, Song C, Shmatikov V (2017) Membership inference attacks against machine learning models. In: 2017 IEEE symposium on security and privacy (SP), pp 3–18
10. Fredrikson M, Jha S, Ristenpart T (2015) Model inversion attacks that exploit confidence information and basic countermeasures. In: Proceedings of the 22nd ACM SIGSAC conference on computer and communications security, pp 1322–1333
11. Lamport L, Shostak R, Pease M (1982) The byzantine generals problem. ACM Trans Prog Lang Syst 4(3):382–401
12. Wang H, Cai Y, Wang J, Ma C, Ge C, Qu X, Zhou L (2024) Voltran: unlocking trust and confidentiality in decentralized federated learning aggregation. IEEE Trans Inf Forensics Secur 19:9744–9759
13. Weng J, Weng J, Zhang J, Li M, Zhang Y, Luo W (2019) Deepchain: auditable and privacy-preserving deep learning with blockchain-based incentive. IEEE Trans Dependable Secure Comput 1:1
14. Chen X, Ji J, Luo C, Liao W, Li P (2018) When machine learning meets blockchain: a decentralized, privacy-preserving and secure design. In: IEEE international conference on big data (BigData), pp 1178–1187

15. Qu Y, Gao L, Luan TH, Xiang Y, Yu S, Li B, Zheng G (2020) Decentralized privacy using blockchain-enabled federated learning in fog computing. IEEE Internet Things J 7(6):5171–5183
16. Ma C, Li J, Shi L, Ding M, Wang T, Han Z, Poor HV (2022) When federated learning meets blockchain: a new distributed learning paradigm. IEEE Comput Intell Mag 17(3):26–33
17. Toyoda K, Zhang AN (2019) Mechanism design for an incentive-aware blockchain-enabled federated learning platform. In: IEEE international conference on big data, pp 395–403
18. Lian Z, Zeng Q, Su C (2022) Privacy-preserving blockchain-based global data sharing for federated learning with non-iid data. In: IEEE 42nd international conference on distributed computing systems workshops, pp 193–198
19. Shayan M, Fung C, Yoon CJM, Beschastnikh I (2021) Biscotti: a blockchain system for private and secure federated learning. IEEE Trans Parallel Distrib Syst 32(7):1513–1525
20. Li H, Sun Y, Yu Y, Li D, Guan Z, Liu J (2023) Privacy-preserving cross-silo federated learning atop blockchain for IoT. IEEE Internet Things J 10(24):21176–21186
21. Tran L, Chari S, Khan MSI, Zachariah A, Patterson S, Seneviratne O (2024) A differentially private blockchain-based approach for vertical federated learning. In: IEEE international conference on decentralized applications and infrastructures, pp 86–92
22. Wu H, Zhu X, Hu W (2024) A blockchain system for clustered federated learning with peer-to-peer knowledge transfer. Proc VLDB Endowment 17(5):966–979
23. W A, Huang, Kandula A, Wang X (2021) A differential-privacy-based blockchain architecture to secure and store electronic health records. In: 2021 The 3rd international conference on blockchain technology, pp 189–194
24. Mukherjee A, Halder R, Chandra J, Shrivastava S (2023) Healthchain: a blockchain-aided federated healthcare management system. In: IEEE international conference on blockchain and cryptocurrency, pp 1–5
25. Hinton G, Vinyals O, Dean J (2015) Distilling the knowledge in a neural network. In: NIPS deep learning and representation learning workshop
26. Liu B, Guo Y, Chen X (2021) Pfa: privacy-preserving federated adaptation for effective model personalization. In: Proceedings of the web conference 2021, Ljubljana, Slovenia, pp 923–934
27. Abadi M, Chu A, Goodfellow I, McMahan HB, Mironov I, Talwar K, Zhang L (2016) Deep learning with differential privacy. In: Proceedings of the 2016 ACM SIGSAC conference on computer and communications security, pp 308–318
28. Elgamal T (1985) A public key cryptosystem and a signature scheme based on discrete logarithms. IEEE Trans Inf Theory 31(4):469–472
29. Charikar M, Chen K, Farach-Colton M (2002) Finding frequent items in data streams. In: International colloquium on automata, languages, and programming, pp 693–703

Chapter 9
Example-Based Explainability in Machine Learning

Ikhtiyor Nematov⊕, Dimitris Sacharidis⊕, Katja Hose⊕, and Tomer Sagi⊕

Abstract Machine learning (ML) systems are increasingly utilized in high-stakes domains, where failures stemming from biases, data quality issues, or model misalignment can lead to severe consequences. To address these risks, explainable AI (XAI) techniques offer valuable insights into ML model decision-making. This chapter provides a comprehensive examination of XAI, with a particular focus on example-based methods, which aim to establish causal links between training data and model behavior, aiding in model debugging and data refinement. We present a taxonomy of XAI techniques, categorizing them into model-based, feature-based, and example-based approaches, and evaluate their strengths and weaknesses in terms of interpretability, adaptability, and robustness. The chapter delves into representative example-based methods, highlighting their key contributions, technical methodologies, and inherent limitations. Furthermore, it critically reviews the state-of-the-art in example-based explainability, assessing how these methods address challenges such as model fidelity, user intent, and data quality. Finally, the chapter identifies open challenges and outlines future research directions, emphasizing the importance of developing user-centered, intent-aware explanations that align more closely with model behavior.

Keywords Machine learning · Explainability · Data attribution · Interpretability · Trustworthy AI

I. Nematov (✉) · D. Sacharidis
Department of Computer & Decision Engineering (CoDE) CP 165/15, Université libre de Bruxelles, Brussels, Belgium
e-mail: ikhtiyor.nematov@ulb.be

D. Sacharidis
e-mail: dimitris.sacharidis@ulb.be

T. Sagi
Department of Computer Science, Aalborg University, Aalborg, Denmark
e-mail: tsagi@cs.aau.dk

K. Hose
Department of Logic and Computation, Vienna University of Technology, Vienna, AT, Austria
e-mail: katja.hose@tuwien.ac.at

G. Dejaegere et al. (eds.), *Data Engineering for Data Science*,
https://doi.org/10.1007/978-3-032-18765-9_9

"

9.1 Introduction

The increasing integration of ML systems into critical domains such as healthcare, finance, law enforcement, and autonomous systems has underscored an urgent need for transparency and accountability in AI decision-making. While these systems demonstrate remarkable predictive capabilities, their inherent complexity often renders them opaque, creating a "black box" effect that obscures how predictions are made. This lack of interpretability raises significant concerns, including eroding trust among stakeholders, impeding widespread adoption, and amplifying risks associated with biases, mislabeled data, and unintentional model behaviors. Moreover, in high-stakes environments, opaque models can lead to ethical dilemmas, legal challenges, and adverse societal impacts if decisions cannot be justified or traced back to understandable causes. XAI emerges as a pivotal solution, offering mechanisms to illuminate the inner workings of ML systems, enabling stakeholders to interpret, trust, and refine these models. Beyond fostering trust, XAI is crucial for tasks such as identifying and mitigating bias, enhancing data quality, debugging model errors, and aligning models with ethical and regulatory standards. Additionally, XAI bridges the gap between AI systems and human decision-making by facilitating collaborative workflows, ensuring accountability, and empowering users with actionable insights. In an era where AI increasingly influences critical decisions, XAI is not merely a tool for understanding-it is a cornerstone for the responsible and ethical deployment of intelligent systems.

The most straightforward way to achieve explainability is to use intrinsically interpretable models with a simple structure, such as linear models or decision trees with reasonable depths. However, in the developing world of AI, to make the most use of the humongous data and solve more sophisticated problems, the capability of such models is not comparable to the complex models with a large number of parameters. This is an unfortunate trade-off between improved quality and transparency. We may be able to observe the set of outputs for a given set of inputs to a model, without knowledge or understanding of its internal workings. Unlike mathematical models that have inherent structure, machine learning models can learn the mapping of inputs to outputs directly from the data. For some models, like decision trees, this mapping is easily discernible. For others, like random forests or deep learning models, it becomes next to impossible to understand how predictions are made. This lack of transparency and understanding can have serious consequences for our trust and adoption of these models. Therefore, post-hoc interpretability is needed to shed light on the decision-making process of these so-called black-box models after they are trained.

9.1.1 Taxonomies

Post-hoc explainability techniques for black-box models are generally divided into global and local approaches based on the interpretability scope they provide.

- **Global Methods**: These techniques attempt to provide an understanding of the model's overall behavior, describing how it tends to make decisions across a range of inputs. Global methods often involve approximating the black-box model with simpler, interpretable models, such as decision trees or rule-based systems. For example, Decision Tree Surrogate Models are frequently used as a global interpretability method by creating a tree that approximates a complex model's decision space.
- **Local Methods**: In contrast, local approaches focus on providing interpretability for individual predictions, explaining why a particular input led to a specific output. This can be particularly useful in cases where understanding a specific decision is more critical than understanding the model's overall behavior. Local Interpretable Model-agnostic Explanations (LIME) [1] is an example of a local method that creates a simple, interpretable model around a single prediction to show which features were most influential.

By their applicability, explainability techniques can further be classified as model-agnostic or model-specific. Model-agnostic methods, like those that compute feature importance scores, are adaptable to any model type, making them versatile across applications. In contrast, model-specific methods are tailored to particular models and leverage unique aspects of their structure, such as specialized methods for neural networks, studying the gradients flowing into the final convolutional layer to produce a coarse localization map.

Various taxonomies have been developed to categorize explainability techniques based on the nature of the explanations they produce. A commonly used classification organizes methods by explanation type: *model-based*, *feature-based*, and *example-based*.

- **Model-Based Methods**: These methods involve building interpretable surrogate models, such as linear models, decision trees, or rule-based models, that approximate the complex model. For example, RuleFit [2] is a method that extracts decision rules from black-box models, making it easier to understand how a model is making decisions across the input space.
- **Feature-Based Methods**: These methods focus on identifying the most critical input features that drive a prediction. Techniques such as LIME [1], SHAP [3], and saliency maps are well-known feature-based approaches that highlight which aspects of the input, like specific words in text or regions in images, have the most influence on a model's output.
- **Example-Based Methods**: Example-based techniques provide interpretability by identifying influential examples from the training dataset or generating synthetic examples. For instance, Influence Functions [4] trace back a model's predictions to

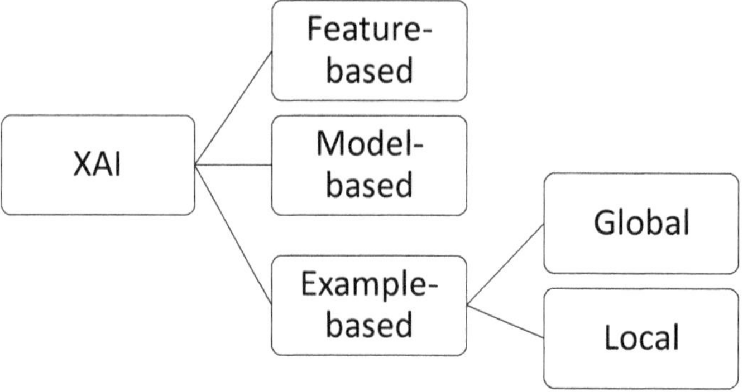

Fig. 9.1 Taxonomy of post-hoc XAI methods depending on the explanation type

the most impactful training data points, helping users understand which examples in the training set most strongly influence a particular prediction. Other example-based methods include Prototypes and Criticisms [5], where representative examples (prototypes) and non-standard examples (criticisms) are used to illustrate the model's understanding of the data.

A taxonomy of the different post-hoc explainability techniques for black-box models is presented in Fig. 9.1.

9.2 Landscape of Example-Based Explainability

Given that machine learning models are fundamentally shaped by their training data, example-based explainability offers numerous advantages, particularly in its ability to link model behavior directly to specific data points. This connection aids in several critical tasks, such as interpreting individual predictions, diagnosing model errors, identifying problematic data points, and detecting potential biases. For instance, consider an image in which the model has classified the contents as a *dog*. A typical feature-based explainer would highlight the area containing the dog as the most influential region in the model's decision, focusing on visual features like fur, shape, or color. However, this form of explanation may be insufficient in cases where a user is interested in understanding why the model ignored other elements in the image, such as a *fish* that is also present.

An example-based explainer, on the other hand, goes beyond highlighting features in the input image by providing influential samples from the training data. These samples reveal patterns or labeling rules learned by the model that contribute to its decision-making process. In this case, the example-based explanation could show the user similar training images where both a dog and a fish appear together and are consistently labeled as *fish*. Such a pattern may indicate that the model has learned

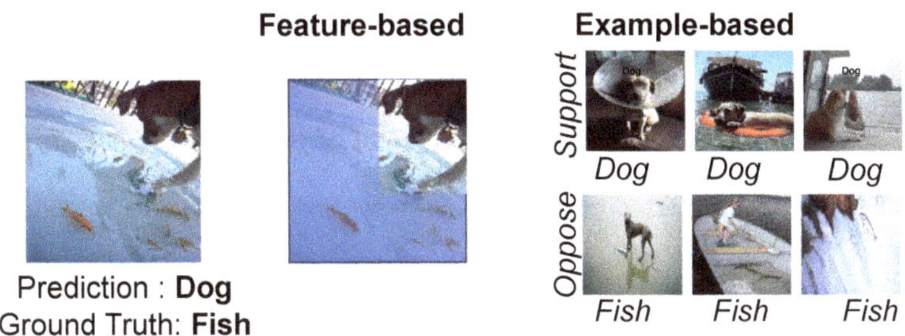

Fig. 9.2 Explanation in image classification (DOG vs FISH)

a labeling rule, for instance,"whenever both a dog and fish are present, classify the image as *fish*." This insight, derived from example-based explanations, allows the user to better understand how specific data correlations or labeling conventions have shaped the model's interpretation, revealing underlying biases or assumptions that a feature-based explanation alone would not disclose.

Figure 9.2 highlights the difference between feature-based and example-based explanations.

Furthermore, in modern machine learning, models are frequently trained through multistage pipelines–incorporating techniques like transfer learning, weak supervision, or data augmentation–which can result in training data that originates from diverse processes or even multiple datasets. In these complex pipelines, example-based methods become particularly valuable by offering traceability, allowing practitioners to link predictions back to fundamental data sources or intermediary datasets. This can provide insight into the influence of specific data characteristics on model behavior, a depth of understanding that is difficult to achieve with other explainability methods.

9.2.1 Local Example-Based Explanations

In contrast to global explanations, *local example-based methods* aim to provide insights into the model's behavior for a specific prediction. Instead of identifying general patterns, these methods identify influential samples that directly impact a single prediction, highlighting how particular examples in the training data affect the model's output for a given input. Fig. 9.3 illustrates this approach.

Local explanations often involve identifying *influential samples*, those training instances whose presence–or absence–would significantly alter the model's prediction for the target sample. This approach, sometimes based on methods like influence functions or similarity-based retrieval, allows users to see the data points most responsible for the model's decision, providing a transparent view of the model's reasoning. These influential samples are especially valuable for debugging misclassifications,

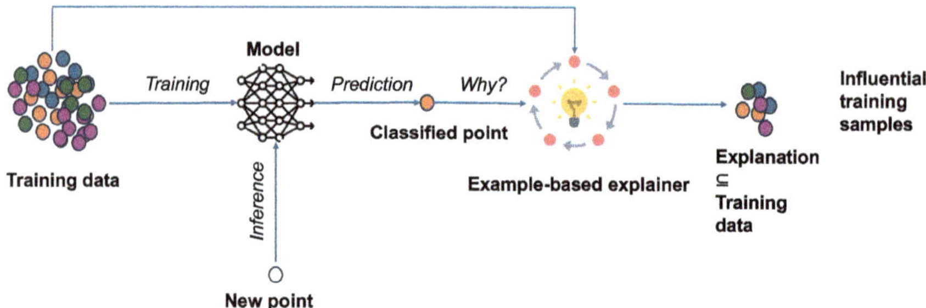

Fig. 9.3 Pipeline of local example-based explainability

identifying biases, and understanding edge cases in the model's behavior, as they help to explain specific decision-making pathways taken by the model.

9.2.2 Global Example-Based Explanations

Most example-based explainability methods operate by selecting a subset of the training data as an explanation. For *global explanations*, this subset typically consists of *prototypes*–a selection of samples that are representative of the training data distribution and serve as exemplars for explaining the model's general behavior. These prototypes help explain how the model interprets the data in a broad sense, enabling users to understand the "themes" or patterns the model has learned. Figure 9.4 illustrates this concept.

Prototypes are representative samples that associate a set of data points with a specific label, allowing for more interpretable classification. Initially used in unsupervised clustering (e.g., k-medoids, which divides data into clusters by identifying central points or prototypes), prototypes have evolved to serve as a foundation for supervised methods that enhance interpretability in machine learning. In supervised settings, prototypes help explain the classification of new data by associating them with prototypes from known classes, inheriting the label of the closest prototypes.

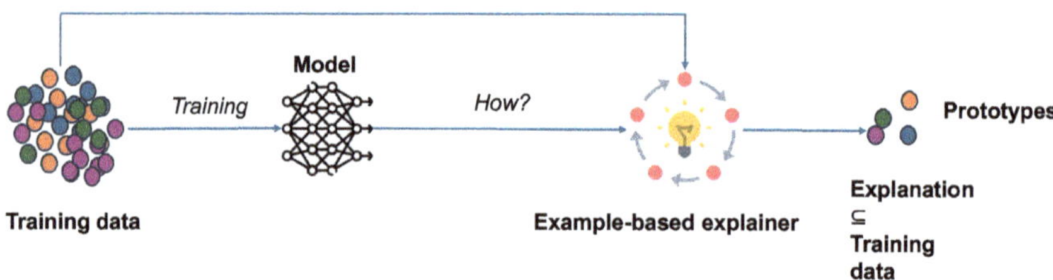

Fig. 9.4 Pipeline of global example-based explainability

This development has led to prototype-based classification methods, in which data distributions are understood through these representative instances.

As deep models have become more prevalent and the need for transparency in high-stakes applications has grown, prototypes have emerged as an effective tool for interpretable classification. Some works advocate that, especially in domains requiring high trust, prototype-based classifiers should be used in place of black-box models due to their inherent explainability. However, prototypes generated independently of a model, used for explaining another model, might not satisfy the faithfulness criterion since identical prototypes can yield explanations that do not align with the distinct internal workings of different models. To address this, some methods propose generating prototypes based on the model's predictions rather than the ground truth, ensuring that the explanation aligns with the specific decision-making process of the model itself. The body of work using prototypes can be split into three categories: data-centric, post-hoc, and by design interpretability.

- **Prototypes for Data-Centric Interpretability**. Prototypes in unsupervised settings, like k-medoids clustering, are used to segment the dataset into distinct clusters. This clustering-based approach selects data points that best represent each cluster, effectively dividing the data into meaningful segments. Over time, data-centric interpretability extended into supervised domains, where prototypes summarize complex distributions and make the model's decision process more transparent
- **Prototype-Based Models Interpretable by Design**. Prototype-based models that are interpretable by design integrate prototype selection into the training process, aiming for both predictive performance and transparency. These models ensure that predictions are closely tied to recognizable instances, making them inherently explainable.
- **Prototypes for Post-Hoc Interpretability**. In post-hoc interpretability, prototypes are used to analyze model predictions by observing the model's performance on representative samples. This approach allows us to validate the model's behavior, especially in regions that are critical for understanding its decisions.

9.2.3 Evaluation Strategies for XAI

The domain of XAI has witnessed significant advancements in the development of various explainability methods. XAI methods are associated with varying goals, scopes, analysis techniques, user groups, and output formats. However, the vast availability of divergent XAI methods introduces the challenge of systematically evaluating and comparing them.

The characteristics of XAI methods suggest that not all of them can be easily measured. Additionally, explanations are inherently subjective, influenced by factors such as user experience, domain knowledge, the purpose of the explanation, the techniques used to generate it, and the scope of the explanation. There are several approaches to assessing the quality of an explanation, which can be grouped into three main categories:

1. **Application-grounded evaluations**: These involve using the machine learning solution within a real-world application, generating explanations for its users, and evaluating the explanation's quality in the context of practical tasks.
2. **Human-grounded evaluations**: These assessments focus on general criteria related to explanation quality by creating simplified tasks that mimic the subject matter of the real-world application. These tasks typically involve participants with less expertise than those engaged in application-grounded evaluations.
3. **Functionally-grounded evaluations**: This category relies on formal definitions of interpretability and quantitative metrics to approximate explanation quality, without involving human participants. This approach is especially useful when time or budget constraints limit human-based experiments or when the explainability technique is still under development.

The former two are **qualitative** approaches, which perform evaluations in the way of user surveys or by just showing anecdotes. This will help users build a mental model to interpret the predictions of the black box. Whereas, the latter is a **quantitative** approach, which determines quantifiable aspects of explanation and proposes metrics to measure them.

9.2.3.1   Qualitative Evaluation

Quantifiable properties critical for assessing explanation quality can be grouped into several classes, each addressing specific aspects of explanations:

- **Consistency and Continuity**: These properties measure the reliability of explanations in producing consistent outcomes for identical or similar samples. In other words, if two samples are similar, their explanations should also be similar. Often referred to as *faithfulness*, this property evaluates how closely an explanation aligns with the actual reasoning used by the model, ensuring that explanations are not only consistent but also predictive of the model's behavior across cases.
- **Contrastivity**: This property allows the model to explain not only why a particular prediction was made but also why other classes were not chosen. This helps users understand the distinctive features that set the prediction class apart, enhancing the interpretability and trustworthiness of the model.
- **Compactness**: Compactness relates to the brevity and sufficiency of the explanation. Ideally, explanations should be concise and free from redundant information, which is particularly crucial for example-based methods. Compactness, often tied to *diversity* metrics, aims to ensure that examples are representative and varied without unnecessary overlap.
- **Contextual Relevance**: Explanations should be user-centric, providing information that aligns with the user's understanding and needs. This property emphasizes

the importance of designing explanations with the end user in mind, considering factors like domain knowledge and decision-making context.

- **Controlled Synthetic Data Checks**: A controlled synthetic data check involves constructing a dataset with specific reasoning embedded into it, ensuring the model's explanations align with this reasoning. The accuracy of explanations is validated by testing whether the model's predictions reflect the same logic embedded in the data, providing a robust mechanism for evaluating *faithfulness* and *reliability*.

9.2.3.2 Qualitative Evaluation

User studies in XAI evaluation have evolved into a rich interdisciplinary field, incorporating perspectives from human-computer interaction (HCI), psychology, and AI to better understand user responses to explanations. Diverse methodologies are employed to assess how explanation formats, system transparency, and individual differences affect user understanding and trust in AI. A recent study by [6] examined the methodologies used by researchers when conducting user studies in XAI applications, identifying four main evaluation dimensions for explainability:

- **Explanation Satisfaction**: Examining whether users find explanations sufficient and helpful in decision-making processes.
- **Transparency and Understanding**: Assessing whether explanations increase users' understanding of the AI model's decision-making process.
- **Trust and Reliability**: Measuring the extent to which explanations enhance user trust and confidence in the model's predictions.
- **User Cognitive Load**: Evaluating the mental effort required to process explanations, which can affect user experience and adoption of the AI system.

These dimensions provide a framework for designing user studies, offering insights into how different explanation characteristics may impact various user groups.

9.3 State of the Art

Research on example-based explainability has advanced rapidly in recent years, producing a diverse range of methods that connect model predictions to training data. This section surveys these developments, distinguishing between local approaches, which explain individual predictions, and global approaches, which characterize the model's behavior more broadly. We also review evaluation strategies that have been proposed to assess the quality and usefulness of such explanations.

9.3.1 Local Example-Based Explainability

Recent research on local example-based explainability has produced several methodological families, each attempting to clarify how individual training samples shape a specific prediction. These approaches differ in how they define and measure the notion of "influence". One of the most prominent and widely studied families is influence-based methods.

9.3.1.1 Influence-Based Methods

The **Influence Function** (IF) is a concept in robust statistics that measures the impact or influence of a single observation on an estimator or statistical model [7]. The intuition behind IF in machine learning is to quantify the change in a model's prediction when a specific training sample is removed. However, removing and retraining the model for each training sample is inefficient. To overcome this problem, one of the foundational works on IF in ML explainability [4] used the first-order Taylor approximation to calculate the change in the loss function. The authors have showcased the effectiveness of IF in identifying influential training points, detecting bias, and identifying mislabeled training samples.

Since this is one of the foundational works for example-based explainability, let us examine it thoroughly. We assume a classification task where a *model* f_θ, defined by *parameters* θ, maps an input $x \in X$ to a predicted class $f_\theta(x) \in Y$. We use the notation $z = (x, y)$ to refer to an input and its actual class pair. Let $S \subseteq X \times Y$ represent a *training set* of size $n = |S|$.

Let $\ell(z, \theta)$ denote the *loss function* of the model for a data point z, and let the *training objective*, defined as the mean loss over S, be represented as $L(S, \theta) = \frac{1}{n} \sum_{z \in S} \ell(z, \theta)$.[1] We denote the parameters that minimize this objective as θ_0^*, i.e., $\theta_0^* = \arg\min_\theta L(S, \theta)$.

The goal is to explain the model's prediction for a specific *test instance* $z_t = (x_t, y_e)$, in terms of the influence each training example $z \in S$ has on the model's prediction $f_\theta(x_t)$, particularly its prediction loss $\ell(z_t, \theta_0^*)$.

Concretely, the *influence* of a training example $z \in S$ on z_t is defined as the change in prediction loss after removing z from the training data [4]. Removing a training example alters the objective and thus the resulting model and parameters. Suppose instead of removing z, we change the weight of its contribution to the objective by a small amount ϵ. We can represent the parameters that minimize this modified objective as a function of ϵ:

$$\theta^*(\epsilon) = \arg\min_\theta \{L(S, \theta) + \epsilon\ell(z, \theta)\}.$$

---

[1] We assume regularization terms are folded into L.

Setting $\epsilon = 0$ retrieves the optimal parameters for the original objective, i.e., $\theta^*(0) = \theta_0^*$. Additionally, setting $\epsilon = -\frac{1}{n}$ corresponds to the parameters minimizing the objective after removing z from the training data.

The *exact influence* of z on z_t is then defined as:

$$I^{\text{exact}}(z, z_t) = \ell(z_t, \theta^*(-1/n)) - \ell(z_t, \theta^*(0)).$$

Calculating exact influence by retraining for each removed training example is prohibitively costly, so we approximate it. Specifically, we view the loss function as a function of ϵ and approximate the exact influence linearly, using the derivative of ℓ at $\epsilon = 0$:

$$I^{\text{exact}}(z, z_t) \approx -\frac{1}{n} \left. \frac{\partial \ell(z_t, \theta^*)}{\partial \epsilon} \right|_{\epsilon=0}.$$

Since $\frac{1}{n}$ is constant across all z, z_t pairs, we define the approximate *influence* [4] as:

$$I(z, z_t) = - \left. \frac{\partial \ell(z_t, \theta^*)}{\partial \epsilon} \right|_{\epsilon=0}.$$

If the influence of z on z_t is *positive*, the loss tends to decrease, indicating that z *supports* the prediction for z_t; otherwise, it *opposes* the prediction.

To compute the derivative of the loss, we use the chain rule, decomposing it into the derivative of the loss with respect to the parameters and the derivative of the parameters with respect to ϵ. This yields:

$$I(z, z_t) = - \left. \nabla_{\theta^*}^{\mathsf{T}} \ell(z_t, \theta^*) \right|_{\theta^*=\theta_0^*} \left. \frac{\partial \theta^*}{\partial \epsilon} \right|_{\epsilon=0},$$

which is the dot product of two row vectors: the gradient of the loss $\nabla_{\theta^*} \ell$ at $\theta^* = \theta^*(0)$ and the derivative of the optimal parameters for the altered objective $\frac{\partial \theta^*}{\partial \epsilon}$ at $\epsilon = 0$.

It can be shown [8] that under certain conditions (e.g., second-order differentiability and convexity of the loss function), the derivative of θ^* is given by:

$$\left. \frac{\partial \theta^*}{\partial \epsilon} \right|_{\epsilon=0} = -H_{\theta^*}^{-1} \left. \nabla_{\theta^*} \ell(z, \theta^*) \right|_{\theta^*=\theta_0^*},$$

where H_{θ^*} is the Hessian matrix of $L(S, \theta)$ evaluated at $\theta^* = \theta_0^*$.

Defining $g(z)$ as the gradient of the loss for z evaluated at $\theta^* = \theta_0^*$, and substituting into Eqs. 9.3.1.1 and 9.3.1.1, we obtain:

$$I(z, z_t) = g^{\mathsf{T}}(z_t) H_{\theta^*}^{-1} g(z).$$

To explain the prediction for z_t, we use Eq. 9.3.1.1 to compute the influence of each training example z, which can be efficiently approximated as suggested in [4]. The influence function (IF) explanation for the prediction of z_t consists of the top-k training examples with the highest influence.

Some subsequent studies have suggested that IF might be non-robust or fragile when applied to deep networks. For instance, [9] argued that this fragility could stem from factors such as the non-convexity of the loss function, the approximation of the Hessian matrix, and the use of weight decay. However, a later study [10] reported contrasting findings, suggesting that IF may not be as fragile as previously thought. The authors argue that evaluating IF by correlating the IF score with the actual change in loss is suboptimal, as the process of removing data points and retraining introduces randomized non-linearities. They further contend that the factors noted by [9] do not make IF inherently fragile in deep models, though they can occasionally lead to semantic dissimilarities or non-relevance between influential instances and the target sample being explained.

The **RelatIF** [11] approach attempts to solve this non-relevance issue by computing a *relative influence* score for each training point, which enables ranking data points by their comparative effect on the loss for a given test instance. RelatIF normalizes these influence scores as follows:

$$\text{RelatIF}(z_i, z_t) = \frac{I_{\text{up, loss}}(z_i, z_t)}{\sum_{j=1}^{n} |I_{\text{up, loss}}(z_j, z_t)|}.$$

This normalization enhances interpretability by representing each sample's proportional influence, allowing for direct influence comparisons across different training points.

TracIn calculates the influence of training points on test predictions by leveraging gradient similarities across multiple training epochs. For a training point z_i and a test point z_t, the influence is computed as:

$$\text{TracIn}(z_i, z_t) = \sum_{j=1}^{N} \langle \nabla_\theta \ell(z_t; \theta_j), \nabla_\theta \ell(z_i; \theta_j) \rangle,$$

where θ_j represents model parameters at training checkpoint j, and $\nabla_\theta \ell$ is the gradient of the loss function.

On the other hand, **Datamodels** [12] is a more empirical approach that selects a test point for explanation and samples multiple subsets from the training set. It trains models on these subsets and then builds a linear model where the input features represent subset encodings and the output corresponds to model performance on the test sample. The resulting weights in the linear model indicate the importance of each training sample. However, the approach requires a large number of intermediate models, which can be computationally expensive. To mitigate this, Trak [13] was introduced as a faster version of Datamodels, maintaining similar accuracy but reducing the computational load.

9.3.1.2 Class Outliers Effect

In [14], the authors demonstrate that the non-relevance problem is not unique to influence function but is also prevalent in other example-based explainability techniques, such as TracIn [15] and Datamodels [12]. Despite their theoretical differences, these methods share conceptual similarities, primarily in their reliance on training data subsets to explain model predictions.

The issue of class outliers–samples that resemble one class but are labeled as another–was identified in [14] as a major cause of the non-relevance problem. These outliers, characterized by high loss and global influence, tend to appear in explanations for many predictions despite being irrelevant. The authors proposed an evaluation framework with three metrics–distinguishability, continuity, and correctness–to highlight the susceptibility of current methods to this issue. Notably, RelatIF, which penalizes explanations by removing high-loss samples, was shown to be suboptimal. Using the correctness metric, the authors demonstrated that removing class outliers from the explanation might not always be desirable. These high-loss samples can be valuable in explaining important user-injected rules in the labeling process, which could lead to higher loss in specific data subsets.

Consecutive work by [16] has proposed **AIDE**, another method for dealing with outliers and data quality issues in general in the context of example-based explainability. They underscore the importance of keeping the outliers in some cases and removing them only when they are irrelevant. For this, they proposed a proximity-based approach which will select explanations solving a weighted sum equation with Influence score, Proximity, and Diversity.

$$\mathcal{E}_q = \arg\max_{\mathcal{E} \subseteq S_q, |\mathcal{E}|=k} \sum_{z \in \mathcal{E}} (\alpha |I(z, z_t)| + \beta P(z, z_t)) + \gamma D(\mathcal{E}), \qquad (9.1)$$

where α, β, γ are weights empirically determined. Similar to other submodular maximization problems [17], they construct $\mathcal{E}_q$ in an incremental way, each time greedily selecting the example that maximizes the objective.

9.3.1.3 User Intent

Example-based methods are rarely adaptable to specific user intents, despite the fact that stakeholders may require different types of explanations depending on their objectives. A critical review [18] identifies misalignments between the design goals of XAI methods and the psychological and cognitive aspects of explainability. A key limitation they identify is that most XAI methods neglect *user intent* during design and evaluation. Although mostly referring to counterfactual explanations, the authors highlight the importance of considering the user's needs and goals while generating explanations. **AIDE** [16] not only addresses the problem of class outliers but also introduces the concept of ***intent*** within the realm of example-based explainability. This framework tailors explanations based on the user's specific goals, enhancing the

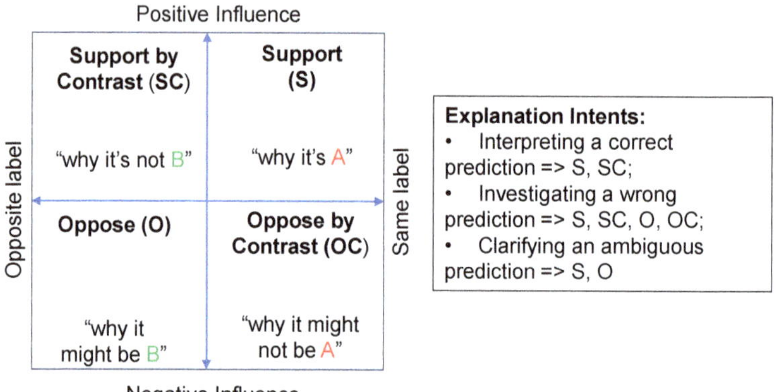

Fig. 9.5 AIDE quadrants and intent

interpretability of machine learning models. AIDE distinguishes three primary types of user intent for local example-based explanations: interpreting a correct prediction, investigating a wrong prediction, and clarifying an ambiguous prediction.

The process begins by classifying influential samples into four quadrants, as shown in Fig. 9.5. These quadrants categorize samples based on their influence and label:

– **Support** examples positively influence the prediction and share the same label as the test instance, helping to confirm the model's decision. – **Support by Contrast** examples also have a positive influence but differ in label, providing contrast and demonstrating why the model did not choose the alternative class. – **Oppose** examples negatively influence the prediction and have a different label, challenging the model's decision and arguing for the test instance's true class. – **Oppose by Contrast** examples also negatively influence the prediction but share the same label as the test instance, showing why the test sample does not belong to the predicted class.

AIDE's approach to interpreting a correct prediction focuses on helping the user understand the rationale behind a correct classification. In this context, *Support* and *Support by Contrast* examples are highlighted: the former explains why the test instance was classified as it was, while the latter illustrates why alternative predictions were less likely.

When investigating a wrong prediction, AIDE helps uncover the causes of errors, which may stem from mislabeled training data or biases within the training set. In the case of mislabeled samples, examining *Supporters* reveals potentially erroneous labels, while *Opposers* provide insight into why an alternative label might be more appropriate. Regarding bias in the training data, AIDE assesses all quadrants. For example, if a model has learned to associate irrelevant features, such as associating snow with wolves (as seen in husky-related data), the bias will be visible in *Supporters* but not in *Supporters by Contrast*. Similarly, *Opposers* will reinforce this bias if they exhibit the biased feature, and *Opposers by Contrast* will show why the model's prediction is incorrect due to the bias.

Finally, for clarifying predictions for ambiguous samples, AIDE explores the model's labeling mechanism. In cases where a test sample contains multiple objects that could influence the classification, the framework identifies training samples with similar ambiguities that helped shape the model's decision. These samples serve as *Supporters*, indicating how the classification rule was learned.

9.3.2 Global Example-Based Explainability

DataShap [19] is a method that utilizes game-theoretic principles to quantify the contribution of individual training samples to a model's performance. By assigning Shapley values to data points, it determines their importance in the training process. However, calculating Shapley values involves evaluating all possible subsets of the dataset, leading to significant computational challenges. Approximation techniques, such as Monte Carlo sampling, have been employed to mitigate this issue, but they remain computationally intensive. To address this, BetaShap [20] was introduced as an enhancement to DataShap, aiming to reduce noise in importance scores. Despite this improvement, BetaShap still faces high computational costs due to the inherent complexity of Shapley value calculations. The MMD-Critic [5] method employs prototypes and criticisms to reflect the data distribution. It selects a subset of data that minimizes the Maximum Mean Discrepancy (MMD) between the empirical distributions of the selected prototypes and the original dataset. By minimizing this discrepancy, prototypes serve as accurate representatives of the data distribution, while criticisms identify areas with poor representation. An advanced version, ProtoDash [21], further refines this approach by providing scores for prototypes, enhancing the selection process. ProtoPNet [22] integrates prototype learning within a convolutional neural network. For each input, it calculates similarity scores to each prototype in a learned embedding space. A fully connected layer then combines these similarity scores to produce the final class prediction. ProtoPNet's architecture supports interpretability, as the prototypes function as reasoning steps within the model's decision-making process. The Bayesian Case Model (BCM) [23] applies probabilistic clustering to create interpretable classifications based on the model's representation of data. This approach generates prototypes that summarize clusters within the data, associating new instances with likely classes based on proximity to learned prototypes. Recent advancements, such as DataModels [12], further refine prototype-based classification by generating prototypes that align with both the model's predictions and its representation of the data. This alignment supports richer, class-wise clustering and ensures that explanations are specific to the model's perspective on the data. The Deep k-Nearest Neighbors (DkNN) [24] method assesses the validity of a neural network's decision by examining nearest-neighbor labels in the intermediate layers. The premise is that a model's decision should align with the nearest neighbors in the hidden layers, adding an interpretability layer to verify whether the network's internal reasoning supports the final decision.

9.3.3 Evaluation Metrics

Evaluating example-based explainability involves distinct challenges due to the nature of explanations that rely on specific data instances to illustrate model behavior. Unlike feature-based explanations that focus on attributing model decisions to input features, example-based approaches seek to make predictions interpretable by referencing training examples that best illustrate the model's reasoning. Consequently, evaluating these explanations requires specialized metrics that go beyond traditional feature importance and consider the nuances of instance-based relevance, faithfulness, and contextual diversity.

Evaluation Properties. In [25] authors introduce a broader evaluation framework with their Co-12 properties, which emphasize the need for multi-dimensional evaluation metrics in XAI. Co-12 includes aspects like correctness, coherence, compactness, and consistency, which collectively assess the quality of an explanation. For example-based explainability, applying these properties means ensuring that explanations are not only accurate (correctness) but also concise and free from redundancy (compactness), thus enhancing user understanding without cognitive overload. Many ways of measuring correctness exist in the literature, some of them perform experiment syntetically inducing data for which the ground truth explanation is known [14, 26, 27], some others check the alignment of explanations with the underlying model [28]. Compactness is usually encoded in diversity and redundancy metrics [29, 30]. In addition, [16] addresses the challenge of class outliers–instances that may not represent typical data patterns but could disproportionately influence explanations. They propose the metric of distinguishability specifically to ensure that explanations maintain contextual relevance across various instances. This metric helps prevent the repeated use of similar instances in explanations for different test cases, which could otherwise reduce an explanation's relevance and informativeness for users. In example-based XAI, distinguishing between globally influential yet unrepresentative examples and locally relevant instances is essential for producing clear, contextually useful explanations.

In [31] the authors posit a reasonable minimal set of desiderata for explanation, such as explicitness—are explanations immediate and understandable, faithfulness—do relevance scores indicate true importance, and stability—are explanations consistent for similar samples. They observe that many existing methods do not satisfy them.

9.4 Open Problems and Future Directions

Despite significant advances, example-based explainability still faces multiple open challenges that hinder its effectiveness, especially in high-stakes applications where transparency and trust are essential.

Model Fidelity and Faithfulness in Explanations. One of the primary challenges is ensuring that explanations are faithful to the model's actual decision-making processes. Current methods often generate prototypes or global example-based explanations independently of the model, leading to explanations that may not accurately reflect the model's internal logic. Developing methods that produce explanations based directly on the model's learned representations, rather than solely the ground truth data distribution, is a critical area for future research. In [32] authors highlighted this issue and demonstrated the unfaithfulness of existing methods and tried to solve it for regression models, however more generic and model-agnostic approach is yet to be proposed.

Handling Outliers and Data Quality Issues. The presence of outliers, mislabeled data, or atypical instances in training data complicates example-based explanations. Current approaches struggle to differentiate between examples that genuinely support the prediction and those that merely coincide with it due to data irregularities. Addressing this issue calls for enhanced data preprocessing and model training techniques that minimize the influence of noisy or irrelevant instances. Additionally, explainability methods that can identify and adjust for outliers, e.g., AIDE [16], are essential for producing robust and meaningful explanations, particularly in domains where data quality may vary.

User-Centered and Intent-Aware Explanations. Another limitation is the lack of adaptability to specific user intents. Explanations that align with the goals and background of individual users–whether they are domain experts, decision-makers, or general users–can improve interpretability and user satisfaction. Future research, pioneered by [16], should explore intent-aware explainability frameworks that can tailor explanations based on user goals. This adaptability could be achieved through interactive explainability systems that allow users to specify their intent, leading to more relevant and informative explanations.

Integrating Global and Local Explainability. Balancing global and local explainability remains an open problem. While global methods provide a high-level understanding of model behavior, they lack the granularity needed to explain specific predictions. Conversely, local methods offer detailed insights but may not capture broader trends in model decision-making. Future research should focus on integrating these approaches, creating hybrid methods that provide both general and instance-specific insights. This integration could bridge the gap between understanding the model's overall tendencies and comprehending its individual predictions.

Evaluating Explainability Techniques. Standardized metrics for evaluating explainability techniques are needed to make meaningful comparisons across methods. Metrics that capture faithfulness, diversity, proximity, and user satisfaction are essential to assess the quality of example-based explanations effectively. Future research should aim to establish comprehensive evaluation frameworks, incorporating both technical and human-centered metrics to provide a balanced assessment of explainability methods in practical applications.

9.5 Conclusion

Example-based explainability methods hold great promise for improving interpretability in complex machine learning models by linking predictions to specific data examples. These methods are particularly valuable in high-stakes settings, where understanding the basis of predictions is critical for building trust and ensuring responsible AI deployment. This chapter reviewed various example-based approaches, including data-centric, post-hoc, and intrinsically interpretable models, highlighting their strengths and limitations.

Despite these advancements, example-based explainability faces challenges such as ensuring faithfulness, handling outliers, adapting to user intent, and integrating global and local insights. Addressing these challenges is essential to improve the robustness, relevance, and usability of explanations. Future research that explores model fidelity, user-centered design, and evaluation frameworks will be crucial for advancing example-based explainability. Ultimately, as machine learning models become increasingly embedded in decision-making processes, the development of effective and trustworthy explainability methods remains a foundational goal for the field of artificial intelligence.

References

1. Ribeiro MT, Singh S, Guestrin C (2016) Why should i trust you?: explaining the predictions of any classifier. In: Proceedings of the 22nd ACM SIGKDD international conference on knowledge discovery and data mining. Association for Computing Machinery, New York, NY, USA, pp 1135–1144
2. Guidotti R, Monreale A, Ruggieri S, Pedreschi D, Turini F, Giannotti F (2021) Local rule-based explanations of black box decision systems. CoRR abs/1805.10820, version 4 update
3. Lundberg SM, Lee SI (2017) A unified approach to interpreting model predictions. In: Guyon I, Luxburg UV, Bengio S, Wallach H, Fergus R, Vishwanathan S, Garnett R (eds) Advances in neural information processing systems, vol 30. Curran Associates Inc, Long Beach, CA, USA, pp 4188–4197
4. Koh PW, Liang P (2017) Understanding black-box predictions via influence functions. In: Precup D, Teh YW (eds) Proceedings of the 34th international conference on machine learning, PMLR, Sydney, Australia, Proceedings of machine learning research, vol 70, pp 1885–1894
5. Kim B, Khanna R, Koyejo OO (2016) Examples are not enough, learn to criticize! criticism for interpretability. In: Lee D, Sugiyama M, Luxburg U, Guyon I, Garnett R (eds) Advances in neural information processing systems, vol 29. Curran Associates Inc, Barcelona, Spain, pp 2280–2288
6. Rong Y, Leemann T, trang Nguyen T, Fiedler L, Qian P, Unhelkar V, Seidel T, Kasneci G, Kasneci E (2023) Towards human-centered explainable AIi: a survey of user studies for model explanations. arXiv:2301.08366
7. Hampel FR (1974) The influence curve and its role in robust estimation. J Am Stat Assoc 69(346):383–393
8. Cook RD, Weisberg S (1982) Residuals and influence in regression. Chapman and Hall, New York

9. Basu S, Pope P, Feizi S (2021) Influence functions in deep learning are fragile. In: International conference on learning representations, Virtual. https://openreview.net/forum?id=kW9K8erRIXW

10. Epifano JR, Ramachandran RP, Masino AJ, Rasool G (2023) Revisiting the fragility of influence functions. Neural Netw 162(C):581–588

11. Barshan E, Thudi A, Khan S, Hashemi M, Bhattacharyya C, John P (2020) Relatif: identifying explanatory training examples via relative influence. In: Daumé H III, Singh A (eds) Proceedings of the 37th international conference on machine learning, PMLR, Virtual, Proceedings of machine learning research, vol 119, pp 658–667

12. Ilyas A, Park SM, Engstrom L, Leclerc G, Madry A (2022) Datamodels: understanding predictions with data and data with predictions. In: Chaudhuri K, Jegelka S, Song L, Szepesvari C, Niu G, Sabato S (eds) Proceedings of the 39th international conference on machine learning, PMLR, Baltimore, Maryland, USA, Proceedings of machine learning research, vol 162, pp 9525–9587

13. Park SM, Georgiev K, Ilyas A, Leclerc G, Madry A (2023) Trak: attributing model behavior at scale. In: Krause A, Brunskill E, Cho K, Engelhardt B, Sabato S, Scarlett J (eds) Proceedings of the 40th international conference on machine learning, PMLR, Honolulu, Hawaii, USA, Proceedings of machine learning research, vol 202, pp 27313–27345

14. Nematov I, Sacharidis D, Sagi T, Hose K (2024) The susceptibility of example-based explainability methods to class outliers. CoRR abs/2407.20678

15. Pruthi G, Liu F, Kale S, Sundararajan M (2020) Estimating training data influence by tracing gradient descent. Adv Neural Inf Process Syst 33:19920–19930. Identical to GLKS20 entry; this is the typical citation format

16. Nematov I, Sacharidis D, Sagi T, Hose K (2024) AIDE: antithetical, intent-based, and diverse example-based explanations. arXiv:2401.09124

17. Gollapudi S, Sharma A (2009) An axiomatic approach for result diversification. In: Proceedings of the 18th international conference on world wide web (WWW '09). ACM, Madrid, Spain, pp 381–390

18. Keane MT, Kenny EM, Delaney E, Smyth B (2021) If only we had better counterfactual explanations: Five key deficits to rectify in the evaluation of counterfactual XAI techniques. In: Zhou ZH (ed) Proceedings of the thirtieth international joint conference on artificial intelligence, IJCAI-21, International joint conferences on artificial intelligence organization, Virtual, pp 4466–4474

19. Ghorbani A, Zou JY (2019) Data shapley: equitable valuation of data for machine learning. In: Chaudhuri K, Salakhutdinov R (eds) Proceedings of the 36th international conference on machine learning, PMLR, Long Beach, CA, USA, Proceedings of machine learning research, vol 97, pp 2242–2251

20. Kwon Y, Zou J (2022) Beta shapley: a unified and noise-reduced data valuation framework for machine learning. In: Camps-Valls G, Ruiz FJR, Valera I (eds) Proceedings of the 25th international conference on artificial intelligence and statistics, PMLR, Virtual, Proceedings of machine learning research, vol 151, pp 6832–6850

21. Gurumoorthy KS, Dhurandhar A, Cecchi GA, Aggarwal CC (2019) Efficient data representation by selecting prototypes with importance weights. In: 2019 IEEE international conference on data mining (ICDM). IEEE, Beijing, China, pp 260–269

22. Li O, Liu H, Chen C, Rudin C (2018) Deep learning for case-based reasoning through prototypes: a neural network that explains its predictions. In: Proceedings of the AAAI conference on artificial intelligence AAAI, Press New Orleans, Louisiana, USA, vol 32, pp 3563–3571

23. Kim B, Rudin C, Shah JA (2014) The bayesian case model: a generative approach for case-based reasoning and prototype classification. In: Proceedings of the 27th international conference on neural information processing systems, vol 2. MIT Press, Montreal, Canada, pp 1952–1960

24. Papernot N, McDaniel P (2018) Deep k-nearest neighbors: towards confident, interpretable and robust deep learning. arXiv:1803.04765

25. Nauta M, Trienes J, Pathak S, Nguyen E, Peters M, Schmitt Y, Schlötterer J, van Keulen M, Seifert C (2023) From anecdotal evidence to quantitative evaluation methods: a systematic review on evaluating explainable AI. ACM Comput Surv 55(9):1–44

26. Adebayo J, Muelly M, Liccardi I, Kim B (2020) Debugging tests for model explanations. In: Larochelle H, Ranzato M, Hadsell R, Balcan M, Lin H (eds) Advances in neural information processing systems, vol 33. Curran Associates Inc, Virtual, pp 700–712
27. Chen J, Song L, Wainwright MJ, Jordan MI (2018) Learning to explain: an information-theoretic perspective on model interpretation. In: Dy J, Krause A (eds) Proceedings of the 35th international conference on machine learning, PMLR, Stockholm, Sweden, Proceedings of machine learning research, vol 80, pp 883–892
28. Adebayo J, Gilmer J, Muelly M, Goodfellow I, Hardt M, Kim B (2018) Sanity checks for saliency maps. In: Bengio S, Wallach H, Larochelle H, Grauman K, Cesa-Bianchi N, Garnett R (eds) Advances in neural information processing systems, vol 31. Curran Associates Inc, Montreal, Canada, pp 921–930
29. Mothilal RK, Sharma A, Tan C (2020) Explaining machine learning classifiers through diverse counterfactual explanations. In: Proceedings of the 2020 conference on fairness, accountability, and transparency, ACM, Barcelona, Spain, pp 607–617
30. Goyal Y, Wu Z, Ernst J, Batra D, Parikh D, Lee S (2019) Counterfactual visual explanations. In: Chaudhuri K, Salakhutdinov R (eds) Proceedings of the conference on machine learning, PMLR, Long Beach, CA, USA, Proceedings of machine learning research, vol 97, pp 2376–2384
31. Alvarez Melis D, Jaakkola TS (2018) Towards robust interpretability with self-explaining neural networks. In: Bengio S, Wallach H, Larochelle H, Grauman K, Cesa-Bianchi N, Garnett R (eds) Advances in neural information processing systems, vol 31. Curran Associates Inc, Montreal, Canada, pp 7775–7784
32. Filho RM, Lacerda AM, Pappa GL (2023) Explainable regression via prototypes. ACM Trans Evol Learn Optim 2(4):1–26

Chapter 10
Table Search in Data Lakes: Methods, Indexing Techniques, and Research Challenges

Ibraheem Taha, Matteo Lissandrini, Alkis Simitsis,
Torben Bach Pedersen, and Yannis Ioannidis

Abstract The exponential growth of data lakes has made efficient table discovery a critical challenge. This chapter explores table search in data lakes, covering methods, indexing techniques, research challenges, and future directions. We break down the complexities of table search into three foundational pillars: table understanding, search mechanisms, and indexing strategies. Table understanding, which includes table annotation and domain discovery, forms the semantic backbone for accurate retrieval. We thoroughly examine three core paradigms of table search, keyword-based, joinable, and unionable searches, each playing a vital role in tasks such as data augmentation and integration. To enhance search efficiency, we analyze advanced indexing techniques, including tree-based, hash-based, graph-based, and quantization-based approaches, emphasizing their scalability and performance benefits. Beyond methodologies, we highlight key open research challenges, such as defining table relatedness, improving semantic search accuracy, and establishing robust benchmarks. By integrating these perspectives, we demonstrate the interconnected nature of table understanding, search, and indexing, showcasing their collective impact in unlocking the analytical and integrative potential of data lakes. This

I. Taha (✉)
Athena Research Center, National and Kapodistrian University of Athens, Athens, Greece
e-mail: ibta@cs.aau.dk

I. Taha · T. B. Pedersen
Aalborg University, Aalborg, Denmark
e-mail: tbp@cs.aau.dk

M. Lissandrini
University of Verona, Verona, Italy
e-mail: matteo.lissandrini@univr.it

A. Simitsis · Y. Ioannidis
Athena Research Center, Athens, Greece
e-mail: alkis@athenarc.gr

Y. Ioannidis
e-mail: yannis@di.uoa.gr

Y. Ioannidis
National and Kapodistrian University of Athens, Athens, Greece

G. Dejaegere et al. (eds.), *Data Engineering for Data Science*,
https://doi.org/10.1007/978-3-032-18765-9_10

chapter not only provides a comprehensive framework for table search but also sets the stage for future advancements in the field.

Keywords Table search · Data lakes · Vector search · Index · Approximate nearest neighbor

10.1 Introduction to Table Search Methods and Indexing Techniques in Data Lakes

Traditional databases and data lakes represent two contrasting paradigms in data management, designed to support distinct workloads and requirements [1]. Traditional databases operate using a *schema-on-write* approach, where data is stored and queried based on predefined schemas. This method ensures consistency, optimization, and strong transactional guarantees, making traditional databases well-suited for operational systems and structured, predictable environments. However, the rigidity of this schema-first design presents challenges in handling heterogeneous and rapidly evolving data types, which are increasingly common in modern applications [2, 3].

Conversely, data lakes adopt a *schema-on-read* philosophy, where raw data, whether structured, semi-structured, or unstructured, can be ingested and stored in its original format without the need for predefined schemas [1]. This flexibility allows data lakes to support diverse use cases, such as big data analytics and exploratory data processing. Data lakes function as scalable, centralized repositories that enable ingestion of multiple datasets, on-demand data integration, and dynamic query execution [3]. These characteristics make data lakes particularly valuable in addressing the variety, volume, and velocity of data generated by contemporary systems and applications [2, 3]. While data lakes excel in scalability and flexibility, they also face challenges that set them apart from traditional databases. Key issues include maintaining robust data governance, ensuring data quality, and achieving efficient query performance [4].

Figure 10.1 illustrates the general architecture of table search and exploration in data lakes. In this chapter, we focus on the three core pillars (highlighted in the gray box): table understanding (Sect. 10.3.1), search methods (Sect. 10.3.2), and indexing techniques (Sect. 10.3.3) for table retrieval. In this chapter, we specifically focus on structured data lakes, where tables serve as the primary data representation. Although structured data lakes benefit from tabular organization, their open and schema-on-read nature introduces complexities in critical processes such as data discovery, an essential task for unlocking the true value of the diverse and extensive data they store [5]. It is worth to note that Fig. 10.2 will serve as the primary reference for examples throughout this chapter.

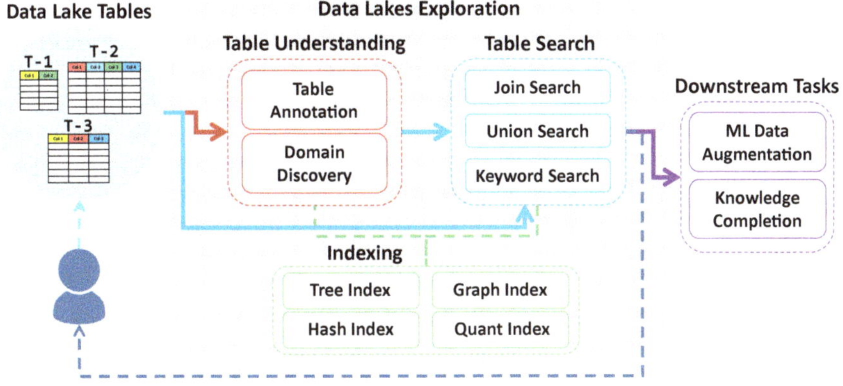

Fig. 10.1 System architecture for table exploration in data lakes

Table (a)

	HQ
Beetle	Germany
Ram	USA
BMW X7	Germany
Porsche 911	Germany

Table (b)

Name	Speed km/h	Weight kg	Lab Location
Beetle	9	0.02	Palestine
Ram	65	110	USA
Jaguar	80	100	Brazil
Hawk	190	1.2	UK
Mustang	48	360	USA

Table (c)

Car	Country
Mercedes	Germany
Lamborghini	Italy
BYD	China
Jaguar	UK

Query Table

Model	Company	Country
Beetle	VW	Germany
Ram	Dodge	USA
Jaguar R75	Jaguar	UK
Mustang	Ford	USA

Fig. 10.2 A running example of table search over a data lake of 3 tables

10.1.1 Table Understanding and Table Search

Indeed, efficient data discovery necessitates robust approaches to generate, enhance, and query metadata, enabling users to efficiently locate datasets that match their analytical needs, such as those containing specific attributes, relationships, or domain-specific information [6]. Methods such as keyword-based search play a vital role by enabling the retrieval of relevant datasets through matching keyword queries with metadata descriptions, as seen in tools like Google Dataset Search [7]. However, metadata often exhibit inconsistency or incompleteness, which hinders their ability to comprehensively describe the datasets they represent. This limitation, in turn, affects keyword-based search approaches, as they rely heavily on metadata and are therefore unable to effectively identify datasets based on specific content within the data itself. To overcome this challenge, alternative methods have been proposed that enable content-based search within datasets, allowing for more accurate and comprehensive discovery [8–18]. In content-based search, queries are performed directly on the actual data values within the tables.

Example 10.1 Consider the data lake tables illustrated in Fig. 10.2. Suppose a user intends to retrieve detailed information about the insect **Beetle** using a keyword-based search. A natural first step might be to input the keyword **"Beetle"**. However, the search results would return both Tables (a) and (b). While Table (b) is relevant, as it pertains to animals and insects, Table (a) is unrelated, as it contains information about cars. This example highlights the limitation of keyword-based search in table retrieval: it often fails to disambiguate between homonyms (words with multiple meanings) or contextually distinct terms. In the case of metadata search, if the query is something like **"What is the speed of the beetle insect"** the most relevant answer would be found in Table (b).

To address the challenges posed by the growing size and complexity of data lakes, table search methods have evolved. Table search is the process of locating and retrieving relevant tables based on criteria like keyword matches, joinable attributes, or unionable schemas to support data integration, augmentation, and analysis. These methods have progressed from simple keyword-based approaches to advanced techniques that incorporate a deep understanding of table structures, including cells, rows, and columns [19–26]. Building on these advancements, recent methods leverage machine learning models such as pretrained language models (PLMs) to encode table content into dense vectors, which are numerical representations with non-zero values across all dimensions, referred to as embeddings. These embeddings capture rich semantic information, enabling retrieval tasks to identify tables based on conceptual similarities rather than mere surface-level matching. Retrieval over these vectors is often performed using similarity metrics, such as cosine similarity, to measure the proximity between query and table embeddings. This paradigm effectively uncovers semantic relationships and contextual alignments, moving beyond traditional syntactic matching to provide more accurate and relevant results.

Recent advancements in table search methods have enhanced two critical tasks for data integration: *joinable table search* and *unionable table search*. Joinable table search identifies tables that share a common attribute with the query table, enabling data enrichment through join operations. This is crucial for tasks such as data augmentation, where additional attributes from related tables provide deeper insights [27–37]. Unionable table search, on the other hand, focuses on locating tables with similar schemas to the query table, facilitating union operations to expand datasets with new rows of compatible data [36–40]. These tasks are critical for enhancing data discovery in data lakes, supporting a variety of downstream applications, including machine learning and knowledge graph construction.

Example 10.2 Given the query table in Fig. 10.2, the goal is to identify the top joinable table(s) based on the *Model* column, ensuring that the retrieved tables are the most suitable for adding columns to the query table. The columns to evaluate for potential joins are the first column in Table (a), *Name* in Table (b), and *Car* in Table (c), as all of these columns share similar values with the query table's *Model* column. The column *Name* in Table (b) has the most overlapping values with the query column. This represents a basic join search that does not account for the context of the table, indicating that Table (b) is related to the context of **Animal** or **Insect**, whereas Table (a) and Table (c) are associated with the context of **Car**.

Example 10.3 Given the query table in Fig. 10.2, the objective is to identify the top unionable table(s) by performing a basic union operation that maximizes the union score between the query table and tables in the data lake. This ensures that the retrieved tables are the most appropriate for adding rows to the query table. Without considering the semantics of the columns or the context of the tables, Table (b) emerges as the top unionable table because the columns *Model* and *Country* in the query table overlap more with *Name* and *Lab Location* in Table (b) than with columns in Tables (a) and (c). Specifically, the number of exact matches between the query table and Table (b) is higher than the matches with other tables in the data lake.

Table understanding is vital for improving table search and retrieval by extracting key semantic and structural information from tables in data lakes [5, 19]. It focuses on tasks such as table annotation, which enriches tables with metadata [20], and domain discovery, which identifies the underlying concepts associated with tables [25, 26]. These methods enhance search accuracy and relevance by providing deeper insights into table content, addressing key challenges in large-scale retrieval, and supporting downstream applications [23, 24, 37].

10.1.2 Table Indexing

Additional challenges in data lake search include the computational resources and time needed for linear searches across massive repositories. These methods become impractical due to the vast size of modern data lakes, which are sourced from diverse domains, including public repositories like Open Data, WikiTables (approximately 154M HTML tables), and GitTables (1M relational tables) [41–47]. To address this, various strategies have been proposed to accelerate table retrieval, with approximate nearest neighbor (ANN) techniques emerging as a key solution. While Nearest Neighbor Search (NNS) aims to find the exact closest data point(s) to a query within a dataset [48], ANN (also referred to as K-Approximate Nearest Neighbor or K-ANN) focuses on efficiently retrieving the top-K approximately similar results to the query. By reducing computation time compared to exhaustive search, ANN becomes a practical solution for large-scale data lakes.

The ANN techniques have been developed to further enhance retrieval efficiency, each tailored to specific data formats and internal structures. Studies on ANN are usually categorized into four different types: tree-based indexes (e.g., M-trees, R-trees) [49–58], hashing-based indexes (e.g., locality-sensitive hashing (LSH)) [48, 59–69], graph-based indexes (e.g., Hierarchical Navigable Small World (HNSW)) [70–77], and quantization-based indexes [78–85]. Such indexing techniques not only reduce query latency but also improve scalability, making them indispensable for handling the ever-growing scale and heterogeneity of data lakes.

This chapter is organized as follows: Sect. 10.2 presents an overview of the landscape, including a general architecture for table exploration and search in data lakes. Section 10.3 provides a detailed discussion of state-of-the-art techniques, covering table understanding in Sect. 10.3.1, table search in data lakes in Sect. 10.3.2, and ANN indexing techniques in Sect. 10.3.3. Research challenges, limitations of these techniques, and potential directions for future work are explored in Sect. 10.4. Finally, Sect. 10.5 concludes the chapter.

10.2   The Landscape of Table Search Methods and Indexing Techniques in Data Lakes

In this section, we present a taxonomy that classifies methods based on their purpose (e.g., table search) and their structure (e.g., indexing). Figure 10.1 illustrates a general architecture for table exploration and search.

Tables from the data lake may first undergo table understanding techniques to enrich their structure and semantics, particularly if they are intended for table search. Alternatively, they can directly proceed to table search methods without prior understanding, depending on the user's specific requirements. In either case, indexing plays a crucial role by supporting both table understanding and table search processes, enhancing their efficiency and scalability. The loop back to the data lake (represented by the dashed line) allows users to review the retrieved tables, refine their queries, or continue exploring the data lake. Additionally, the retrieved tables can be utilized in downstream tasks such as data augmentation for machine learning or knowledge completion. The system is usually supported by an ANN index to optimize the table retrieval process. This discussion focuses on the three main components highlighted in the gray box in the figure: table understanding, table search, and indexing.

We recognize an established categorization of table understanding methods into table annotation and domain discovery, as summarized in the taxonomy presented in Table 10.1. Table annotation methods involve techniques that assign semantic labels or annotations to various components of a table, enhancing both understanding and usability. These methods are categorized based on the specific components they

Table 10.1 A taxonomy table of table understanding techniques

Main categories	Sub categories	Techniques
Table annotation	Cell filling	Limaye et al. [20]
		TURL [22]
	Column relations	TABLE [21]
		TURL [22]
	Column type	Limaye et al. [20]
		TABLE [21]
		TURL [22]
		Sherlock [23]
		SATO [24]
Domain discovery	PLMs	Starmie [37]
		DeepJoin [31]
	Domain relations	C^4 [25]
		D4 [26]

annotate, such as columns type, or entity linking. Domain discovery methods focus on identifying the domains and topics associated with tables, providing valuable contextual information to enhance table search and integration. This improves the organization, retrieval, and overall understanding of tables. These methods are further classified based on the approaches used to determine the associated domain.

The taxonomy presented in Table 10.2 reflects our proposed categorization, dividing table search methods into three primary categories: keyword search, join search, and union search. These techniques are categorized into different search methods, which are further classified based on the specific approach used for table retrieval, as shown in Table 10.2. Finally, indexing techniques in the literature are classified based on their structure, as shown in Table 10.3. These techniques are grouped into four categories: tree, hash, graph, and quantization methods.

Each category encompasses various approaches that optimize data retrieval and table indexing. Trees and hashing techniques are further classified based on the specific approaches used for data indexing, while graph-based methods are categorized according to the focus area of each technique. Lastly, quantization indexes are divided into basic techniques and advanced, more advanced methods (see Table 10.3).

10.3 State of the Art

In this section, we provide a detailed exploration of the topics presented in Sect. 10.2, offering an in-depth analysis of the current research landscape, methodologies, and advancements in the areas of table understanding, table search, and ANN indexing.

Table 10.2 A taxonomy table of table search techniques

Main categories	Sub categories	Techniques
Keyword search	Schema (Metadata)	GDS [7]
		DBXplorer [10]
		MeanKS [14]
		Query Annotator [15]
	Content	Discover [8]
		Banks [9]
		WWT [18]
	Semantic relations	Précis [11–13]
		Octopus [16]
		Aurum [17]
Join search	Values	Juneau [28]
		JOSIE [29]
		DICE [32]
		COCOA [33]
	Hashing	LSH-Ensemble [27]
		MATE [30]
	PLMs	DeepJoin [31]
		YADL [34]
		D^3L [36]
		Starmie [37]
Union search	Values	TUS [38]
		SANTOS [39]
	PLMs	D^3L [36]
		Starmie [37]
		AUTOTUS [40]

Each topic is examined to showcase the state-of-the-art techniques, their foundational principles, and their contributions to addressing key challenges in the field.

10.3.1 Table Understanding

Table understanding refers to the process and methodologies aimed at enabling the effective search and retrieval of tables from repositories [19]. It addresses diverse search tasks [5], including keyword search, join search and table search, each of which will be explored in Sects. 10.3.2.1, 10.3.2.2, 10.3.2.3, respectively. This field involves a variety of techniques designed to extract and interpret essential information about tables. We present the most recent techniques used in table search. These methods include table annotation (Sect. 10.3.1.1), and domain discovery (Sect. 10.3.1.2).

Table 10.3 A taxonomy table of indexing techniques

Main categories	Sub categories	Techniques
Tree	Partition	KD-Tree [51]
		RPTs [55]
		Cover Tree [54]
	Metric	R-Tree [50]
		M-Tree [52]
		SR-Tree [53]
	Learning	FLANN [56]
		JTR [57]
Hashing	Partition	LSH [48]
		P-Stable LSH [59]
		Lei et al. [63]
		VHP [65]
		SRS [66]
		LSH-APG [86]
	Collision	DCC [60]
		Query Aware [62]
		DB-LSH [68]
	Learning	IDEC [61]
		Li et al. [64]
		PM-LSH [69]
Graph	Performance	HNSW [70]
		NSW [75]
		NSG [73]
		τ-MG [76]
	Scalability	DiskANN [71]
		SSG [72]
		SPANN [77]
Quantization	Fundamental	PQ [82]
		OPQ [80]
		AQ [81]
	Advances	RaBitQ [78]
		ITQ [79]
		AVQ [83]
		VAQ [84]
		PQBF [85]

10.3.1.1   Table Annotation

Table annotation (TA) involves completing missing data and/or enriching existing data with additional details to the input table (see Example 10.4). TA includes supporting various downstream tasks, including cell filling (completing missing values based on context), relation extraction (identifying relationships between entities in a table), and column annotation (assigning semantic labels to table columns). Limaye et al. [20] proposed an early work on table discovery that leverages various forms of annotations, including entity annotations for identifying specific entities from an ontology (YAGO catalog [87]), type annotations to assign semantic types, and relation annotations to capture binary relationships. These annotations enable more precise table searches by allowing queries to leverage the semantic meaning of table content. For instance, type annotations align columns with predefined semantic categories, and relation annotations connect columns through meaningful relationships.

TABLE [21] presents a system for recovering the semantics of web tables by annotating them with two key types of information: *entity class* labels and *binary relationship* labels. Entity class labels describe the semantic class of a column's contents. Binary relationship labels capture relationships between pairs of columns. These annotations enable meaningful table search by aligning tables with the semantics of user queries, rather than relying solely on text-based matching. The system ranks and retrieves tables based on their semantic relevance to the query, leveraging annotated class labels, binary relationships, schema tokens, and contextual signals.

As machine learning and neural networks continue to gain prominence, state-of-the-art research has utilized deep learning techniques for semantic type annotation, also known as column classification. **Sherlock** [23] proposes a framework for identifying the semantic types of individual columns, leveraging both column values and associated metadata. The framework extracts four categories of features: value embeddings, global statistical features, character distributions, and the raw column values. These features are combined and fed into a model that classifies columns into one of 78 predefined semantic categories. Expanding on this foundation, **SATO** [24] extends Sherlock's approach by incorporating inter-column dependencies, enabling a more structured analysis of tables.

Pre-trained language models (PLMs) have recently emerged as powerful tools for table understanding, facilitating a variety of tasks. A state-of-the-art notable framework in this area is **TURL** [22]. It offers a new approach to table understanding by utilizing pre-training and fine-tuning techniques to model relational web tables. The framework employs a structure-aware transformer encoder, which incorporates a visibility matrix to model the row-column structure of tables. This design ensures that each element aggregates information only from relevant table components, such as related rows, columns, or metadata. TURL introduces a primary approach during pre-training, *Masked Entity Recovery (MER)*, which focuses on recovering masked entities based on surrounding table content and metadata to encode factual knowledge. The PLM is subsequently fine-tuned for a variety of downstream tasks, including entity linking, column type annotation, relation extraction, row population, cell filling, and schema augmentation. State-of-the-art methods prioritize semantic type annotation for columns, as this approach has proven effective for both table under-standing and table search, even compared to other table annotation techniques.

Example 10.4 Considering the first column in Table (a) in Fig. 10.2, the goal is to determine the domain to which this column belongs. By applying table annotation techniques, more particularly, *domain annotation*, the results would be ***Car*** or ***Car Model***. To provide additional details about the *HQ* column, the results could include information such as *Country* or *Location*.

10.3.1.2 Domain Discovery

Domain discovery is the process of finding groups of terms or dataset attributes that share a common meaning or belong to the same category. A domain represents a set of values related to a specific concept in an application. Instead of focusing on matching individual columns to specific types, domain discovery brings together all relevant values across the dataset that fit the same concept.

C^4 [25] is discovering concepts from relational tables in spreadsheet datasets by constructing concept hierarchies. The approach begins with a deep clustering tree, where values are grouped based on co-occurrence statistics across spreadsheet columns. Entities with high co-occurrence scores are iteratively merged in a bottom-up manner to form fine-grained clusters organized into a hierarchical structure. In the second step, a tree reduction process selects a subset of nodes from the deep clustering tree to produce a reduced concept hierarchy. This reduction is guided by an optimization problem that evaluates the ability of nodes to cover the original dataset, quantified using Jaccard similarity between cluster nodes and table columns. The resulting reduced tree is concise, interpretable, and captures meaningful concepts.

D4 [26] begins by generating context signatures, which capture how terms co-occur across different columns in datasets, i.e., tables. These signatures are then refined to remove noise, keeping only the most relevant terms. To handle incom-plete columns, D4 uses an iterative column expansion process, adding terms that are strongly supported by the refined signatures, ensuring more complete domains. Next, D4 performs graph-based clustering on the expanded columns, grouping terms into local domains based on their co-occurrence patterns. Finally, these local domains are filtered and merged into strong domains by assessing their similarity and support across multiple columns, ensuring high-quality domain discovery.

An alternative methodology employs PLMs as encoders to generate embeddings, which are vectorized representations of table elements (cells, rows, and columns), metadata, and their relationships. These embeddings enable table search by measur-ing the distance (based on a distance metric, e.g., cosine) between a query vector and vectors stored in a data lake, where the closest vectors indicate the most simi-lar content. Examples of systems adopting this approach include **Starmie** [37], and **DeepJoin** [31]. A detailed discussion of these systems is provided in Sects. 10.3.2.2 and 10.3.2.3. Additionally, clustering methods can group similar vectors in the data lake, helping identify tables that can be joined or combined based on their shared domain.

Example 10.5 For the last column in Tables (a), (b), and (c), labeled as *HQ*, *Lab Location*, and *Country*, respectively, domain discovery techniques identify them as belonging to the same domain, potentially *Location* or *Country*. This information aids in understanding the tables as a whole and enhances the effectiveness of subsequent table search techniques. Unlike Example 10.4, where domains are predefined by annotations, domain discovery automatically identifies domains by analyzing the current datasets, grouping terms or attributes that share a common meaning without relying on external knowledge or predefined labels. This approach is particularly useful for datasets lacking metadata or annotations.

10.3.2 Table Search

The work on table search in data lakes can be categorized into the following approaches: (a) *keyword* search (KS), detailed in Sect. 10.3.2.1, which retrieves tables based on matching user-specified keywords to metadata or content; (b) *join* search (JS), discussed in Sect. 10.3.2.2, where the objective is to identify tables that can be joined with a query table based on common keys or attributes; and (c) *union* search (US), outlined in Sect. 10.3.2.3, focusing on locating tables with similar schemas to facilitate union operations [5]. These techniques collectively aim to bridge the gap between raw table repositories and actionable insights, thereby enhancing the utility of data lakes in diverse analytical tasks.

10.3.2.1 Keyword Search

Keyword search involves entering a query composed of a finite set of keywords to retrieve data that fulfills the user's informational needs [88] (refer to Example 10.1, where the user issues a simple keyword query to retrieve tables matching the keyword). A result is considered relevant if it provides information perceived as valuable and aligned with the user's requirements [88]. Identifying datasets relevant to a specific task or information need is a core challenge in data discovery. Initial research addressed this issue within relational databases, with systems like **Discover** [8], **Banks** [9], **DBXplorer** [10], **Précis** [11–13], and **MeanKS** [14] providing solutions tailored to structured data retrieval. For example, the **Query Annotator** [15] introduces a probabilistic algorithm for annotating parts of keyword queries with table names, attribute names, and selection predicates on product catalog datasets. The method involves structured annotations by mapping keywords to tables and columns, generating all valid annotations using a tokenizer and tagger, and scoring them probabilistically to identify the most plausible matches.

Over time, the focus broadened to include web tables and, more recently, data lakes, reflecting the increasing diversity and complexity of modern data repositories. For instance, **Octopus** [16] employs keyword search to identify web tables and determines correlations between keywords and table cell values, enabling the grouping of tables with similar schemas. Similarly, **Google Dataset Search (GDS)** [7] emphasizes the standardization and enrichment of table metadata while constructing relationships between datasets using metadata-derived knowledge graphs, improving the accuracy of table search. Systems like **Aurum** [17] enhance this approach by representing dataset relationships as a knowledge graph, enabling users to navigate, search, and explore connections between datasets. These graphs allow for refined searches using keyword queries and uncover relationships derived from metadata. **WWT** [18] is a search engine designed for finding web tables using multi-column keyword queries. It indexes web tables based on headers, context, and content to retrieve candidates that match the input keywords describing the desired table's columns. Through graphical models, it maps query columns to table columns by analyzing keyword matches in headers, context, and content overlap, ensuring accurate and relevant table retrieval. Together, these systems illustrate the evolution of data discovery techniques, transitioning from structured database queries to comprehensive solutions for navigating vast and heterogeneous data repositories.

10.3.2.2   Join Search

The term *joinable tables* originates from the relational join operation. These are tables that can be combined into a new table or record set by adding additional columns while retaining all or a subset of the original records. **Finding Related Tables** [89] defines joinable tables as schema-complements. Schema matching is performed using attribute names, types, and values to identify joinable attributes. Non-matching attributes are used to augment the query.

Several notable works have addressed the challenges of identifying joinable tables. **LSH-Ensemble** [27] tackles the issue of multi-column joinability, particularly under skewed column value distributions, which are common in data lakes where certain domains dominate. To mitigate this skewness, LSH-Ensemble partitions datasets into buckets based on value distributions, ensuring a balanced application of LSH and enhancing joinability detection.

Juneau [28] adopts a holistic approach by profiling datasets based on intrinsic data values and extrinsic user interactions, such as co-occurrence patterns derived from data science notebooks. This dual perspective enhances the identification of joinable tables by incorporating behavioral insights alongside traditional value-based metrics. Other approaches focus on set-based techniques and super key generation. For instance, **JOSIE** [29] utilizes set intersections on table columns to determine joinability, returning the top-k joinable tables for a given query table. **MATE** [30] introduces XASH, a technique that aggregates row values into hash values, generating super keys to streamline the search for joinable tables across multiple columns.

DICE [32] facilitates interactive data discovery in data lakes by leveraging user-provided examples to uncover relevant datasets through join path discovery, where a join path represents a sequence of join operations. DICE iteratively refines its results based on user feedback and employs indexing and pruning techniques to ensure scalability and efficiency. This system simplifies complex queries by focusing on relational joins and semantic data augmentation. Moreover, **COCOA** [33] identifies and integrates external tables with high correlation to a target column. It employs an inverted index for efficient searching of joinable tables, leveraging overlapping attribute values, and uses a Join Mapper to execute lightweight, virtual joins, avoiding the overhead of materialized joins. Post-join, COCOA calculates non-linear correlation coefficients (e.g., Spearman's rank) to filter and retain the most correlated features, ensuring scalability and effective data augmentation.

DeepJoin [31] is an advanced method for discovering equi-joinable tables using PLMs. It employs seven column-to-text transformation strategies, leveraging metadata and contextual information to generate embeddings that effectively represent each column as a vector. During the search phase, the query column is encoded into a vector, which is then compared against pre-computed column embeddings using similarity metrics such as cosine similarity. This embedding-based approach ensures accurate and efficient identification of joinable tables, even in large-scale data lakes.

Recent end-to-end systems further advance joinability techniques. **YADL** [34] integrates three phases: retrieval, merging, and prediction, with retrieval of candidate joinable tables being its core. YADL employs techniques such as Jaccard containment and contrastive learning to streamline the augmentation process with scalability and precision. Similarly, **ARDA** [35] identifies joinable tables by combining precise and approximate join key discovery with optimized join plans. It addresses complex joins through pre-aggregation, resampling, and interpolation techniques, enabling efficient integration of external tables for predictive modeling.

D^{3}L [36] and **Starmie** [37] are frameworks capable of supporting both join search and union search. However, their primary application lies in union table search, which will be discussed further in Sect. 10.3.2.3.

Example 10.6 Revisiting **Example 10.2**, where the goal is to identify the top joinable table for the column *Model* in the query table, a context-aware approach would rank Table (a) as the most suitable. The first column in Table (a) shares the most related values with the query column, has the closest semantic meaning, and both tables share a common context.

10.3.2.3   Union Search

Another approach to data augmentation involves enriching query tables by adding additional data points (records) that share the same set of attributes (columns or features). The literature refers to such tables as *unionable tables*, relating to the relational union operation. However, a strict formal definition of unionability remains absent, as many systems adopt their own interpretations of what constitutes unionable tables [5]. For instance, **Finding Related Tables** [89] defines unionable tables as entity complements that share a subject attribute and exhibit schema similarity. These tables provide complementary subsets of entities, enabling their union to form a coherent and comprehensive table. **Starmie** [37], on the other hand, identifies unionability by maximizing the similarity score between the column representations (embeddings) of the query table and those in the data lake.

TUS [38], one of the foundational works in this area, introduces union search criteria based on three statistical models: set domains (value-based model), semantic domains (ontology-based model), and natural language domains. To enhance column correlations, TUS assigns separate scores to columns under each model and computes an aggregate union search score from these. **SANTOS** [39] adopts a relational approach to define semantic connections between columns within a table. This framework leverages both an established knowledge base and a synthetic knowledge base derived directly from the data lake. SANTOS constructs semantic graphs for tables, where columns are nodes and relationships are edges. To assess unionability, the algorithm evaluates the match strength between column relationships and the degree of query tree alignment within the semantic graph when aligning a query table with a target column (intent column).

D³L [36] identifies unionable and joinable tables using five metrics: word embedding strategies, column name similarity, column value overlap, format representation similarity (using regular expressions for text columns), and domain distribution similarity (for numeric columns). This approach employs a word embedding model (WEM) [90] to capture nuanced semantic contexts within column values.

Starmie [37] is an integrated framework for unionable and joinable table search, leveraging the PLMs model RoBERTa [91] to evaluate similarities between table contents. Operating in an unsupervised mode, Starmie employs multi-column contrastive learning to capture contextual semantic relationships within table values. Columns from the entire data lake are then represented in a unified vector space.

AUTOTUS [40] extends this field by introducing a framework for table union search that utilizes a multi-stage self-supervised approach powered by PLMs such as BERT [92] and RoBERTa [91]. It generates contextualized embeddings of column relations, iteratively refining them through adaptive clustering to uncover latent relational features and using pseudo-label classification to enhance clustering purity. To address real-world challenges such as noisy column names and values, AUTOTUS incorporates a table noise generator that introduces realistic perturbations during training, thereby improving the model's generalization capabilities.

Example 10.7 Revisiting **Example 10.3**, the objective is to identify the top unionable table for the query. Using a context-aware approach, Table (a) emerges as the top candidate in the data lake. Table (a), which represents car models, shares the same context as the query table. Furthermore, the union score between its columns and the query table's columns *Model* and *Country* achieves the maximum value, as defined by Starmie [37].

10.3.3 Indexing Techniques

In this section, we explore various indexing techniques, categorized based on their internal structures, which play a crucial role in accelerating the process of finding related tables during table search. These techniques include tree-based indexing (Sect. 10.3.3.1), hash-based indexing (Sect. 10.3.3.2), graph-based indexing (Sect. 10.3.3.3), and quantization-based indexing (Sect. 10.3.3.4). Each of these methods leverages distinct principles and structures to optimize the search and retrieval process. Furthermore, in Sect. 10.4, we provide an in-depth discussion of the challenges and limitations associated with each of these indexing approaches, highlighting areas for potential improvement and future research.

Early studies on ANN typically focused on raw data points. For instance, LSH-Ensemble [27] employed a hash-based technique using LSH on column values. However, with the advent of state-of-the-art table search methods leveraging machine learning models and PLMs, as well as their demonstrated success, recent ANN techniques have shifted towards vector representations (dense retrieval). Consequently, our discussion will primarily focus on these vector-based ANN approaches.

Example 10.8 The tables in the data lake, illustrated in Fig. 10.2, are commonly represented as dense vectors. These vectors are indexed as data points. As a use case, Starmie [37] encodes each column in the data lake into a vector, resulting in 8 vectors (data points) to be indexed. By considering the context of the table, vectors corresponding to related columns are positioned closer together in the proximity space, reflecting a higher likelihood of similarity. For instance, the vectors representing the columns *HQ*, *Lab Location*, and *Country* would be near each other, with *HQ* and *Country* being even closer due to their shared context.

10.3.3.1 Tree Indexes

Tree-based indexes, originally designed for structured and spatial data, can support table search by enabling efficient partitioning, especially when combined with other indexing techniques to mitigate high-dimensional challenges. The origins of tree-based indexes can be traced back to early data structures like **B-trees** [49] and **R-trees** [50], designed for efficient management of ordered and spatial data, respectively. B-trees provided mechanisms for maintaining balanced trees with logarithmic access times, while R-trees extended this to spatial data by organizing objects using Minimum Bounding Rectangles (MBRs). R-trees proved effective for range and intersection queries, making them vital in geographic information systems. These foundational structures inspired the development of **kd-trees** [51], which partition data using axis-aligned hyperplanes and excel in low-dimensional spaces. However, as data dimensionality increases, these structures face challenges due to the "curse of dimensionality," where their partitioning strategy became less efficient and query performance degraded. This limitation highlighted the need for new techniques tailored to handle the complexities of high-dimensional data.

To address the challenges of high-dimensionality, researchers developed more sophisticated metric-based indexes like the **M-tree** [52] and hybrid models such as the **SR-tree** [53]. The M-tree was specifically designed for similarity search in metric spaces, utilizing dynamic clustering to organize data into nodes that optimize for distance-based queries. These techniques are particularly relevant in table search scenarios where tables are represented as dense vectors in high-dimensional spaces.

This structure allowed pruning of irrelevant nodes during queries, ensuring scalability. SR-trees combined features of R-trees and bounding spheres, enabling efficient nearest-neighbor searches even in high-dimensional settings. These advancements formed the basis for early ANN techniques by enabling sub-linear search times, a critical improvement over brute-force methods. Additionally, **cover trees** [54] extended these principles by organizing data hierarchically using cover sets, achieving logarithmic scaling with the dataset size under certain conditions. Cover trees provided a robust solution for embedding data into hierarchical structures.

In the modern era, the focus has shifted towards integrating probabilistic and learning-based methods with tree-based indexes, as exemplified by **Random Projection Trees** (RPTs) [55] and scalable libraries like **FLANN** [56]. RPTs leverage random projections to reduce dimensionality while maintaining essential structures, enabling efficient ANN search on low-dimensional manifolds embedded within high-dimensional spaces. FLANN further optimized ANN by adaptively selecting the most suitable indexing algorithm for a given dataset and query workload, combining tree-based and clustering techniques. Recent innovations, such as **Joint optimization of TRee-based index (JTR)** [57], have taken these concepts further by jointly optimizing tree-based indexes and query encoders through end-to-end training. By employing unified contrastive learning loss and overlapped clustering, JTR effectively balances efficiency and retrieval quality, demonstrating the evolving adaptability of tree-based indexes in ANN applications. These developments highlight the enduring relevance of tree-based methods for dense retrieval and large-scale search.

The table understanding method C^4 [25] incorporates a bottom-up hierarchical tree-based index, specifically constructing a deep clustering tree to group distinct values from spreadsheet tables based on their statistical co-occurrence. Despite this application, tree indexes are less commonly used in table search compared to other indexing techniques. They are less efficient than hash indexes for point look-ups, struggle with handling complex relationship queries compared to graph indexes, and perform poorly in high-dimensional or approximate searches where quantization indexes are more effective. While tree indexes are well-suited for scalar data and range queries, they lack optimization for specialized data structures and specific query types. Tree indexes can be highly efficient and effective when combined with other indexing structures, leveraging the logarithmic properties of trees to enhance search performance. For instance, LSH with tree indexes can address the limitations of each approach: LSH mitigates the curse of dimensionality inherent in tree structures, while tree indexes reduce the additional memory overhead introduced by hash tables by efficiently partitioning the data.

Example 10.9 Tree-based indexes, such as KD-trees or B-trees, organize vectors into hierarchical structures based on geometric partitions. In this example, the 8 vectors (representing columns such as *HQ*, *Lab Location*, and *Country*) would be recursively divided into smaller regions of the vector space. During a query, the tree navigates through these partitions to find the nearest neighbors efficiently. Columns with similar contexts (e.g., *HQ* and *Country*) would likely be placed in adjacent branches of the tree, making their retrieval faster when querying for related columns.

10.3.3.2 Hash Indexes

The development of ANN algorithms has been instrumental in mitigating the curse of dimensionality inherent in high-dimensional data processing. **LSH** [48] laid the foundation for ANN by partitioning the data space probabilistically, ensuring that similar points are mapped to the same bucket with high probability. This approach enabled sub-linear query times, revolutionizing search in domains like image retrieval and document similarity. Extensions to LSH incorporated **p-stable distributions** [59], tailoring hash functions for Euclidean distance, improving its applicability to continuous metric spaces. Further advancements included a **dynamic collision counting** mechanism [60], which dynamically adjusts hash functions based on collision distributions, offering a better balance between query accuracy and efficiency.

Beyond LSH, novel indexing techniques have emerged to address diverse ANN requirements. **Indexable Distance Estimating Codes (IDEC)** [61] approximate pairwise distances through compact vector representations, ensuring efficient nearest neighbor retrieval while maintaining accuracy. A **query-aware hashing mechanism** [62] adapts the hashing process to the query workload, optimizing precision for query-specific scenarios. Additionally, an LSH scheme leveraging **longest circular co-substrings** [63] enhances ANN performance for sequence-based datasets. These advancements showcase the adaptability of LSH and related techniques to various data types and application needs.

Recent innovations focus on integrating advanced partitioning strategies and machine learning methods to improve ANN search. **Learning-to-hash methods** [64] dynamically optimize hash functions for complex datasets using deep learning. **Virtual Hypersphere Partitioning (VHP)** [65] organizes data into hyperspheres, enabling efficient query processing in high-dimensional spaces. The **SRS index** [66] provides a lightweight structure for solving c-ANN queries, particularly suited to memory-constrained environments. In the **LSB-tree** [67], a framework influenced by B-trees, combining pruning techniques and tight bounds ensures efficient and accurate nearest neighbor and closest pair searches in high-dimensional spaces.

Further refinements on dynamic query-specific optimizations, **DB-LSH** [68] integrates query-based dynamic bucketing into LSH, adapting the indexing structure to query distributions for improved accuracy and efficiency. **PM-LSH** [69] enhances LSH with multi-probe techniques and optimized parameters, achieving scalability for large-scale datasets.

Hashing indexes have been utilized in various table search methods. For example, LSH is employed in systems such as Aurum [17], LSH-Ensemble [27], D^3L [36], Starmie [37], and TUS [38]. Additionally, methods like YADL [34] and DICE [32] leverage MinHash to accelerate the table retrieval process. Although hashing indexes remain in use across some approaches, they introduce complexity, particularly in dense retrieval tasks. Achieving greater effectiveness often requires employing more hashing functions to reduce false positives, which increases both computational complexity and memory usage. Recent advancements, such as **LSH-APG** [86], combine hashing indexes with graph indexes (discussed next in Sect. 10.3.3.3) to address these limitations. However, the application of such techniques in table search methods still requires further experimentation and evaluation.

Example 10.10 LSH uses hash functions to map similar vectors into the same buckets. In this scenario, the 8 column vectors would be hashed into buckets based on their proximity in the vector space. Columns like *HQ*, *Lab Location*, and *Country*, which are close in the space, would likely fall into the same or nearby hash buckets. This enables fast approximate lookups, as queries for one column (e.g. *Country* column in the query table) would immediately retrieve others from the same bucket without scanning the entire dataset.

10.3.3.3 Graph Indexes

Graph-based indexing methods have emerged as powerful solutions for ANN search, leveraging the structure of graphs to efficiently navigate high-dimensional spaces. Navigable Small World (**NSW**) graphs are a foundational approach in the ANN search, characterized by logarithmic scalability and small-world navigation properties [75]. NSW constructs a graph where vertices represent data points connected by edges optimized for efficient traversal using greedy search algorithms. By maintaining a balance of local and long-range connections, NSW ensures effective search performance, even in high-dimensional spaces. Hierarchical Navigable Small World (**HNSW**) graphs [70] represent a prominent example, constructing multi-layered graphs where each layer connects points at different granularity levels. This hierarchical approach allows for efficient pruning of the search space by navigating from coarse-grained to fine-grained layers, achieving both high accuracy and fast query times. **DiskANN** [71] extends this idea by optimizing graph-based search for disk-based storage, enabling billion-scale ANN queries on a single machine while maintaining competitive speed and accuracy.

To further enhance scalability and flexibility, methods like the Satellite System Graph (**SSG**) [72] and the Navigating Spreading-Out Graph (**NSG**) [73] refine graph construction and traversal strategies. SSG employs a divide-and-conquer approach to manage high-dimensional similarity searches, ensuring efficient scalability without requiring pre-indexed queries. NSG improves upon traditional graph structures by reducing redundant connections and employing a spreading-out strategy during graph construction, optimizing memory usage and query speed. These systems demonstrate the adaptability of graph-based indexes for various dataset characteristics and computational constraints, making them integral to modern ANN search systems.

Recent advancements focus on billion-scale data and hybrid architectures. **SPANN** [77] introduces a highly efficient solution for billion-scale ANN search, combining graph-based traversal with optimized hierarchical storage and computation techniques. Additionally, experiments and analyses [74] have shed light on practical considerations for ANN graph structures, highlighting the trade-offs between precision, scalability, and memory usage. These developments, complemented by algorithms like navigable small world graphs [75] and efficient graph-based multi-dimensional search [76], underline the critical role of graph-based indexing in addressing the challenges of high-dimensional data retrieval in large-scale systems.

Few methods for table search currently utilize graph-based indexes, despite their competitive efficiency and effectiveness [93, 94]. Recent advancements in table join search, such as DeepJoin [31], and table union search, such as Starmie [37], have effectively employed HNSW graph index to accelerate the retrieval of column representations. Graph indexes are not only efficient but also versatile, often serving as a powerful ANN indexing technique. For instance, the state-of-the-art LSH-APG approach [86] combines graph-based indexes with hashing techniques to improve the performance and accuracy of ANN retrieval. This integration highlights the adaptability and robustness of graph indexes in modern data search frameworks.

However, the adoption of graph-based indexes is not without its challenges. Key limitations include difficulties in dynamically updating the graph, high consumption of main memory resources, and reduced time efficiency when relying on disk storage. These issues are critical and will be discussed in greater detail in Sect. 10.4. Exploring methods to address these constraints, such as integrating graph indexes with quantization-based techniques or other complementary approaches, holds promise for improving their applicability and scalability.

Example 10.11 Graph indexes, such as HNSW [70], represent vectors as nodes in a graph, with edges connecting nearest neighbors. Here, the 8 vectors from the data lake would form a graph, where connections reflect their proximity. For example, the columns *HQ* and *Country* would have a direct edge due to their close semantic relationship, and *Lab Location* would also connect to these nodes. When querying for a related column, the graph structure allows efficient traversal to nearby neighbors, leveraging both local and global connections.

10.3.3.4 Quantization Indexes

Quantization-based methods have become essential for ANN search, offering efficient encoding of high-dimensional vectors into compact representations. **Product Quantization (PQ)** [82] is a foundational approach that splits the data space into subspaces and quantizes each subspace independently, reducing storage and computation requirements for large-scale datasets. This method is effective for minimizing quantization errors, making it a standard choice for vector compression. Extensions like **Optimized Product Quantization** [80] improve upon PQ by refining subspace assignments and optimizing codebooks, thereby enhancing retrieval accuracy. Other refinements, such as **Additive Quantization** [81], address extreme vector compression by combining multiple quantized vectors to approximate original data points more accurately, further improving search precision while maintaining scalability.

Advanced quantization techniques incorporate domain-specific optimizations to adapt to diverse datasets. **Iterative Quantization** [79] introduces a Procrustean rotation mechanism to minimize quantization errors, providing an efficient solution for binary code generation in large-scale image retrieval systems. **Anisotropic Vector Quantization (AVQ)** [83] focuses on enhancing large-scale inference by adjusting the quantization process based on the anisotropy of vector distributions, achieving better performance in imbalanced datasets. Similarly, **Variance-Aware Quantization** [84] dynamically adapts the quantization process based on the variance of data, ensuring improved similarity search performance. **PQBF** [85] combines Product Quantization with disk-based storage systems and further enhances I/O efficiency through data-aware buffering techniques, making it highly effective for high-throughput environments. Leveraging a B+-tree structure, PQBF minimizes disk access and highlights opportunities for optimizing disk-based quantization in hybrid storage systems and real-time massive data retrieval.

The most recent contribution, **RaBitQ** [78], represents a leap forward in quantization-based ANN search. RaBitQ introduces a theoretically grounded approach to quantizing high-dimensional vectors with tight error bounds, ensuring both retrieval accuracy and computational efficiency. Unlike traditional methods, RaBitQ employs a hierarchical structure for encoding, which enables it to maintain low quantization errors even as dataset scales increase. Additionally, the method incorporates adaptive quantization techniques that dynamically adjust to varying data distributions, making it versatile across different applications.

Unfortunately, the table search methods have not yet incorporated the use of the quantization indexes. However, DeepJoin [31] has referenced its use for encoding data vectors along with the graph index HNSW [70]. A key limitation of quantization in the table search domain is the lack of concrete evaluation for its effectiveness. Moreover, a general drawback of this technique is that all data points must be available during the index construction process.

Example 10.12 Quantization indexes, such as PQ [82], represent vectors in compact forms by partitioning the space into smaller subspaces. In this example, the 8 column vectors are grouped into clusters, with similar vectors (e.g., *HQ*, *Lab Location*, and *Country*) placed in the same or nearby clusters. For a query, such as the column *Country* from the query table, the approach identifies the nearest cluster to the query vector and retrieves approximate neighbors from within that cluster.

10.4 Research Challenges and Future Directions

Firstly, a significant gap exists in *defining key concepts* for table search in data lakes. This involves examining the varying definitions of "relatedness" between tables employed by different methods, identifying commonalities and gaps, and addressing what is fundamentally missing. Specifically, it requires establishing a unified and comprehensive framework for defining "relatedness" in the context of table union, along with robust metrics or methodologies to quantify it effectively for achieving specific objectives. Secondly, while semantic and context-aware search methods like Starmie [37] show promise, there remains a need to capture the complex relationships and similarities between tables in data lakes. Additionally, these methods should be capable of generalizing their learning across multiple and diverse data lakes beyond the ones they are initially trained on. Although several table search benchmarks for data lakes exist, including initiatives like LakeBench [95], few focus on the ultimate objectives of table search, such as data augmentation for downstream machine learning tasks [96]. Existing benchmarks often fall short in providing adequate table diversity to support various domains of downstream tasks and fail to reflect the complex, real-world challenges inherent in data lakes. This gap restricts researchers and practitioners from effectively developing, testing, and comparing new methods for table search and indexing.

ANN indexing methods—tree, hashing, graph, and quantization indexes—face notable challenges both broadly and within the context of table search. Tree-based indexes are limited by the curse of dimensionality, which restricts their scalability in handling higher-dimensional data, particularly with the growing adoption of PLMs, thus significantly reducing the effectiveness of table search. Hashing methods like LSH are efficient but struggle with capturing semantic relationships when applied to raw data and require numerous hash functions for indexing dense embedding vectors, reducing efficiency. Graph-based indexes inherently model relationships and structures when applied to raw data, making them more effective than other methods at capturing semantic meaning. They also provide high precision but demand substantial memory, especially for dense vectors, as all dimensions of the vectors must be stored in main memory [70]. Quantization reduces memory and storage overhead by reducing the size of data point, but it comes at the cost of reduced effectiveness. Therefore, to overcome these limitations, future research could focus on hybrid approaches that combine the strengths of multiple methods, such as LSH-APG [86]. A promising direction is the integration of modified quantization with graph-based techniques to enhance performance. Finally, optimizing these approaches for distributed and large-scale table search systems, such as those handling billions of tables in data lakes, remains a critical area for future research.

10.5 Conclusion

This chapter explores table search and indexing techniques in data lakes, highlighting their role in facilitating data discovery, knowledge integration, and supporting scalable table retrieval. As research has transitioned from traditional keyword-based approaches to semantic-rich methods, including dense vector embeddings and pre-trained language models, this chapter emphasizes the importance of understanding table content and context for tasks like join and union search. Furthermore, the discussion on indexing techniques, such as tree, hash, graph, and quantization-based indexes, reveals their individual strengths and limitations, emphasizing the necessity for hybrid solutions to balance efficiency, scalability, and retrieval accuracy.

The research challenges identified in the chapter focus on the lack of standardized definitions and benchmarks for table search, as well as the need for innovative methods to address tables (and their content) scalability and heterogeneity. The integration of semantic understanding into indexing and search methods, coupled with the development of hybrid and distributed frameworks, is suggested as a promising direction for future advancements.

References

1. Hai R, Koutras C, Quix C, Jarke M (2023) Data lakes: a survey of functions and systems. IEEE Trans Knowl Data Eng 35(12):12571–12590
2. Stonebraker M, Hellerstein JM (2005) What goes around comes around. MIT Press
3. Armbrust M, Ghodsi A, Xin R, Zaharia M (2021) Lakehouse: a new generation of open platforms that unify data warehousing and advanced analytics. In: Conference on innovative data systems research (CIDR), conference on innovative data systems research (CIDR), Online
4. Sawadogo PN, Darmont J (2021) Benchmarking data lakes featuring structured and unstructured data with dlbench. In: Golfarelli M, Wrembel R, Kotsis G, Tjoa AM, Khalil I (eds) Big data analytics and knowledge discovery. Springer International Publishing, Cham, pp 15–26
5. Fan G, Wang J, Li Y, Miller RJ (2023) Table discovery in data lakes: state-of-the-art and future directions. In: Companion of the 2023 international conference on management of data. Association for Computing Machinery, New York, NY, USA, pp 69–75
6. Nargesian F, Zhu E, Miller RJ, Pu KQ, Arocena PC (2019) Data lake management: challenges and opportunities. Proc VLDB Endow 12(12):1986–1989
7. Brickley D, Burgess M, Noy N (2019) Google dataset search: building a search engine for datasets in an open web ecosystem. In: The world wide web conference. Association for Computing Machinery, New York, NY, USA, pp 1365–1375
8. Hristidis V, Papakonstantinou Y (2002) Discover: keyword search in relational databases. In: Proceedings of the 28th international conference on very large data bases. PVLDB, Hong Kong, China, pp 670–681
9. Bhalotia G, Hulgeri A, Nakhe C, Chakrabarti S, Sudarshan S (2002) Keyword searching and browsing in databases using BANKS. In: Agrawal R, Dittrich KR (eds) Proceedings of the 18th international conference on data engineering. IEEE Computer Society, San Jose, CA, USA, pp 431–440
10. Agrawal S, Chaudhuri S, Das G (2002) Dbxplorer: A system for keyword-based search over relational databases. In: Agrawal R, Dittrich KR (eds) Proceedings of the 18th international conference on data engineering. IEEE Computer Society, San Jose, CA, USA, pp 5–16

11. Koutrika G, Simitsis A, Ioannidis YE (2006) Précis: the essence of a query answer. In: Liu L, Reuter A, Whang K, Zhang J (eds) Proceedings of the 22nd international conference on data engineering, ICDE 2006. IEEE Computer Society, Atlanta, GA, USA, pp 69–78

12. Simitsis A, Koutrika G, Ioannidis YE (2007) Generalized précis queries for logical database subset creation. In: Proceedings of the 23rd international conference on data engineering, ICDE 2007. The Marmara Hotel, Istanbul, Turkey, IEEE Computer Society, Istanbul, Turkey, pp 1382–1386

13. Simitsis A, Koutrika G, Ioannidis YE (2008) Précis: from unstructured keywords as queries to structured databases as answers. VLDB J 17(1):117–149

14. Kargar M, An A, Cercone N, Godfrey P, Szlichta J, Yu X (2014) Meanks: meaningful keyword search in relational databases with complex schema. In: Proceedings of the 2014 ACM SIG-MOD international conference on management of data. Association for Computing Machinery, New York, NY, USA, pp 905–908

15. Sarkas N, Paparizos S, Tsaparas P (2010) Structured annotations of web queries. In: Proceedings of the 2010 ACM SIGMOD international conference on management of data. Association for Computing Machinery, New York, NY, USA, pp 771–782

16. Cafarella MJ, Halevy A, Khoussainova N (2009) Data integration for the relational web. PVLDB 2(1):1090–1101

17. Fernandez RC, Abedjan Z, Koko F, Yuan G, Madden S, Stonebraker M (2018) Aurum: a data discovery system. In: 2018 IEEE 34th international conference on data engineering (ICDE), pp 1001–1012

18. Pimplikar R, Sarawagi S (2012) Answering table queries on the web using column keywords. Proc VLDB Endow 5(10):908–919

19. Shigarov A (2023) Table understanding: problem overview. WIREs Data Min Knowl Discov 13(1):e1482

20. Limaye G, Sarawagi S, Chakrabarti S (2010) Annotating and searching web tables using entities, types and relationships. Proc VLDB Endow 3(1–2):1338–1347

21. Venetis P, Halevy A, Madhavan J, Paşca M, Shen W, Wu F, Miao G, Wu C (2011) Recovering semantics of tables on the web. Proc VLDB Endow 4(9):528–538

22. Deng X, Sun H, Lees A, Wu Y, Yu C (2020) Turl: table understanding through representation learning. PVLDB 14(3):307–319

23. Hulsebos M, Hu K, Bakker M, Zgraggen E, Satyanarayan A, Kraska T, Demiralp C, Hidalgo C (2019) Sherlock: a deep learning approach to semantic data type detection. In: Proceedings of the 25th ACM SIGKDD international conference on knowledge discovery and data mining. Association for Computing Machinery, New York, NY, USA, pp 1500–1508

24. Zhang D, Hulsebos M, Suhara Y, Demiralp C, Li J, Tan WC (2020) Sato: contextual semantic type detection in tables. PVLDB 13(12):1835–1848

25. Li K, He Y, Ganjam K (2017) Discovering enterprise concepts using spreadsheet tables. In: Proceedings of the 23rd ACM SIGKDD international conference on knowledge discovery and data mining. Association for Computing Machinery, New York, NY, USA, pp 1873–1882

26. Ota M, Müller H, Freire J, Srivastava D (2020) Data-driven domain discovery for structured datasets. Proc VLDB Endow 13(7):953–967

27. Zhu E, Nargesian F, Pu KQ, Miller RJ (2016) Lsh ensemble Internet-scale domain search. PVLDB 9(12):1185–1196

28. Zhang Y, Ives ZG (2020) Finding related tables in data lakes for interactive data science. In: Proceedings of the 2020 ACM SIGMOD international conference on management of data. Association for Computing Machinery, New York, NY, USA, pp 1951–1966

29. Zhu E, Deng D, Nargesian F, Miller RJ (2019) JOSIE: Overlap set similarity search for finding joinable tables in data lakes. In: Proceedings of the 2019 international conference on management of data. Association for Computing Machinery, New York, NY, USA, pp 847–864

30. Esmailoghli M, Quiané-Ruiz JA, Abedjan Z (2022) MATE: multi-attribute table extraction. PVLDB 15(8):1684–1696

31. Dong Y, Xiao C, Nozawa T, Enomoto M, Oyamada M (2023) DeepJoin: joinable table discovery with pre-trained language models. PVLDB 16(10):2458–2470

32. Rezig EK, Bhandari A, Fariha A, Price B, Vanterpool A, Gadepally V, Stonebraker M (2021) DICE: data discovery by example. Proc VLDB Endow 14(12):2819–2822
33. Esmailoghli M, Quiané-Ruiz JA, Abedjan Z (2021) COCOA: correlation coefficient-aware data augmentation. Konstanz, Germany: OpenProceedings.org, University of Konstanz, University Library
34. Cappuzzo R, Coelho A, Lefebvre F, Papotti P, Varoquaux G (2025) Retrieve, merge, predict: augmenting tables with data lakes. arxiv:abs/2402.06282
35. Chepurko N, Marcus R, Zgraggen E, Fernandez RC, Kraska T, Karger D (2020) ARDA: automatic relational data augmentation for machine learning. Proc VLDB Endow 13(9):1373–1387
36. Bogatu A, Fernandes AAA, Paton NW, Konstantinou N (2020) Dataset discovery in data lakes. In: 2020 IEEE 36th international conference on data engineering (ICDE). IEEE, Dallas, TX, USA, pp 709–720
37. Fan G, Wang J, Li Y, Zhang D, Miller RJ (2023) Semantics-aware dataset discovery from data lakes with contextualized column-based representation learning. PVLDB 16(7):1726–1739
38. Nargesian F, Zhu E, Pu KQ, Miller RJ (2018) Table union search on open data. PVLDB 11(7):813–825
39. Khatiwada A, Fan G, Shraga R, Chen Z, Gatterbauer W, Miller RJ, Riedewald M (2023) SANTOS: relationship-based semantic table union search. Proc ACM Manag Data 1(1):1–25
40. Hu X, Wang S, Qin X, Lei C, Shen Z, Faloutsos C, Katsifodimos A, Karypis G, Wen L, Yu PS (2023) Automatic table union search with tabular representation learning. In: Rogers A, Boyd-Graber J, Okazaki N (eds) Findings of the association for computational linguistics: ACL 2023. Association for Computational Linguistics, Toronto, Canada, pp 3786–3800
41. Lehmberg O, Ritze D, Meusel R, Bizer C (2016) A large public corpus of web tables containing time and context metadata. In: Proceedings of the 25th international conference companion on world wide web, international world wide web conferences steering committee. Republic and Canton of Geneva, CHE, pp 75–76
42. Lehmberg O, Ritze D, Meusel R, Bizer C (2025) Web data commons—web table corpora. https://webdatacommons.org/webtables/
43. Hulsebos M, Demiralp c, Groth P, (2023) Gittables: a large-scale corpus of relational tables. Proc ACM Manag Data 1(1):1–17
44. Hulsebos M, Demiralp c, Groth P (2025) Gittables corpora. https://gittables.github.io/
45. Bhagavatula CS, Noraset T, Downey D (2015) Tabel: Entity linking in web tables. In: The semantic web—ISWC 2015: 14th international semantic web conference. Bethlehem, PA, USA, October 11-15, 2015, Proceedings, Part I, Springer, Berlin, Heidelberg, pp 425–441
46. Bhagavatula CS, Noraset T, Downey D (2025) Wikitables corpora. http://websail-fe.cs.northwestern.edu/TabEL/
47. Government of Canada (2025) Open Data portal. https://open.canada.ca/en
48. Indyk P, Motwani R (1998) Approximate nearest neighbors: towards removing the curse of dimensionality. In: Proceedings of the thirtieth annual ACM symposium on theory of computing. Association for Computing Machinery, New York, NY, USA, pp 604–613
49. Bayer R, McCreight E (1970) Organization and maintenance of large ordered indices. In: Proceedings of the 1970 ACM SIGFIDET (Now SIGMOD) workshop on data description, access and control. Association for Computing Machinery, New York, NY, USA, pp 107–141
50. Guttman A (1984) R-trees: a dynamic index structure for spatial searching. In: Proceedings of the 1984 ACM SIGMOD international conference on management of data. Association for Computing Machinery, New York, NY, USA, pp 47–57
51. Bentley JL (1975) Multidimensional binary search trees used for associative searching. Commun ACM 18(9):509–517
52. Ciaccia P, Patella M, Zezula P (1997) M-tree: an efficient access method for similarity search in metric spaces. In: Proceedings of the 23rd international conference on very large data bases. Morgan Kaufmann Publishers Inc., San Francisco, CA, USA, pp 426–435
53. Katayama N, Satoh S (1997) The sr-tree: an index structure for high-dimensional nearest neighbor queries. In: Proceedings of the 1997 ACM SIGMOD international conference on

management of data. Association for Computing Machinery, New York, NY, USA, pp 369–380

54. Beygelzimer A, Kakade S, Langford J (2006) Cover trees for nearest neighbor. In: Proceedings of the 23rd international conference on machine learning. Association for Computing Machinery, New York, NY, USA, pp 97–104

55. Dasgupta S, Freund Y (2008) Random projection trees and low dimensional manifolds. In: Proceedings of the fortieth annual ACM symposium on theory of computing. Association for Computing Machinery, New York, NY, USA, pp 537–546

56. Muja M, Lowe DG (2014) Scalable nearest neighbor algorithms for high dimensional data. IEEE Trans Pattern Anal Mach Intell 36(11):2227–2240

57. Li H, Ai Q, Zhan J, Mao J, Liu Y, Liu Z, Cao Z (2023) Constructing tree-based index for efficient and effective dense retrieval. In: Proceedings of the 46th international ACM SIGIR conference on research and development in information retrieval. Association for Computing Machinery, New York, NY, USA, pp 131–140

58. Hellerstein JM, Naughton JF, Pfeffer A (1995) Generalized search trees for database systems. In: Proceedings of the 21st international conference on very large data bases. Morgan Kaufmann Publishers Inc., San Francisco, CA, USA, pp 562–573

59. Datar M, Immorlica N, Indyk P, Mirrokni VS (2004) Locality-sensitive hashing scheme based on p-stable distributions. In: Proceedings of the twentieth annual symposium on computational geometry. Association for Computing Machinery, New York, NY, USA, pp 253–262

60. Gan J, Feng J, Fang Q, Ng W (2012) Locality-sensitive hashing scheme based on dynamic collision counting. In: Proceedings of the 2012 ACM SIGMOD international conference on management of data. Association for Computing Machinery, New York, NY, USA, pp 541–552

61. Gong L, Wang H, Ogihara M, Xu J (2020) IDEC: indexable distance estimating codes for approximate nearest neighbor search. Proc VLDB Endow 13(9):1483–1497

62. Huang Q, Feng J, Zhang Y, Fang Q, Ng W (2015) Query-aware locality-sensitive hashing for approximate nearest neighbor search. Proc VLDB Endow 9(1):1–12

63. Lei Y, Huang Q, Kankanhalli M, Tung AKH (2020) Locality-sensitive hashing scheme based on longest circular co-substring. In: Proceedings of the 2020 ACM SIGMOD international conference on management of data. Association for Computing Machinery, New York, NY, USA, pp 2589–2599

64. Li J, Yan X, Zhang J, Xu A, Cheng J, Liu J, Ng KKW, Cheng TC (2018) A general and efficient querying method for learning to hash. In: Proceedings of the 2018 international conference on management of data. Association for Computing Machinery, New York, NY, USA, pp 1333–1347

65. Lu K, Wang H, Wang W, Kudo M (2020) VHP: approximate nearest neighbor search via virtual hypersphere partitioning. Proc VLDB Endow 13(9):1443–1455

66. Sun Y, Wang W, Qin J, Zhang Y, Lin X (2014) SRS: solving c-approximate nearest neighbor queries in high dimensional Euclidean space with a tiny index. Proc VLDB Endow 8(1):1–12

67. Tao Y, Yi K, Sheng C, Kalnis P (2010) Efficient and accurate nearest neighbor and closest pair search in high-dimensional space. ACM Trans Database Syst 35(3):1–46

68. Tian Y, Zhao X, Zhou X (2024) DB-LSH 2.0: Locality-sensitive hashing with query-based dynamic bucketing. IEEE Trans Knowl Data Eng 36(3):1000–1015

69. Zheng B, Zhao X, Weng L, Hung NQV, Liu H, Jensen CS (2020) PM-LSH: a fast and accurate lsh framework for high-dimensional approximate NN search. Proc VLDB Endow 13(5):643–655

70. Malkov YA, Yashunin DA (2020) Efficient and robust approximate nearest neighbor search using hierarchical navigable small world graphs. IEEE Trans Pattern Anal Mach Intell 42(4):824–836

71. Jayaram Subramanya S, Devvrit F, Simhadri HV, Krishnawamy R, Kadekodi R (2019) DiskANN: fast accurate billion-point nearest neighbor search on a single node. In: Wallach H, Larochelle H, Beygelzimer A, d'Alché-Buc F, Fox E, Garnett R (eds) Advances in neural information processing systems. Curran Associates Inc., Vancouver, Canada

72. Fu C, Wang C, Cai D (2022) High dimensional similarity search with satellite system graph: efficiency, scalability, and unindexed query compatibility. IEEE Trans Pattern Anal Mach Intell 44(8):4139–4150

73. Fu C, Xiang C, Wang C, Cai D (2019) Fast approximate nearest neighbor search with the navigating spreading-out graph. Proc VLDB Endow 12(5):461–474

74. Li W, Zhang Y, Sun Y, Wang W, Li M, Zhang W, Lin X (2020) Approximate nearest neighbor search on high dimensional data—experiments, analyses, and improvement. IEEE Trans Knowl Data Eng 32(8):1475–1488

75. Malkov Y, Ponomarenko A, Logvinov A, Krylov V (2014) Approximate nearest neighbor algorithm based on navigable small world graphs. Inf Syst 45:61–68

76. Peng Y, Choi B, Chan TN, Yang J, Xu J (2023) Efficient approximate nearest neighbor search in multi-dimensional databases. Proc ACM Manag Data 1(1):1–27

77. Chen Q, Zhao B, Wang H, Li M, Liu C, Li Z, Yang M, Wang J (2021) SPANN: highly-efficient billion-scale approximate nearest neighbor search. In: Advances in Neural Information Processing Systems 34 (NeurIPS 2021), pp 5199–5212. Curran Associates, Inc

78. Gao J, Long C (2024) RaBitQ: quantizing high-dimensional vectors with a theoretical error bound for approximate nearest neighbor search. Proc ACM Manag Data 2(3):1–27

79. Gong Y, Lazebnik S, Gordo A, Perronnin F (2013) Iterative quantization: a procrustean approach to learning binary codes for large-scale image retrieval. IEEE Trans Pattern Anal Mach Intell 35(12):2916–2929

80. Ge T, He K, Ke Q, Sun J (2013) Optimized product quantization for approximate nearest neighbor search. In: Proceedings of the 2013 IEEE conference on computer vision and pattern recognition. IEEE Computer Society, USA, pp 2946–2953

81. Babenko A, Lempitsky V (2014) Additive quantization for extreme vector compression. In: 2014 IEEE conference on computer vision and pattern recognition. IEEE, Columbus, OH, USA, pp 931–938

82. Jégou H, Douze M, Schmid C (2011) Product quantization for nearest neighbor search. IEEE Trans Pattern Anal Mach Intell 33(1):117–128

83. Guo R, Sun P, Lindgren E, Geng Q, Simcha D, Chern F, Kumar S (2020) Accelerating large-scale inference with anisotropic vector quantization. In: Proceedings of the 37th international conference on machine learning. JMLR.org, Online, pp 1–10

84. Paparrizos J, Edian I, Liu C, Elmore AJ, Franklin MJ (2022) Fast adaptive similarity search through variance-aware quantization. In: 2022 IEEE 38th international conference on data engineering (ICDE). IEEE, Kuala Lumpur, Malaysia, pp 2969–2983

85. Liu Y, Cheng H, Cui J (2017) PQBF: I/O-efficient approximate nearest neighbor search by product quantization. In: Proceedings of the 2017 ACM on conference on information and knowledge management. Association for Computing Machinery, New York, NY, USA, pp 667–676

86. Zhao X, Tian Y, Huang K, Zheng B, Zhou X (2023) Towards efficient index construction and approximate nearest neighbor search in high-dimensional spaces. Proc VLDB Endow 16(8):1979–1991

87. Suchanek FM, Kasneci G, Weikum G (2007) YAGO: a core of semantic knowledge. In: Proceedings of the 16th international conference on world wide web. Association for Computing Machinery, New York, NY, USA, pp 697–706

88. Manning CD, Raghavan P, Schütze H (2008) Introduction to information retrieval. Cambridge University Press

89. Das Sarma A, Fang L, Gupta N, Halevy A, Lee H, Wu F, Xin R, Yu C (2012) Finding related tables. In: Proceedings of the 2012 ACM SIGMOD international conference on management of data. Association for Computing Machinery, New York, NY, USA, pp 817–828

90. Joulin A, Grave E, Bojanowski P, Mikolov T (2017) Bag of tricks for efficient text classification. In: Lapata M, Blunsom P, Koller A (eds) Proceedings of the 15th conference of the european chapter of the association for computational linguistics: volume 2, Short Papers. Association for Computational Linguistics, Valencia, Spain, pp 427–431

91. Liu Y, Ott M, Goyal N, Du J, Joshi M, Chen D, Levy O, Lewis M, Zettlemoyer L, Stoyanov V (2019) ROBERT: a robustly optimized BERT pretraining approach. arxiv:abs/1907.11692
92. Devlin J, Chang MW, Lee K, Toutanova K (2019) BERT: pre-training of deep bidirectional transformers for language understanding. In: Burstein J, Doran C, Solorio T (eds) Proceedings of the 2019 conference of the North American chapter of the association for computational linguistics: human language technologies, Volume 1 (Long and Short Papers). Association for Computational Linguistics, Minneapolis, Minnesota, pp 4171–4186
93. Taha I, Lissandrini M, Simitsis A, Ioannidis Y (2024) A study on efficient indexing for table search in data lakes. In: 2024 IEEE 18th international conference on semantic computing (ICSC). IEEE, Laguna Hills, CA, USA, pp 245–252
94. Taha I, Lissandrini M, Simitsis A, Ioannidis Y (2025) Comparative analysis of indexing techniques for table search in data lakes. Int J Semantic Comput 19(2):173-196; https://doi.org/10.1142/S1793351X25420024
95. Deng Y, Chai C, Cao L, Yuan Q, Chen S, Yu Y, Sun Z, Wang J, Li J, Cao Z, Jin K, Zhang C, Jiang Y, Zhang Y, Wang Y, Yuan Y, Wang G, Tang N (2024) Lakebench: a benchmark for discovering joinable and unionable tables in data lakes. Proc VLDB Endow 17(8):1925–1938
96. Leventidis A, Christensen MP, Lissandrini M, Di Rocco L, Hose K, Miller RJ (2024) A large scale test corpus for semantic table search. In: Proceedings of the 47th international ACM SIGIR conference on research and development in information retrieval. Association for Computing Machinery, New York, NY, USA, pp 1142–1151

Part IV
Analysis

Chapter 11
Adversarial Learning for Fraud Detection

Daniele Lunghi, **Alkis Simitsis**, and **Gianluca Bontempi**

Abstract Over the last decade, machine learning security has become an essential field of study. Researchers developed attacks aimed at compromising data-driven systems' integrity, availability, and privacy and used them to assess the security of machine learning systems. However, most of these attacks have been studied in the domain of image recognition, and their applicability in security applications remains unclear. Despite its economic relevance, the research community has largely ignored credit card fraud detection, also known as fraud detection, and only a few solutions have been developed in this field. This work aims to provide a guide to understanding adversarial attacks in the context of fraud detection. We begin with an overview of the challenges posed by credit card fraud detection and adversarial machine learning, introducing the concept of threat modeling as a framework for constructing a taxonomy of adversarial strategies and declining it to the fraud detection context. Next, we review the main types of attacks in the literature, explore how these techniques can be applied in the context of fraud detection, and examine the solutions proposed to bridge the gap. Finally, we discuss promising research directions, drawing inspiration from the literature on other threat detection systems to discuss possible approaches to better understanding the domain.

Keywords Adversarial machine learning · Fraud detection · Threat modeling · Evasion attack

D. Lunghi (✉)
Université libre de Bruxelles, Athena Research Center, National and Kapodistrian University of Athens, Athens, Greece
e-mail: daniele.lunghi@ulb.be

A. Simitsis
Athena Research Center, Athens, Greece
e-mail: alkis@athenarc.gr

G. Bontempi
Université libre de Bruxelles, Brussels, Belgium
e-mail: gianluca.bontempi@ulb.be

G. Dejaegere et al. (eds.), *Data Engineering for Data Science*,
https://doi.org/10.1007/978-3-032-18765-9_11

11.1 Overview

The impact of machine learning on our lives can hardly be overestimated. From recommender systems [1] to automated stock trading [2], most applications have been revolutionized by the power of deep learning, incorporating it into their systems. Machine learning has been used to make our technology more effective and efficient. As an increasing part of our life has moved online, the need to protect us from malicious agents led to an arms race between defenders and attackers in fields like spam [3], malware [4], intrusion [5], and credit card fraud detection [6, 7]. In all these fields, machine learning has proven to be a handy tool to improve defense quality.

However, the increasing use of machine learning for security led researchers to focus on the vulnerabilities of machine learning algorithms. If the algorithms become vulnerable, the security of existing systems may severely deteriorate. This growing concern has given rise to Adversarial Machine Learning (AML), a field dedicated to studying attacks explicitly targeting data-driven systems [40]. In the last decade, state-of-the-art algorithms have been proven extremely vulnerable to the malicious introduction of noise in the data [9–11], changes in their training set [12, 13], and privacy leakages [14].

The first adversarial machine learning works focused on image recognition [16], showing how images may be modified with changes that, while imperceptible to humans, completely subvert the prediction of machine learning classifiers [16]. Eventually, researchers showed that even changing one pixel could lead to successful attacks [17]. Such attacks are, however, hard to apply to other domains. Challenges such as the presence of sparse feature engineering process [18], syntactic and semantic constraints on the data [19], and the different threat models among applications severely limit the applicability of image-based attacks outside of the image domain. In financial fraud detection, data class imbalance, semantic constraints on the features, and editability constraints severely hinder the performance and usability of state-of-the-art attacks [20]. Later attacks have been adapted to work in domains such as spam detection [21], malware detection [15, 22], and intrusion detection [23, 24]. While this research significantly enhanced our understanding of machine learning security, other security applications have gone largely ignored.

In this work, we focus on *credit card fraud detection* (referred to in the rest of this work also simply as *fraud detection*), a problem that, despite its economic relevance, has received limited attention in adversarial machine learning research. Specifically, we focus on frauds performed in online payment systems, where fraudsters clone or steal a card from a legitimate cardholder and use it to conduct illicit transactions on a set of terminals. Notably, fraud detection heavily relies on machine learning, and understanding whether fraudsters can bypass it through adversarial attacks is paramount to guarantee its security. This is not a trivial task, as this domain presents multiple challenges. First, fraud detection engines often rely on sparse aggregated features derived from transactional data over time, as noted in [18]. These aggregated features pose significant challenges in designing valid attacks [18] and in terms of required knowledge for attackers [25]. Moreover, fraud detection systems often involve humans in the loop, where manual reviews or human interventions play a crucial role in validating flagged transactions. This process can introduce delayed classification feedback [25]. Such delays complicate the attacker's ability to evaluate the success of their actions in real-time. Despite these challenges, exploring adversarial approaches in such security-critical domains is essential, as it can lead to more robust and resilient fraud detection systems capable of withstanding sophisticated attacks.

This chapter is intended to bridge the concepts of adversarial machine learning and credit card fraud detection. Specifically, we aim to provide readers with an understanding of the main principles behind machine learning security and adversarial machine learning and how they apply to the relatively unexplored domain of credit card fraud detection. Furthermore, we showcase how significant open challenges exist when applying adversarial machine learning principles to largely unexplored domains, using card fraud detection as a primary example.

This work is organized as follows. First, we give the background in Sect. 11.2, showing how credit card fraud detection works, how to understand adversarial attacks in such context, and explaining the concept of threat model [26], which is necessary to understand and organize attacks. Section 11.3 describes the primary adversarial attacks in the literature, starting from the most common ones and moving to the adaptations to security domains like malware detection, ending with the analysis of the few existing solution for fraud detection. Section 11.4 critically analyzes the open challenges, discusses which strategies have proved effective in fraud detection or similar domains, and how we can expect research to evolve. Finally, we conclude this work in Sect. 11.5.

11.2 Background

In this section, we review the elementary concepts that form the basis of our study. We first introduce data-driven approaches to fraud detection, which have become the standard methodology in financial applications. We then present the main ideas of adversarial machine learning.

11.2.1 Data-Driven Fraud Detection

Fraud detection has increasingly shifted from static rule-based systems to data-driven methods. By leveraging large volumes of transaction records, these approaches can capture patterns that distinguish legitimate from fraudulent behavior and adapt to evolving attack strategies. At the same time, challenges such as data imbalance, scarcity of fraud examples, and the cost of false positives remain central concerns.

11.2.1.1 System Overview

Credit card fraud can generally be categorized into two main groups: *card-present* and *card-not-present* frauds [27, 28]. In cases of card-present fraud, the criminal typically gains unauthorized access to the physical card through theft and uses it directly at physical points of sale or ATMs. Card-not-present, instead, fraud refers to situations where the fraudulent transaction is executed without using the physical card in online or remote purchases where only the card details are required. Over recent years, as e-commerce and digital transactions have expanded rapidly, card-not-present fraud has emerged as the dominant form of credit card fraud, posing significant challenges for both consumers and financial institutions.[1] Therefore, this study will primarily analyze card-not-present fraud and online payment systems.

At a high level, fraud detection works as follows [7, 28, 29]. Cardholders own one or more cards. They then use a card they own to contact a merchant, select a terminal, and perform a payment. For example, when buying on an e-commerce platform, the merchant is the platform, and it may allocate different virtual terminals for different payment methods. Payments comprise an amount, a timestamp, other transaction-specific features, and the terminal and card identifiers. Cards and terminals are also characterized by other features, such as the country of the terminal or the type of card. Notably, since terminals and cards are provided with an identifier, they can be analyzed through aggregations, where all transactions performed using a terminal or card are combined to build aggregate features, such as the average amount of transactions conducted at that specific terminal in the last seven days.

This system is subject to a relevant threat; fraudsters may clone cards and perform frauds. Complex fraud detection systems are employed and continuously upgraded to fight this phenomenon. These systems are, in turn, composed of multiple layers [7] (see Fig. 11.1). First, transactions are checked at terminal level, where transactions may be blocked due to a wrong PIN, excessive number of attempts, and wrong card status (active or blocked) [6]. At a second layer, expert-driven transaction-blocking rules block transactions that follow known suspicious patterns. Traditionally, this is the only part performed online, with other steps of the fraud detection process performed after the transaction has been accepted [7]. Machine learning classification is instead performed offline and used to block the card *after the transaction has been accepted*, blocking a card only after it has successfully conducted one or multiple frauds. However, with machine learning being increasingly employed in all fields of security [30], it is reasonable to expect it will be used more and more in online fraud detection to not only block the card but also block transactions in real-time, hence reducing the number of successful frauds.

---

[1] See ECB Sixth report on card fraud.

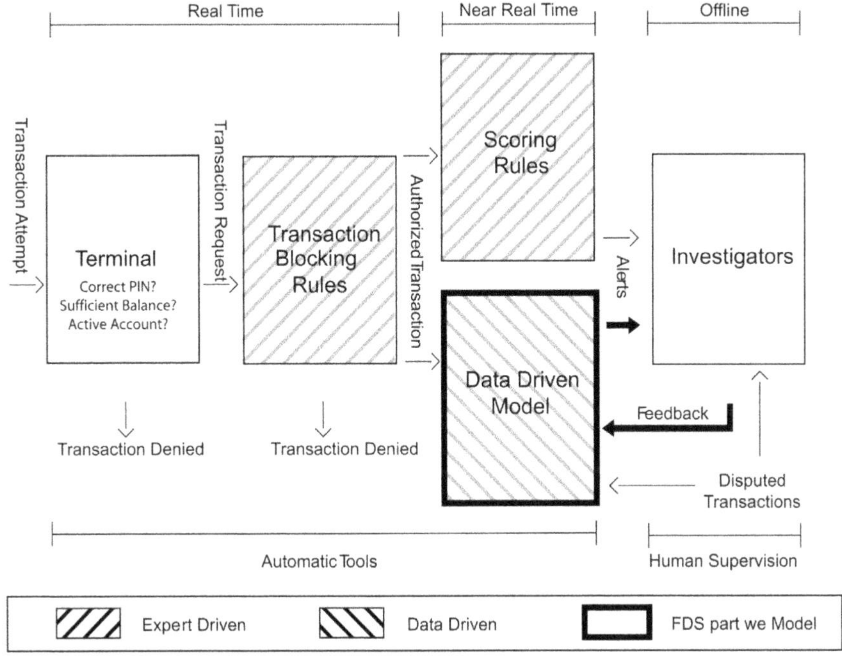

Fig. 11.1 Fraud detection system, reproduced from [28]

11.2.1.2 Machine Learning for Fraud Detection

Machine learning for fraud detection is a complex and widely studied problem, but we can use the following as a baseline formulation. Notably, fraud detection is mainly performed through supervised methods [7], as unsupervised approaches struggle with covering all possible scenarios of legitimate transaction activities [31]. As we have seen, a transaction can be expressed as an amount paid to a merchant by a cardholder at a particular time [28]. Fraud detection systems can then evaluate the transactions using not only its time and amount but also the characteristics of the terminal and card involved. Furthermore, aggregate features over the customer [32] and terminal [33] can also be used. Consequently, credit card fraud detection systems divide each transaction into three groups of features [6, 34]:

- Transactions features, such as the amount and time;
- Card features. This group includes both card characteristics as the card brand and card-based aggregations;
- Terminal features. This group includes terminal characteristics such as the country brand and terminal-based aggregations.

In formal terms, data-driven fraud detection can be expressed as follows. First, we assume to have access to a training set and a test set, called D_{train} and D_{test} respectively. Both datasets are composed of sets of transactions $x_1, x_2 \ldots$, each associated to a label $y(x_i) = \{0, 1\}$, where $y(x_i) = 1$ if the transaction is a fraud, and $y(x_i) = 0$ otherwise. The goal is to train a classification function f capable of modeling the relationship between x and $y(x)$ and that generalizes well on the test set.

Notably, fraud detection classifiers cannot be measured only using accuracy, given the high data imbalances [25, 28]. Instead, fraud detection systems should be evaluated in terms of Area under the Precision-Recall Curve (PRAUC) [35, 36] and precision top K, expressed as:

$$\text{Precision}@K = \frac{TP_k}{k} \tag{11.1}$$

where TP_k indicates the number of transactions correctly classified as frauds (true positives) in the top k transactions (ranked by predicted probability of being a fraud). Furthermore, it should be noted that, once in production, fraud detection systems operate in an ever-changing environment, where regular fraudsters continuously change their behavior [29]. For this reason, designing a model capable of adapting to evolving situations is necessary, and the test set should be created from data that directly succeeds the training period, simulating a real-world deployment where future data is unseen during training.

Finally, the corresponding credit card is blocked once a transaction is labeled as fraudulent. Considering multiple transactions from the same card in the metrics measurement may lead to overestimating the accuracy. To avoid this issue, a proposed solution is grouping by card [33], i.e., considering only the most suspicious transaction for each card.

11.2.2 Adversarial Machine Learning

Adversarial machine learning is firmly based on an accurate understanding of plausible threats. Poor risk modeling may have catastrophic consequences, from a false sense of security to designing defenses that tackle the wrong risks. In computer security, risk modeling is generally obtained through a careful design of the *threat model* [37], defined as using abstract models to find and identify security problems [38]. The central concept behind the threat model is to jointly model the system's functioning and security requirements, focusing on "what could possibly go wrong" [38]. At a high level, the threat model is the basis of all security choices in our lives. For instance, when we decide whether to install a security system in our house, we are balancing the cost, the probability of getting robbed, and the value of the things we have at home. This balance between costs and benefits is called the security-performance tradeoff, and it has significant applications in computer security [39].

An effective way to model adversarial security is through a game-theoretic approach [26], where both attacker and defender are represented as game players. Attackers, in particular, are modeled based on three main characteristics: *goal, knowl-*

edge, and *capability* [40]. Furthermore, machine learning in production is generally employed in complex pipelines. Understanding which parts are vulnerable is crucial to proper threat modeling [41].

11.2.2.1 Attackers Model

Attack Time

We begin defining the life cycle of a standard machine learning application. First, during the data collection phase, the scientists collect and label the data. The second stage is feature engineering, where the data representation is modified better to suit the needs of the machine learning model. Finally, the scientists train a machine learning model on the resulting data, test it, and put it in production, where it may be subject to different interactions with the external world. In some applications, the borders among the various phases are fuzzy. For instance, active and online learning algorithms may loop between the data collection and the training phase [42, 43], and foundational models in deep learning complicate the relationship between feature engineering and model training [44, 45].

Malicious agents can attack the model at different process stages, depending on their goals and capabilities. First, if they can alter the training set of the algorithm, they can modify or insert some points in the training set. Such attacks, called *poisoning* [12, 13], aim to change the model's decision boundaries by targeting its training data [40]. Attackers can also target different parts of the machine-learning pipeline. For instance, they target the learning process when outsourced to untrusted sources [46, 47].

If attackers have no control over the algorithm's training process, they can still attack at inference time to fool the classifier and to find existing errors in its decision boundaries. These so-called *evasion attacks* require weaker assumptions on the attackers' capabilities and are a more realistic threat in those fields where the test set can be safely assumed to be protected [25].

Fraud detection engines rely on intricate pipelines designed to analyze transactions iteratively, enabling them to determine the most effective boundaries between legitimate and fraudulent activities [29]. Despite their sophistication, these pipelines are challenging for fraudsters to manipulate. First, customers can report fraudulent transactions performed with their cards, ensuring that the labels used in training data are generally accurate. Additionally, companies handle the storage and processing of fraud detection data internally, eliminating the possibility of attacks targeting the training phase of the models.

The primary challenge in credit card fraud detection lies in evasion tactics [18, 25]. In particular, attackers may steal or clone credit cards and use them repeatedly to commit fraud during the system's production phase. Finally, fraud detection is often performed online [29]. Online attacks require the attacker to select the best moment to perform the attack, and the decision, when taken, is irrevocable [48]. Choosing the right moment to conduct an attack is a further complication for the attacker [25].

Let us consider a fraud detection engine that aggregates features over time intervals, such as the average number of transactions over the previous week. Old transactions gradually leave the system, and new ones are introduced, potentially altering the detection dynamics. Additionally, time is a critical constraint, as customers may report card compromises, causing the card to be blocked. This creates a complex decision-making problem, where attackers must carefully weigh the timing of their actions, considering both the evolving data and the limited window before detection occurs.

Goal

In a system-centered view, the first question when designing threat models is which property of our system is at risk, i.e., the likely security violations we can face [49]. A helpful acronym is that of CIA: Confidentiality, Integrity, and Availability [50]. Systems must not leak information to unauthorized users (Confidentiality), must not allow unauthorized information modification (Integrity), and must be able to serve genuine users without interruptions (Availability). The same principles can be applied to machine learning systems and are the first step to define most threat models [18, 26, 40, 41, 49]. Specifically, we can group the attackers according to their desired security violations. For security applications, where the target model is often a detector (for malware, intrusions, frauds …), compromising the system's integrity means having some or all malicious points constantly bypass the system. Usually, this means attackers aim at reducing the classifier detection power for a malicious class (*targeted attacks*).

Availability attacks may instead play on false positives, i.e., the fraction of observations wrongly predicted as fraudulent. As an example, if a fraud detection classifier wrongly predicted the majority of transactions as fraudulent, it would become unsustainable, forcing the company to turn off the detector, thus renouncing a significant defense. This type of attack is common among poisoning attacks, where attackers mistrain the classifier to achieve their goal [51]. Instead, it is fairly uncommon in evasion, where the attacker does not influence the target model.

Another common distinction is that between *Untargeted* attacks, which aim to generally reduce the model's accuracy, and *targeted* attacks, which target a particular class. However, this distinction overlaps with the previous one for binary detectors. Targeted attacks target the positive class to let malicious points bypass the detector, reducing its integrity. In contrast, untargeted attacks reduce accuracy, creating false predictions and hindering the system's availability.

To maximize their revenue, attackers must steal some cards and use them to perform transactions until they are blocked. Cards may be blocked for multiple reasons, including detected fraud and the customer noticing suspicious movements in their account [25]. Hence, fraudsters need to maximize the number of successful frauds from the beginning through an exploration-exploitation tradeoff [52]. Notably, however, credit card fraud detection systems undergo periodic retraining [7]. This retraining process introduces a constantly evolving classifier, meaning that even

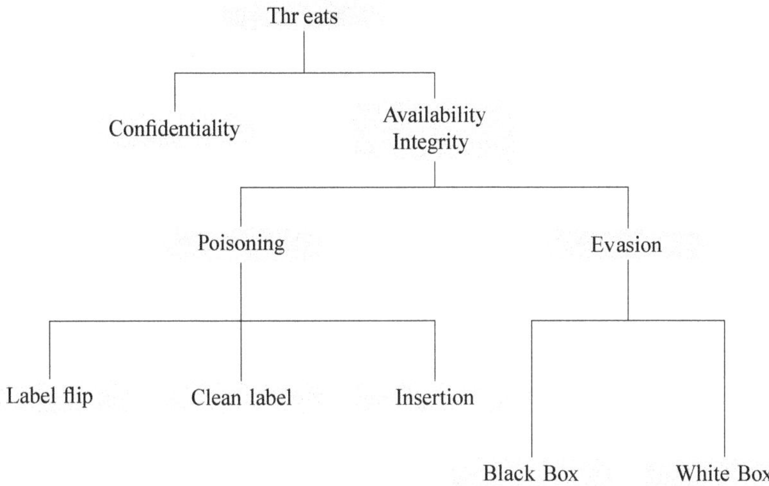

Fig. 11.2 Hierarchical representation of threats posed by adversarial attacks in machine learning. In bold, the threats relevant for fraud detection

if attackers identify successful patterns, those patterns may become obsolete and ineffective after the model has been updated. While these transformations may be too slow to effectively block attacks, given the speed at which they could be performed, fraudsters could find that the attack patterns they identified no longer work once the model has been updated. A hierarchical representation of threats posed by adversarial attacks in machine learning is represented in Fig. 11.2.

Knowledge

A key factor in threat modeling is the attacker's level of knowledge [26, 53], typically categorized as complete (White Box), partial (Grey Box), or none (Black Box) [18, 23]. Attackers may know some or all of the training set. Surrogate model attacks [54], for example, use partial training data to build a copy of the target classifier. Attackers may also know details about the classifier itself. In White Box settings, they know the model and its parameters, enabling gradient-based attacks. In Grey Box, they know only the model type. In Black Box, they have no internal knowledge and rely solely on output labels. Most production systems expose only hard, limiting attacks to hard-label strategies. For instance, in credit card fraud detection, the system simply accepts or rejects a transaction without exposing model confidence. Thus, attacks must rely on hard-label black-box methods [55]. Moreover, attackers cannot query the system directly-they can only observe whether a fraud attempt succeeds. Feature knowledge is also limited. Aggregated behavioral features are usually hidden [25], and while Trojans could reveal them [18], this approach is costly and uncommon. Finally, fraud detection involves delayed feedback: human investigators review suspicious transactions before cards are blocked [29]. Therefore, attackers cannot immediately infer if a transaction was truly successful, further complicating attack design.

Capabilities

Evasion attacks can be grouped by the way they craft their attacks. First, attacks can be generated by modifying existing observations with the goal of making the target classifier misclassify them, or they can create new observations from scratch. In attacks against image recognition systems, for example, attackers can either modify an image to force the system to misclassify it, or they can create a new image ex novo that, despite belonging to one class, is classified as belonging to a different one. In image recognition, the primary threat model concerns update attacks, where attackers modify an image to bypass a classifier while being imperceptible to human observers [10, 16].

Notably, update attacks assume that the updated observation must be somewhat "similar" to the original one. Similarity can be assessed using some distance in the feature space, such as L_0, L_1, L_2, and L_∞, through indistinguishability for human eyes, or through application-specific losses such as preserving important semantical characteristics of the original data [56]. Since most attacks work by solving optimization problems, they introduce the distance measure in the loss. For this purpose, L_0, L_1, L_2, and L_∞ are the most used.

In credit card fraud detection, only a small fraction of transactions are ever reviewed manually by human investigators. Typically, these investigators focus solely on transactions that have already been flagged as fraudulent by the detection system. As a result, successful fraud attempts, i.e., those that evade detection by the automated system, are rarely analyzed by humans. This significantly reduces the need for attackers to craft their actions to be imperceptible to human observers. Relaxing the imperceptibility constraint, sophisticated black-box adversarial attack methods such as Boundary and HopSkipJump have proven less effective than straightforward attacks in that they do not bind to the point where simply random sampling the data space [55] produces valid attacks with fewer queries than them.

Then, they can easily decide the value of the transaction features like time and amount. Instead, they cannot modify fixed card's features as its brand. Finally, in principle, they may find changes in the data spaces that correspond to the desired changes in the aggregated features [18]. In practice, this would mean finding a sequence of transactions that, when the aggregated features are extracted, result in valid attacks in the resulting feature space. However, this would require a complete knowledge of the feature generation process and the previous transactions performed with the used card and terminal. A second significant limitation attackers face is the impossibility of querying the model before performing any attack. Fraudsters can only interact with the system by performing transactions on which they collect rewards.

Attack effectiveness also depends on whether attackers operate in the feature space, i.e., the input to the classifier, or in the data space, when features are unknown

Table 11.1 Notation

Symbol	Description
$x \in \mathbb{R}^D$	Original observation
$r(x) \in \mathbb{R}^D$	Adversarial perturbation applied to x
$x^{adv} = x + r(x)$	Adversarial attack
$y(x)$	True label of x
$f(x)$	Classifier decision for input x
$\bar{(y)}$	Target class
$L(f(x), y')$	Classifier loss over input x and class y'
$d(x_1, x_2)$	Distance metric between two observations
B	Max perturbation allowed
N	Size of the dataset used for evaluation

[57]. In domains like malware detection, where features are hand-crafted, operating directly in the data space may be infeasible [58].

11.2.2.2 Problem Formulation

We describe here the classical problem formulation for evasion attacks, i.e., attacks that do not interfere with the classifier's training set but aim to *evade* it at test time, i.e., when the trained model is deployed in the real world. The main notations used are described in Table 11.1. Adversarial attacks were historically designed in the context of image recognition. Typically, the problem is formulated as follows: First, attackers are given a set of N images $x_1, x_2 \ldots x_N$ in the test set, each belonging to a class $y(x_1), y(x_2) \ldots y(x_N)$. The target algorithm is a trained classifier f which, given an observation x, outputs a class $f(x)$. Attackers apply a perturbation $r(x)$ to x, obtaining the adversarial observation $x^{adv} = x + r(x)$. The attacker's goal depends on whether the attack is targeted or not. Targeted attacks aim to obtain a specific prediction, i.e., $f(x^{adv}) = \bar{y}$, with $\bar{y}$ being the target class. Untargeted attacks instead aim to cause any classification error, i.e., $f(x^{adv}) \neq y(x)$. To keep the attacks invisible, attackers usually want to minimize the magnitude of the perturbation $r(x)$. Finally, attacks can be White Box or Black Box, with White Box attacks knowing the weights and structure of the classifier. In contrast, Black Box attacks require estimating the model using various strategies, typically involving querying the model through attacks.

11.2.2.3 Metrics

We evaluate attackers' success through objective metrics that balance the effectiveness of the attacks and the cost and feasibility of performing them. We report the metrics for the untargeted setting, repeat them for the targeted one, and treat those specific to black box attacks in the end.

Untargeted Metrics

Untargeted attacks aim to obtain any misclassification, i.e., obtain $f(x^{adv}) \neq y(x)$. The most straightforward metric is the Success Rate, the percentage of successful attacks among the attempted ones [8]. Given a set of observations $x_1, x_2 \ldots x_N$, we measure the success rate as:

$$SR = \frac{\text{COUNT}_i^N(f(x_i + r(x_i)) \neq y(x_i))}{N} \tag{11.2}$$

Since most threat models assume attacks cannot modify observations over a certain perturbation budget B, attacks are generally measured through their success rate inside a certain perturbation budget.

$$SR = \frac{\text{COUNT}_i^N(f(x_i + r(x_i)) \neq y(x_i))}{N} \tag{11.3}$$
$$s.t. \quad d(r(x_i)) < B \quad \forall i \in 1, 2, \ldots N$$

with $d(r(x_i)) = d(x_i^{adv}, x_i)$ being the distance metric used. Instead, other works measure the minimal perturbation size to craft successful attacks. The resulting metric, Empirical Robustness [59], is measured as follows

$$ER = \min_{i \in 1..N}(d(r(x_i))) \tag{11.4}$$
$$s.t. \quad f(x_i + r(x_i)) \neq y(x_i)$$

Targeted Metrics

While untargeted attacks aim to obtain any type of classification, i.e., $f(x^{adv}) \neq y(x)$, targeted attacks want the classifier to target a specific pre-determined class $\bar{y}$, hence obtaining $f(x^{adv}) = \bar{y}$ with $\bar{y} \neq y(x)$ being the target class. We need to adapt the aforementioned metrics to this setting. Specifically, we can express the Success Rate as:

$$SR = \frac{\text{COUNT}_i^N(f(x_i + r(x_i)) = \bar{y})}{N} \tag{11.5}$$

where $\bar{y}$ is the target class. Similarly, the constrained Success Rate becomes

$$SR = \frac{\text{COUNT}_i^N \left(f(x_i + r(x_i)) = \bar{y} \right)}{N}$$

$$\text{s.t.} \quad d(r(x_i)) < B \quad \forall i \in \{1, 2, \dots, N\} \tag{11.6}$$

Notably, the budget B is at the level of individual perturbation. This is consistent with the traditional threat model of image recognition, where each image should be as close as possible to the original one.

Finally, we can express the Empirical Robustness as follows

$$ER = \min_{i \in 1..N} \left(d(r(x_i)) \right)$$

$$s.t. \quad f(x_i + r(x_i)) = \bar{y} \tag{11.7}$$

In the case of binary classification, the two sets of metrics usually overlap. In fraud detection, for instance, attackers generate fraudulent transactions and want them to be evaluated as genuine transactions. We say $y(x) = 1$ if a transaction is a fraud, and $y(x) = 0$ otherwise. In this case, if the attacker targets only observations x s.t. $y(x) = 1$, the only misclassification possible happens if $f(x + r(x)) = 0$, with 0 being both the target class $\bar{y}$ and the only possible misclassification. We can then express the constrained success rate as:

$$SR = \frac{\text{COUNT}_i^N \left(f(x_i + r(x_i)) = 0 \right)}{N}$$

$$\text{s.t.} \quad d(r(x_i)) < B \quad \forall i \in \{1, 2, \dots, N\} \tag{11.8}$$

while the adversarial robustness becomes:

$$ER = \min(d(r(x_i)))$$

$$s.t. \quad f(x_i + r(x_i)) = 1. \tag{11.9}$$

Black Box Metrics

Notably, black-box attacks assume no prior knowledge of the target classifier and need to estimate the function f to craft successful frauds. To measure the cost of obtaining the required information, metrics such as the total number of queries (for query-based attacks) [60] and training set properties (for mimicry and transfer learning can be employed.

11.3 State of the Art: Evasion Attacks

In the following, we present a review of the main AML approaches found in the literature. For simplicity, we focus primarily on White Box and Black Box attacks, which are the most commonly used. For both groups, we present only a minority of attacks, focusing on the most important and the ones that best help understand the traditional functioning of adversarial attacks. The techniques presented here can not be directly applied to fraud detection. Yet, they serve to illustrate the fundamental principles behind adversarial machine learning.

White-box attacks assume that the attacker has full knowledge of the target classifier and can therefore utilize information such as the gradient of the loss. Instead, black-box attacks require the attacker to infer insights about the target. This is generally done through two families of approaches. Query-based attacks such as ZOO [11], Boundary [61], and HopSkipJump [60] query the classifier to get insights about local properties of the model, such as its gradient's loss and decision boundaries, and use it to iteratively improve the quality of their attacks. Alternatively, if the attacker can observe a dataset resembling the one used to train the target classifier, various surrogate model techniques are possible.

11.3.1 White-Box

Fast Gradient Sign Method

Fast Gradient Sign Method (FGSM) [10] is one of the first adversarial attacks designed and one of the simplest. Originally designed to minimize the L_∞ distance, it can also be adapted to minimize the L_2 distance. The attack is white-box, requiring the attacker to know the trained classifier f. The main idea of the algorithm is to maximize the loss of the target classifier using gradient ascent. When used in a single iteration, the attack can be defined as:

$$x^{adv} = x + \epsilon \cdot sign\big(\nabla L(f(x), y(x))\big) \tag{11.10}$$

where x is the initial observation, x^{adv} the resulting adversarial observation, $\epsilon > 0$ controls the perturbation magnitude, and $\nabla L(f(x), y)$ is the gradient of the loss. Due to the gradient-ascent procedure, FGSM necessitates a gradient-based algorithm.

Projected Gradient Descent

Projected Gradient Descent (PGD) [62] is an extension of FGSM, where the amount of the perturbation iteratively increases to cause the misclassification. Starting from $x^{adv}(0) = x$, each iteration is computed as:

$$x^{adv}(t+1) = Proj[x^{adv}(t) + \epsilon \cdot sign(\nabla L(f(x^{adv}(t), y(x))))] \qquad (11.11)$$

where the projection function $Proj$ projects adversarial examples into the $e - ball$ of acceptable changes. Overall, PGD is a slower but more effective variant of FGSM.

Carlini and Wagner

One of the most effective attacks in White Box settings is the Carlini and Wagner attack [9]. The main idea of the attack is that attacks can be thought of as a constrained minimization problem, where attackers minimize the perturbation size $r(x)$ under the constraint of changing the classification label:

$$\min_{r(x)}(r(x))$$
$$s.t. \quad f(x + r(x)) \neq f(x) \qquad (11.12)$$

Directly optimizing this problem is difficult, so Carlini and Wagner reformulate it as an *unconstrained minimization* problem by introducing a loss function $J(x^{adv})$ that penalizes inputs not classified as the desired target class $\bar{y}$:

$$\min_{r(x)}(r(x) + c \cdot J(x^{adv})) \qquad (11.13)$$

where $c > 0$ balances perturbation size and classification success.
A common choice for $J(x^{adv})$ is:

$$J(x^{adv}) = \max \left(\max_{i \neq t} Z(x^{adv})_i - Z(x^{adv})_t, -k \right), \qquad (11.14)$$

where $Z(x^{adv})_i$ is the logit (pre-softmax score) for class i, t is the target class, and k is a confidence parameter that enforces a margin between the target class and all other classes. Lower values of k produce smaller perturbations, while higher values increase attack confidence.

This framework is flexible: any differentiable proxy loss J' can be substituted as long as $f(x^{adv}) = \bar{y} \Rightarrow J'(x^{adv}) \leq 0$. The C&W attack can also be adapted to minimize perturbations under different norms, including L_0, L_2, and L_∞.

11.3.2 Black-Box Attacks

ZOO

Zeroth Order Optimization Based Black-box Attack (ZOO) [11] is a black box attack designed against neural networks, which relies on the confidence scores of the model

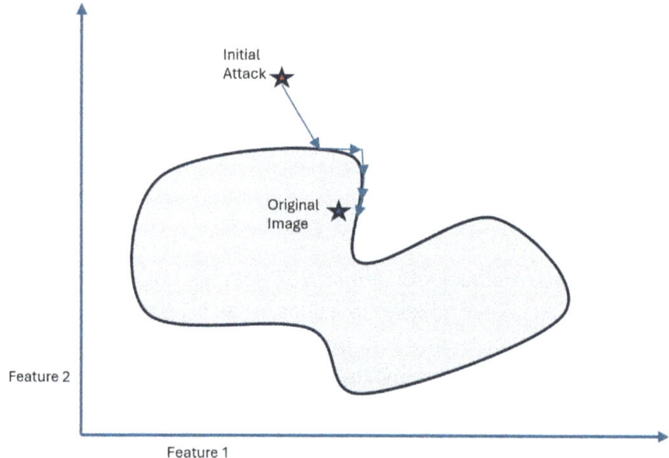

Fig. 11.3 Illustration of the boundary attack

to estimate its behavior. The main idea of the attack is to use the Zeroth Order Optimization to locally estimate the gradient of the classifier from the confidence scores and combine it with the Carlini & Wagner attack to obtain a powerful black box attack. They employ dimension reduction, hierarchical attack, and importance sampling techniques to limit the number of queries.

Boundary

BoundaryAttack [61] is a decision-based adversarial attack effective against any classifier. The main idea of the attack is the following. First, it randomly samples the data space until it finds an observation $x^{adv}(0)$ such that $f(x^{adv}(0)) = \bar{y}$, regardless of far it is from x. Then, it performs a binary search to move the adversarial observation toward the class boundary. Once it finds it, it iteratively creates a random perturbation, finds a new valid attack over the border, and makes a small move toward the original point. An illustration of this process can be seen in Fig. 11.3. The result effectively moves across the class's decision boundary to get close to the original observation but relies on many queries.

HopSkipJump

HopSkipJump [60] is another decision-based black box attack combining some Boundary and ZOO principles. The main idea is the same as Boundary, i.e., move across the decision boundary to get closer to the original observation. The main difference is that instead of performing the orthogonal step of Boundary, it locally estimates the gradient using Zeroth Order Approximation. It then makes a small step toward the gradient and reaches the decision border again.

Surrogate Model

The idea behind surrogate models [23, 54, 63] is straightforward. Since querying the target classifier may not always be feasible or efficient, if the attacker can access a labeled training set resembling the one used to train the target classifier, they can train a surrogate model whose properties should mimic those of the target. Attackers then test their attack on the resulting classifier, also known as a surrogate, and utilize attack transferability [54] to break the target classifier. Attacks have proven to transfer over different datasets and models [64]. Empirically, complex models have proved particularly vulnerable to such attacks [54].

11.3.3 Adversarial Attacks in Security Applications

Adversarial machine learning has recently attracted increasing attention in the security domain. In these settings, the attacker's goal is to manipulate input data in order to evade detection while respecting domain-specific constraints. Below, we review two prominent application areas: intrusion detection and malware detection.

11.3.3.1 Intrusion Detection

Intrusion detection systems typically employ anomaly detection tools based on the host (HIDS) or in the network (NIDS) [65]. Intrusion detection attacks need to consider semantic and syntactic constraints [19], the presence of unmodifiable features [23]. Traditional attacks designed for image recognition fail when applied to intrusion detection because they lead to unfeasible observations [19]. Applying manually selected perturbations on a limited number of features can effectively solve these issues [66]. However, the resulting attacks may still struggle with the need for many queries. If the target classifier employs query detection systems [67], this may lead to an attack failure. Surrogate model attacks seem promising in this scenario, assuming training data can fit the surrogate model [23, 68].

11.3.3.2 Malware Detection

Since researchers started focusing on adversarial attacks research in security applications, reinforcement learning has gained traction as a valid tool for designing attacks. For instance, successful applications of reinforcement learning have been developed for cross-site scripting [69] and malware detection [70, 71]. In malware detection, for instance, attacks can iteratively perform pre-defined changes to the code of malware until they breach the target classifier [70]. Notably, these attacks can also be performed in a fully black-box manner, where only the final, hard-label decision is disclosed [71].

These works, however, are domain-specific and cannot be directly applied to fraud detection. First, these attacks aim to alter malicious code while preserving its functionality. As such, they necessitate a starting point (e.g., the original malware) along with a range of potential transformations to apply to it. In credit card fraud detection, however, attackers can freely choose the values of the features they control without predefined values to modify. More importantly, malware can be copied cost-free to target all devices defended by the breached malware detection tool. Therefore, given its inherent scalability, attackers are highly rewarded for any successful attack. Credit card fraudsters, instead, need to steal cards to perform their frauds and can only perform a limited number before cards are blocked.

11.3.4 Fraud Detection and Adversarial Machine Learning

In this section, we examine the limited number of solutions that have been proposed to address the problem of adversarial attacks in machine learning. Although these solutions are relatively few, we highlight how they differ from previously discussed techniques in both their approach and objectives. Each method attempts to tackle the unique challenges posed by adversarial scenarios. We present only attacks, as, to the best of our knowledge, no specific defense has been proposed.

11.3.4.1 Challenges

In fraud detection, attackers typically have access to a limited number of payment cards and aim to maximize their profits before these cards are detected and blocked. Unlike malware attacks, which persist over time, fraudsters can only operate until they have exhausted their available cards, i.e., the cards they have compromised. Each card is unique, and stolen cards are costly to acquire. Furthermore, if a consistent attack pattern emerges, it is unlikely to remain exploitable for long due to the model's adaptation to tackle concept drift in the data. As a result, fraud attacks are often brief, opportunistic, and highly tailored to exploit vulnerabilities swiftly before countermeasures can be deployed.

Most adversarial attacks face significant challenges when applied to credit card fraud detection. Traditional adversarial approaches are constrained by the need to be imperceptible to humans, thus applying only small perturbations to the original data. In fraud detection, however, any transaction that respects the domain's semantics and constraints can potentially constitute a valid attack. This difference is not merely theoretical: when the number of allowed queries is restricted, a reasonable constraint in fraud detection, the performance of state-of-the-art attacks often degrades to levels worse than random sampling [55].

Finally, attacks in the fraud detection setting must operate under black-box constraints, relying solely on hard labels to infer the decision boundaries of the model. Since the only way to interact with the classifier is by executing real transactions, each query comes at a tangible cost, as it may lead to one stolen card being blocked. Hard-label attacks such as Boundary and HopSkipJump typically require extensive queries for each successful attack, making them impractical in this domain.

11.3.4.2 Attacks

To our knowledge, only a handful of attacks have been designed to detect credit card fraud. Interestingly, these attacks follow three different threat models and approaches, allowing us to compare the strengths and weaknesses and learn the best approaches for future work. First, Mimicry has been employed [72]. The main idea behind the attack is that fraud detection engines often use a user-centric approach, working on a setting of cardholder-based aggregated features. If the attacker can approximate genuine users' behavior, they will likely bypass the detector successfully. In this work, the authors assume that the attacker can obtain a database of genuine users' transactions. Moreover, having infected the cardholder's device, they can access the cardholder's past transactions. The main challenge in this setting is to generate transactions such that the resulting aggregated behavior resembles that in the training set. This is a form of Mimicry, and can be formulated as a constrained maximization problem. The goal is to maximize the reward, expressed as the total money amount of successful frauds, i.e.,

$$\sum_{i}^{N}(1 - f(x_i))AMT_i \tag{11.15}$$

where x_i is the attempted fraud, $y(x_i)$ the label returned by the classifier, and AMT_i the amount of x_i. Transactions are designed to resemble the genuine patterns and respect the domain constraints. A second work proposed by the same authors [18] extended the idea to the situation where the fraud detection engine can be unknown to the attacker, hence operating in a black-box setting. Unlike previous work, the authors use a small dataset to generate a substitute model called Oracle, which they employ to develop their attack in a white-box manner, selecting the amount and the timestamp that maximize the chance of bypassing the classifier. The time is tokenized to simplify the computation, and the success is measured in terms of reward.

Regarding applicability, both attacks effectively tackle some of the constraints that credit card fraud detection poses. First, they consider the existence of aggregated features and opt for the realistic choice of working at the data level. Moreover, they do not require attackers to perform queries for each attack. Furthermore, the substitute model attack can be performed in a fully black-box manner. However, they work under the assumption that the attacker can infer the customers' past transactions and that they either know the target classifier or can access a sufficiently large training set to build a surrogate model. While this assumption makes these attacks highly effective, it also requires attackers to access significant resources and information, which may limit the attacks' scalability.

Finally, recent works have tried to tackle adversarial attacks using reinforcement learning. The resulting algorithm, employing Proximal Policy Optimization [73] as a reinforcement learning algorithm to find the best fraudulent pattern, tackles aspects as the presence of unknown features, but the approach is still in an early stage, and it requires further validation.

11.4 Open Problems and Future Directions

While research has made significant advancements in designing realistic and effective attacks for fraud detection systems, numerous open challenges remain, limiting the state of adversarial robustness in this domain. One of the most critical gaps is the limited attention given to the issue of aggregated features. These features, which aggregate the information of multiple transactions or behaviors over time, are central to real-world fraud detection systems. However, very few studies have addressed how these features may impact the feasibility of adversarial attacks. For instance, the sole existing work on this topic [18] assumes that fraudsters possess complete knowledge of the data aggregation mechanisms and the previous transactions performed with the card. As we have seen, such a scenario is unrealistic in practical settings, where attackers typically operate under significant uncertainty and limited information. In addition to the challenge posed by aggregated features, fraud detection data also suffers from concept drift, a phenomenon where the statistical properties of the data change over time. This dynamic nature of the data can alter the decision boundaries of the target classifiers, leading to a decrease in the long-term effectiveness of static attack strategies, as well as creating new openings for new attacks that may exploit vulnerabilities that appear when the system is updated.

Nevertheless, the threat posed by these attacks remains significant. Both mimicry and substitute model techniques have demonstrated the ability to operate effectively in a Black-Box setting, enabling attackers to generate fraudulent transactions that closely resemble genuine user behavior. By leveraging previously acquired transaction datasets or training a substitute model (the Oracle) attackers can craft frauds capable of evading detection systems. These approaches are constrained by substantial limitations, such as requiring access to transaction data to construct the surrogate models and the complex requirement of injecting banking Trojans into victims' devices. Still, they reveal that attackers can design methods that bypass many of the challenges typically posed by modern fraud detection systems with sufficient planning and sophistication. As such, although not easily scalable, these techniques prove that design challenges in fraud detection attacks can be overcome with careful design.

Another type of attack that should be studied is that of reinforcement learning. These attacks have already proved effective in malware detection, and answer well the problem of designing attacks operating in the data space and not in the features space [58]. Specifically, one could model the adversary as a reinforcement learning (RL) agent operating in an environment with limited knowledge: the attacker does not have access to past transaction data or internal model details, and must make decisions based solely on the current transaction's observable features. These features may include controllable variables (like transaction amount), fixed attributes (like card type), and hidden aggregated patterns that the attacker cannot access. In this setup, the task becomes an RL problem where the agent is rewarded for crafting undetectable fraudulent transactions. Such approaches could be adapted to incorporate categorical features, handle delayed feedback, and test defensive applications.

Finally, research should focus on the robustness of fraud detection. Classical adversarial defenses such as distillation [74], input gradient regularization (penalizing large gradients of the model's output with respect to its input) [75], and randomized smoothing [76] are not appropriate in a setting where attackers may change the values of some features in an unconstrained way. In fraud detection, attackers have a great degree of control over specific transaction attributes, such as the time and amount, while they do not control other features. This violates the assumptions behind many traditional defenses, which typically rely on bounded, norm-constrained perturbations. Instead, adversarial training [62], i.e., training a classifier using adversarially generated data to improve its robustness, may be used. This, however, requires us first to understand the threat, as adversarial training may be partially attack-specific and costly in terms of resources and lost accuracy over non-adversarial data and model generalization capabilities [77], and applying for attacks that do not represent an actual threat may be a waste of resources and hinder the classifier's performance without any advantage.

11.5 Conclusions

This chapter aimed to discuss the problem of adversarial machine learning for credit card fraud detection. Specifically, we introduced in Sect. 11.2 the concepts of fraud detection and adversarial machine learning. First, we discussed how online payment systems work and how fraud detection engines are structured. Then, we introduced the concept of threat modeling, showing how threats to machine learning systems can be understood according to an attacker-centric model based on the attack time and the attackers' goal, knowledge, and capability. We also declined it to the problem of fraud detection. We showed how evasion attacks are the most serious risk for fraud detection systems and how limited card budgets, feature knowledge, and delayed feedback present challenges for fraudsters.

Then, we used in Sect. 11.3 this analysis as a blueprint to discuss existing attacks in the literature. We started with traditional attacks against image recognition systems such as FGSM and ZOO, and we introduced different approaches, such as the Surrogate Model attacks and Mimicry, as well as the applications of adversarial attacks in similar security domains, such as intrusion detection and malware detection. Then, we discussed the promising approaches in the literature, explaining their threat and current limitations.

Specifically, we highlighted the risk posed by Mimicry, Substitute Model attacks, and reinforcement learning. We then briefly discussed how traditional defenses work and showed how understanding adversarial attacks in the context of credit card fraud detection is a crucial step to improving their robustness.

Understanding the gaps in the literature has significant implications for designing the proper defense. Designing a poor threat model or assessing security through attacks not fit for the task may lead to serious challenges when assessing the risk adversarial attacks pose to fraud detection and designing defenses accordingly. First, misunderstanding why existing attacks do not work may lead to a false sense of security and do not allow for developing effective defenses that tackle realistic threats. On the flip side, overestimating the risk may lead to an excessive focus on the system's robustness at the expense of accuracy. Suppose a system employs a defense designed for an attack that, in practice, can not be performed due to the discussed constraints. This will result in an ongoing cost in terms of model accuracy, translating into an excessive number of false alarms or too many regular frauds that are not detected. While this trade-off is familiar to any security application, the lack of an adequate understanding of the threats makes it significantly more dangerous. Defending against attacks is an open problem even in incredibly well-studied applications such as image recognition, where attacks have been studied for years. Improving our understanding of adversarial attacks in fraud detection is therefore paramount if we do not want to risk severe consequences.

References

1. Portugal I, Alencar P, Cowan D (2018) The use of machine learning algorithms in recommender systems: a systematic review. Expert Syst Appl 97:205–227. https://doi.org/10.1016/j.eswa.2017.12.020, https://linkinghub.elsevier.com/retrieve/pii/S0957417417308333
2. Yang H, Liu XY, Zhong S, Walid A (2020) Deep reinforcement learning for automated stock trading: an ensemble strategy. In: Proceedings of the first ACM international conference on AI in Finance, ACM, New York New York, pp 1–8. https://doi.org/10.1145/3383455.3422540
3. Crawford M, Khoshgoftaar TM, Prusa JD, Richter AN, Al Najada H (2015) Survey of review spam detection using machine learning techniques. J Big Data 2(1):23. https://doi.org/10.1186/s40537-015-0029-9, http://www.journalofbigdata.com/content/2/1/23
4. Liu K, Xu S, Xu G, Zhang M, Sun D, Liu H (2020) A Review of Android Malware Detection Approaches Based on Machine Learning. IEEE Access 8:124579–124607. https://doi.org/10.1109/ACCESS.2020.3006143, https://ieeexplore.ieee.org/document/9130686/
5. Demontis A, Melis M, Pintor M, Jagielski M, Biggio B, Oprea A, Nita-Rotaru C, Roli F (2019) Why do adversarial attacks transfer? explaining transferability of evasion and poisoning attacks pp 321–338, https://www.usenix.org/conference/usenixsecurity19/presentation/demontis
6. Van Vlasselaer V, Bravo C, Caelen O, Eliassi-Rad T, Akoglu L, Snoeck M, Baesens B (2015) Apate: a novel approach for automated credit card transaction fraud detection using network-based extensions. Decis Support Syst 75:38–48

7. Dal Pozzolo A, Boracchi G, Caelen O, Alippi C, Bontempi G (2018) Credit card fraud detection: a realistic modeling and a novel learning strategy. IEEE Trans Neural Netw Learn Syst 29(8):3784–3797. https://doi.org/10.1109/TNNLS.2017.2736643, https://ieeexplore.ieee.org/document/8038008/

8. Wang J, Tao Y, Zhang Y, Liu W, Kong Y, Tan S, Yan R, Liu X (2024) Adversarial examples against wifi fingerprint-based localization in the physical world, vol 19, pp 8457–8471. https://doi.org/10.1109/TIFS.2024.3453041, https://ieeexplore.ieee.org/document/10662949/

9. Carlini N, Wagner D (2017) Towards evaluating the robustness of neural networks. In: 2017 IEEE symposium on security and privacy (SP), IEEE, San Jose, CA, USA, pp 39–57. https://doi.org/10.1109/SP.2017.49, http://ieeexplore.ieee.org/document/7958570/

10. Moosavi-Dezfooli SM, Fawzi A, Fawzi O, Frossard P (2017) Universal adversarial perturbations pp 1765–1773

11. Chen PY, Zhang H, Sharma Y, Yi J, Hsieh CJ (2017) ZOO: zeroth order optimization based black-box attacks to deep neural networks without training substitute models. In: Proceedings of the 10th ACM workshop on artificial intelligence and security, ACM, Dallas Texas USA, pp 15–26. https://doi.org/10.1145/3128572.3140448, https://dl.acm.org/doi/10.1145/3128572.3140448

12. Biggio B, Nelson B, Laskov P (2012) Poisoning attacks against support vector machines. In: Proceedings of the 29th international conference on international conference on machine learning, Omni Press, Madison, WI, USA, ICML'12, pp 1467–1474

13. Shafahi A, Huang WR, Najibi M, Suciu O, Studer C, Dumitras T, Goldstein T (2018) Poison frogs! targeted clean-label poisoning attacks on neural networks, pp 6106–6116. https://proceedings.neurips.cc/paper/2018/hash/22722a343513ed45f14905eb07621686-Abstract.html

14. Rigaki M, Garcia S (2024) A survey of privacy attacks in machine learning. ACM Comput Surv 56(4):1–34. https://doi.org/10.1145/3624010

15. Aryal K, Gupta M, Abdelsalam M, Kunwar P, Thuraisingham B (2025) A survey on adversarial attacks for malware analysis. IEEE Access 13:428–459. https://doi.org/10.1109/ACCESS.2024.3519524, https://ieeexplore.ieee.org/document/10806701/

16. Szegedy C, Zaremba W, Sutskever I, Bruna J, Erhan D, Goodfellow I, Fergus R (2013) Intriguing properties of neural networks. arXiv:1312.6199

17. Su J, Vargas DV, Sakurai K (2019) One pixel attack for fooling deep neural networks. IEEE Trans Evol Comput 23(5):828–841. https://doi.org/10.1109/TEVC.2019.2890858, https://ieeexplore.ieee.org/document/8601309/

18. Carminati M, Santini L, Polino M, Zanero S (2020) Evasion attacks against banking fraud detection systems. In: Egele M, Bilge L (eds) 23rd international symposium on research in attacks, intrusions and defenses, RAID 2020, San Sebastian, Spain, October 14–15, 2020, USENIX Association, pp 285–300. https://www.usenix.org/conference/raid2020/presentation/carminati

19. Merzouk MA, Cuppens F, Boulahia-Cuppens N, Yaich R (2022) Investigating the practicality of adversarial evasion attacks on network intrusion detection. Ann Telecommun 77(11-12):763–775. https://doi.org/10.1007/s12243-022-00910-1

20. Cartella F, Anunciação O, Funabiki Y, Yamaguchi D, Akishita T, Elshocht O (2021) Adversarial attacks for tabular data: application to fraud detection and imbalanced data 2808. https://ceur-ws.org/Vol-2808/Paper_4.pdf

21. Imam NH, Vassilakis VG (2019) A survey of attacks against twitter spam detectors in an adversarial environment. Robotics 8(3):50. https://doi.org/10.3390/robotics8030050, https://www.mdpi.com/2218-6581/8/3/50

22. Li D, Li Q, Ye YF, Xu S (2023) Arms race in adversarial malware detection: a survey. ACM Comput Surv 55(1):1–35. https://doi.org/10.1145/3484491

23. Debicha I, Cochez B, Kenaza T, Debatty T, Dricot JM, Mees W (2023) Adv-bot: realistic adversarial botnet attacks against network intrusion detection systems. Comput Sec 129:103176. https://doi.org/10.1016/j.cose.2023.103176, https://linkinghub.elsevier.com/retrieve/pii/S016740482300086X

24. He K, Kim DD, Asghar MR (2023) Adversarial machine learning for network intrusion detection systems: a comprehensive survey. IEEE Commun Surv Tutor 25(1):538–566. https://doi.org/10.1109/COMST.2022.3233793, https://ieeexplore.ieee.org/document/10005100/
25. Lunghi D, Simitsis A, Caelen O, Bontempi G (2023) Adversarial learning in real-world fraud detection: challenges and perspectives. In: Proceedings of the second ACM Data economy workshop, ACM, Seattle WA USA, pp 27–33. https://doi.org/10.1145/3600046.3600051
26. Huang L, Joseph AD, Nelson B, Rubinstein BI, Tygar JD (2011) Adversarial machine learning. In: Proceedings of the 4th ACM workshop on security and artificial intelligence, pp 43–58
27. Kou Y, Lu CT, Sirwongwattana S, Huang YP (2004) Survey of fraud detection techniques. IEEE international conference on networking, sensing and control, 2004. IEEE, Taipei, pp 749–754
28. Le Borgne YA, Siblini W, Lebichot B, Bontempi G (2022) Reproducible machine learning for credit card fraud detection - practical handbook. Université Libre de Bruxelles
29. Carcillo F, Dal Pozzolo A, Le Borgne YA, Caelen O, Mazzer Y, Bontempi G (2018) SCARFF : A scalable framework for streaming credit card fraud detection with spark. Inf Fusion 41:182–194. https://doi.org/10.1016/j.inffus.2017.09.005, https://linkinghub.elsevier.com/retrieve/pii/S1566253517305444
30. Ozkan-Okay M, Akin E, Aslan m, Kosunalp S, Iliev T, Stoyanov I, Beloev I (2024) A Comprehensive survey: evaluating the efficiency of artificial intelligence and machine learning techniques on cyber security solutions. IEEE Access 12:12229–12256. https://doi.org/10.1109/ACCESS.2024.3355547, https://ieeexplore.ieee.org/document/10403908/
31. Carcillo F, Le Borgne YA, Caelen O, Kessaci Y, Oblé F, Bontempi G (2021) Combining unsupervised and supervised learning in credit card fraud detection. Inf Sci 557:317–331. https://doi.org/10.1016/j.ins.2019.05.042, https://linkinghub.elsevier.com/retrieve/pii/S0020025519304451
32. Carminati M, Caron R, Maggi F, Epifani I, Zanero S (2015) BankSealer: a decision support system for online banking fraud analysis and investigation. Comput Sec 53:175–186. https://doi.org/10.1016/j.cose.2015.04.002, https://linkinghub.elsevier.com/retrieve/pii/S0167404815000437
33. Lunghi D, Paldino GM, Caelen O, Bontempi G (2023) An adversary model of fraudsters' behavior to improve oversampling in credit card fraud detection. IEEE Access 11:136666–136679. https://doi.org/10.1109/ACCESS.2023.3337635, https://ieeexplore.ieee.org/document/10332176/
34. Adewumi AO, Akinyelu AA (2017) A survey of machine-learning and nature-inspired based credit card fraud detection techniques. Int J Syst Assur Eng Manag 8:937–953
35. Dal Pozzolo A (2015) Adaptive machine learning for credit card fraud detection
36. Fawcett T (2006) An introduction to ROC analysis. Pattern Recognit Lett 27(8):861–874. https://doi.org/10.1016/j.patrec.2005.10.010, https://linkinghub.elsevier.com/retrieve/pii/S016786550500303X
37. Torr P (2005) Demystifying the threat-modeling process. IEEE Sec Privacy Mag 3(5):66–70. https://doi.org/10.1109/MSP.2005.119, http://ieeexplore.ieee.org/document/1514406/
38. Shostack A (2014) Threat modeling: designing for security. Wiley
39. Wolter K, Reinecke P (2010) Performance and security tradeoff. International school on formal methods for the design of computer, Communication and software systems, pp 135–167
40. Biggio B, Roli F (2018) Wild patterns: Ten years after the rise of adversarial machine learning. Pattern Recognit 84:317–331. https://doi.org/10.1016/j.patcog.2018.07.023, https://linkinghub.elsevier.com/retrieve/pii/S0031320318302565
41. Cinà AE, Grosse K, Demontis A, Vascon S, Zellinger W, Moser BA, Oprea A, Biggio B, Pelillo M, Roli F (2023) Wild patterns reloaded: a survey of machine learning security against training data poisoning. ACM Comput Surv 55(13s):1–39. https://doi.org/10.1145/3585385
42. Hoi SC, Sahoo D, Lu J, Zhao P (2021) Online learning: a comprehensive survey. Neurocomputing 459:249–289. https://doi.org/10.1016/j.neucom.2021.04.112, https://linkinghub.elsevier.com/retrieve/pii/S0925231221006706

43. Settles B (2009) Active learning literature survey. https://api.semanticscholar.org/CorpusID: 324600
44. Devlin J, Chang M, Lee K, Toutanova K (2019) BERT: pre-training of deep bidirectional transformers for language understanding. In: Burstein J, Doran C, Solorio T (eds) Proceedings of the 2019 conference of the North American chapter of the association for computational linguistics: human language technologies, NAACL-HLT 2019, June 2-7, Volume 1 (Long and Short Papers), Association for computational linguistics, Minneapolis, MN, USA, pp 4171– 4186. https://doi.org/10.18653/V1/N19-1423
45. Bommasani R, Hudson DA, Adeli E, Altman R, Arora S, von Arx S, Bernstein MS, Bohg J, Bosselut A, Brunskill E, et al. (2021) On the opportunities and risks of foundation models. arXiv:2108.07258
46. Gu T, Liu K, Dolan-Gavitt B, Garg S (2019) Badnets: evaluating backdooring attacks on deep neural networks. IEEE Access 7:47230–47244
47. Nguyen TA, Tran AT (2021) Wanet - imperceptible warping-based backdoor attack. https:// openreview.net/forum?id=eEn8KTtJOx
48. Mladenovic A, Bose AJ, Berard H, Hamilton WL, Lacoste-Julien S, Vincent P, Gidel G (2022) Online adversarial attacks. In: The Tenth International Conference on Learning Representations, ICLR 2022, Virtual Event, April 25–29, 2022, OpenReview.net. https://openreview.net/ forum?id=bYGSzbCM_i
49. Barreno M, Nelson B, Sears R, Joseph AD, Tygar JD (2006) Can machine learning be secure? In: Proceedings of the 2006 ACM Symposium on Information, computer and communications security, ACM, Taipei Taiwan, pp 16–25. https://doi.org/10.1145/1128817.1128824
50. Samonas S, Coss DL (2014) The cia strikes back: Redefining confidentiality, integrity and availability in security. https://api.semanticscholar.org/CorpusID:215838643
51. Barreno M, Nelson B, Joseph AD, Tygar JD (2010) The security of machine learning. Mach Learn 81(2):121–148. https://doi.org/10.1007/S10994-010-5188-5
52. Macready W, Wolpert D (1998) Bandit problems and the exploration/exploitation tradeoff. IEEE Trans Evol Comput 2(1):2–22. https://doi.org/10.1109/4235.728210
53. Biggio B, Corona I, Maiorca D, Nelson B, Srndic N, Laskov P, Giacinto G, Roli F (2013) Evasion attacks against machine learning at test time. In: Blockeel H, Kersting K, Nijssen S, Zelezný F (eds) Machine Learning and Knowledge Discovery in Databases - European Conference, ECML PKDD 2013, Prague, Czech Republic, September 23–27, 2013, Proceedings, Part III, Springer, Lecture Notes in Computer Science, vol 8190, pp 387–402. https://doi.org/ 10.1007/978-3-642-40994-3_25
54. Demontis A, Melis M, Pintor M, Jagielski M, Biggio B, Oprea A, Nita-Rotaru C, Roli F (2019) Why do adversarial attacks transfer? explaining transferability of evasion and poisoning attacks. In: 28th USENIX security symposium (USENIX security 19), pp 321–338
55. Lunghi D, Simitsis A, Bontempi G (2024) Assessing adversarial attacks in real-world fraud detection. In: 2024 IEEE international conference on web services (ICWS), IEEE, Shenzhen, China, pp 27–34. https://doi.org/10.1109/ICWS62655.2024.00021, https://ieeexplore. ieee.org/document/10707487/
56. Engstrom L, Tsipras D, Schmidt L, Madry A (2017) A rotation and a translation suffice: fooling CNNs with simple transformations. CoRR abs/1712.02779
57. Suciu O, Mărginean R, Kaya Y, Daumé H, Dumitraş T (2018) When does machine learning fail? generalized transferability for evasion and poisoning attacks. In: Proceedings of the 27th USENIX conference on security symposium, USENIX Association, USA, SEC'18, pp 1299– 1316
58. Pierazzi F, Pendlebury F, Cortellazzi J, Cavallaro L (2020) Intriguing properties of adversarial ML attacks in the problem space. In: 2020 IEEE symposium on security and privacy, SP 2020, San Francisco, CA, USA, May 18–21, 2020, IEEE, pp 1332–1349. https://doi.org/10.1109/ SP40000.2020.00073
59. Moosavi-Dezfooli SM, Fawzi A, Frossard P (2016) DeepFool: a simple and accurate method to fool deep neural networks. In: 2016 IEEE conference on computer vision and pattern recognition (CVPR), IEEE, Las Vegas, NV, USA, pp 2574–2582. https://doi.org/10.1109/CVPR. 2016.282, URL http://ieeexplore.ieee.org/document/7780651/

60. Chen J, Jordan MI, Wainwright MJ (2020) HopSkipJumpAttack: a query-efficient decision-based attack. In: 2020 IEEE symposium on security and privacy (SP), IEEE, San Francisco, CA, USA, pp 1277–1294. https://doi.org/10.1109/SP40000.2020.00045, https://ieeexplore. ieee.org/document/9152788/

61. Brendel W, Rauber J, Bethge M (2018) Decision-based adversarial attacks: reliable attacks against black-box machine learning models. https://openreview.net/forum?id=SyZI0GWCZ

62. Madry A, Makelov A, Schmidt L, Tsipras D, Vladu A (2018) Towards deep learning models resistant to adversarial attacks. https://openreview.net/forum?id=rJzIBfZAb

63. Papernot N, McDaniel P, Goodfellow I, Jha S, Celik ZB, Swami A (2017) Practical black-box attacks against machine learning. In: Proceedings of the 2017 ACM on Asia conference on computer and communications security, ACM, Abu Dhabi United Arab Emirates, pp 506–519. https://doi.org/10.1145/3052973.3053009

64. Papernot N, McDaniel PD, Goodfellow IJ (2016) Transferability in machine learning: from phenomena to black-box attacks using adversarial samples. CoRR abs/1605.07277, http://arxiv. org/abs/1605.07277

65. Debicha I, Cochez B, Kenaza T, Debatty T, Dricot JM, Mees W (2023) Review on the Feasibility of Adversarial Evasion Attacks and Defenses for Network Intrusion Detection Systems DOI https://doi.org/10.48550/ARXIV.2303.07003, URL https://arxiv.org/abs/2303.07003, version Number: 1

66. Apruzzese G, Andreolini M, Ferretti L, Marchetti M, Colajanni M (2022) Modeling realistic adversarial attacks against network intrusion detection systems. Digital Threats: Res Pract 3(3):1–19. https://doi.org/10.1145/3469659

67. Zhang C, Costa-Perez X, Patras P (2020) Tiki-Taka: attacking and defending deep learning-based intrusion detection systems. In: Proceedings of the 2020 ACM SIGSAC conference on cloud computing security workshop, ACM, Virtual Event USA, pp 27–39. https://doi.org/10. 1145/3411495.3421359

68. Roshan K, Zafar A (2024) Black-box adversarial transferability: an empirical study in cyber-security perspective. Comput Sec 141:103853. https://doi.org/10.1016/j.cose.2024.103853, https://linkinghub.elsevier.com/retrieve/pii/S0167404824001548

69. Fang Y, Huang C, Xu Y, Li Y (2019) Rlxss: optimizing xss detection model to defend against adversarial attacks based on reinforcement learning. Future Internet 11(8):177

70. Hyrum SA, Anant K, Bobby F, David E, Phil R (2018) Learning to evade static PE machine learning malware models via reinforcement learning. CoRR abs/1801.08917

71. Tsingenopoulos I, Shafiei AM, Desmet L, Preuveneers D, Joosen W (2022) Adaptive malware control: decision-based attacks in the problem space of dynamic analysis. In: Proceedings of the 1st workshop on robust malware analysis, ACM, Nagasaki Japan, pp 3–14, https://doi.org/ 10.1145/3494110.3528243

72. Carminati M, Polino M, Continella A, Lanzi A, Maggi F, Zanero S (2018) Security evaluation of a banking fraud analysis system. ACM Trans Privacy Sec (TOPS) 21(3):1–31

73. Schulman J, Wolski F, Dhariwal P, Radford A, Klimov O (2017) Proximal policy optimization algorithms. CoRR abs/1707.06347, http://arxiv.org/abs/1707.06347, 1707.06347

74. Papernot N, McDaniel P, Wu X, Jha S, Swami A (2016) Distillation as a defense to adversarial perturbations against deep neural networks. In: 2016 IEEE symposium on security and privacy (SP), IEEE, San Jose, CA, pp 582–597, https://doi.org/10.1109/SP.2016.41, http://ieeexplore. ieee.org/document/7546524/

75. Ross A, Doshi-Velez F (2018) Improving the adversarial robustness and interpretability of deep neural networks by regularizing their input gradients. Proc AAAI Conf Artif Intell 32(1). https:// doi.org/10.1609/aaai.v32i1.11504, https://ojs.aaai.org/index.php/AAAI/article/view/11504

76. Cohen J, Rosenfeld E, Kolter JZ (2019) Certified adversarial robustness via randomized smoothing. In: Chaudhuri K, Salakhutdinov R (eds) Proceedings of the 36th international conference on machine learning, ICML 2019, 9–15 June 2019, PMLR, Long Beach, California, USA, Proceedings of machine learning research, vol 97, pp 1310–1320, http://proceedings. mlr.press/v97/cohen19c.html

77. Raghunathan A, Xie SM, Yang F, Duchi JC, Liang P (2019) Adversarial training can hurt generalization. CoRR abs/1906.06032, http://arxiv.org/abs/1906.06032, 1906.06032

Chapter 12
Approximate and Adaptive Methods for Inference Queries

Christos C. Papadopoulos, Alkis Simitsis⬭, and Torben Bach Pedersen⬭

Abstract With the emergence of Artificial Intelligence technologies like machine and deep learning, data analysis is increasingly associated with executing so-called *Inference Queries* over unstructured data such as text, images, and video, which are an important aspect of modern data analytics. Executing such queries involves the evaluation of predicates based on user-defined functions (UDF) containing a so-called *oracle* in the form of packaged deep models or other procedures, such as labeling data through human effort. For example, querying large video datasets for given objects like humans or wheelchairs requires the invocation of a deep object detector on every frame of the video. The main aim of this chapter is to introduce the reader to the problem of unstructured data querying in the realm of big data analytics. Starting from the relevant problem formulations, we will introduce the challenges associated with integrating AI models and other oracle types (like human inference) in big data analytics workloads. We will then proceed with introducing the main research directions in the data management community in order to mitigate the impact of those issues. The chapter includes representative use cases, examples, and a literature review of the state of the art, highlighting current and future research directions in the area.

Keywords Inference queries · Unstructured data · Big data analytics · Approximate query processing · Index-based inference

C. C. Papadopoulos (✉) · T. B. Pedersen
Aalborg University, Aalborg, Denmark
e-mail: ccpa@cs.aau.dk

T. B. Pedersen
e-mail: tbp@cs.aau.dk

A. Simitsis
Athena Research Center, Athens, Greece
e-mail: alkis@athenarc.gr

© The Author(s) 2026

G. Dejaegere et al. (eds.), *Data Engineering for Data Science*,
https://doi.org/10.1007/978-3-032-18765-9_12

12.1 Overview

Inference queries over unstructured data such as text, images, and video are an important aspect of modern data analytics. Executing such queries involves the evaluation of predicates based on user-defined functions (UDFs) containing a so-called *oracle* in the form of packaged deep models or other procedures, such as labeling data through human effort. Optimizing the execution of these queries poses a unique challenge, as oracles are often computationally expensive and their execution over the entire dataset ends up dominating the total query as each data point needs to be processed by the oracle. Listing 12.1 shows a representative example of how an oracle may be integrated into a query for two different domains. In (a), the oracle identifies video frames that contain a car through the corresponding HasCar() UDF. In (b), IsAgg() determines whether the natural language question selected corresponds to an aggregate query.

```
SELECT frameId                          SELECT question
FROM Frames                             FROM WikiSql
WHERE HasCar(frame)=True;               WHERE IsAgg(question)=True;
```

Listing 12.1: (left) Invoking an object detection model through the use of the HasCar UDF (right) Evaluating whether a given question corresponds to an aggregate query through the IsAgg UDF.

There is currently an ongoing effort to create systems and methods that accelerate the execution of such queries over data at scale by minimizing the number of oracle invocations needed in order to execute the query, and research efforts have focused on optimizing model inference costs by using *proxy models*, computationally cheap variants of the main oracle model that can execute orders of magnitude faster and typically require fewer computational resources to do so. Systems supporting proxy models are *query-driven* in the sense that proxy models have to be trained and deployed according to the given query workload and may not be useful for future queries [1–3]. In fact, proxy model performance is often restricted to the task at hand; a proxy model trained for recognizing blue cars will be useless for the task of recognizing red cars. Thus, sharing of proxy models between queries may not be feasible at all. Furthermore, training a proxy model may lead to reduced gains in performance since they often require a high amount of data to be labeled, in order to achieve a satisfactory approximation of the oracle model [3]. Most notably, when injected in the query plan, proxy models still require a full scan of the data with the oracle to derive results, leading to high costs on large datasets [4].

A different line of work follows the observation that, when data becomes large enough, similarity between groups of data points, i.e., being close with respect to some distance measure, can be exploited as highly similar groups of data points only require one *representative point* to be labeled in order to derive their predicate evaluation results [5]. Therefore, the execution of the oracle model (or a proxy variant) over the entire dataset is superfluous. In order to leverage this observation, a class of systems constructs *data similarity indexes* with the aim of grouping together similar data [5–7]. In the case of inference queries, these indexes are then used to generate *representative sets* of data points that accelerate query execution by selectively applying the predicate to specific data points that are chosen as *representatives* for each data group in the index.

12.1.1 Example Use Cases

To better understand the importance of fast and efficient large-scale data analysis when it comes to using oracles such as Artificial Intelligence models, we present two example use cases.

Example 1: Video Analytics. An engineer could easily implement the predicate containing an object detector as shown in the query presented in Fig. 12.1a, by packaging the target model inside a UDF and executing the UDF predicate over the entire dataset during query processing. However, while easy to implement, the above query is highly inefficient; the UDF containing the car detection model will have to be applied to every frame in the dataset and accurate Deep Neural Networks (DNNs) have expensive inference operations [8, 9]. In fact, even using newer, faster deep learning architectures can be prohibitively expensive as the available data grows in size. It should be noted that cheap object detectors deployed over video datasets with sizes in the order of hundreds of thousands to a few million frames can take up to 9 h to scan the whole video [4]. What is more, typical query optimization methods such as predicate push down cannot be directly applied here since the optimizer does not have access to the internals of the UDF [10, 11]. As a result, the evaluation time of the UDF operator can be orders of magnitude slower than the other query operations, effectively bottlenecking the performance of the query at hand.

Example 2: Large Language Models. A data analyst working for a large technology company wants to perform a large-scale data analysis on collected e-mail threads in order to find threads that indicate fraudulent intent. In order to do so, they decide to use a pre-trained Large Language Model (LLM) that will process each e-mail thread and indicate whether it should be flagged or not. Large Language models have made remarkable progress and currently hold the state of the art in numerous natural language applications. However, such models like versions of the llama model [12] typically feature billions of parameters, making them expensive to deploy both in computational resources and latency. In our example, in order to deploy the model, the analyst is required to get access to a shared large-scale compute cluster that can support such intensive workloads, which is often the case among practitioners and researchers. Usage of computational resources is distributed between different analytics teams, and each team cannot exceed its allocated quota. Therefore, in this situation, the required inference workload must fit the computational resource budget even though there may be no latency-related constraints. Given that approximating the Large Language Model for inference requires additional resources to train and deploy a smaller model, there is a challenge associated with using the available budget to efficiently and effectively replicate the results of a full-scale inference by the original model.

12.1.2 Background

The emergence of capable AI models has led to an increased need to integrate them in large-scale analytics workloads. Throughout this chapter, we shall refer to such workloads as "inference queries", database queries that incorporate the usage of a deployed model in order to produce additional insights about the available data. To better understand the basis of our task at hand, we shall first define the notion of an oracle.

Oracle: An oracle is any black-box function, that is, a function that only allows access to its outputs given some input, and the output is always correct with respect to some accuracy metric.

The definition is broad on purpose. AI models such as LLMs and object detectors are by-design oracles once deployed, e.g. through an API or packaged inside a UDF. However, oracles need not be AI or ML models, or even models at all. Large analytics workloads may require annotation through human effort, where the oracle is an actual human being. Even a UDF that does not integrate an AI model can be considered an oracle, given the fact that there are no statistics or other insights about it, which would allow traditional query optimization techniques to be applied. We also note that oracles are mostly treated as binary functions. However, categorical oracle outcomes are not prohibitive and most methods presented in this chapter can be easily generalized to more than two possible outcomes. The key issue concerning all types of oracles is their poor scalability as they involve expensive inference operations, and the definition of the oracle implies that every data point needs to be processed by a given oracle in order to derive the result which can then be utilized in order to execute the rest of the query.

The unique nature of an oracle and the challenges it presents give rise to a new paradigm in query processing, *inference queries*. Now that we have defined oracles, we can move on to define inference queries.

Inference Queries: An inference query is a query that contains an operator integrating the inference results of an oracle and requires the invocation of the oracle to be resolved.

As stated previously, the main challenge of processing inference queries is the dominating cost associated with executing the oracle in order to get inference results for all the associated data. Hence, in this chapter, we will review the efforts made by the data management community to make inference queries more efficient and scalable with regard to increasing data sizes. These efforts revolve around optimizing oracle inference by introducing an *approximation to the full oracle results*. As we will see in Sect. 12.2, these approximations can either provide strong empirical results, directly replacing the operator integrating the oracle without suffering significantly in terms of result quality, or, additionally, provide guarantees on the resulting approximation following the popular Approximate Query Processing (AQP) template [13].

12.1.3 Contributions

In summary, this chapter makes the following contributions:

1. We provide an introduction to the emerging paradigm Inference Query Processing that is becoming more and more prevalent in modern data analytics workloads.
2. We introduce a detailed classification of works associated with efficient Inference Query Processing, and describe in detail the main methods introduced by each associated works reported.
3. We present important next steps and challenges associated with Inference Query Processing.

12.2 The Landscape of Approximate Inference Query Processing

In this chapter we will present the main methods associated with approximating oracles in inference queries. Regardless of the approach, all approximation methods have the same goal: Minimize the number of calls to the oracle, or replace the oracle-based predicate with its approximate counterpart. We note here that we are interested in minimizing the number of calls to the oracle and not the overall time or computational resources required to execute an inference query. The number of oracle calls is a measure that is invariant to the type of oracle, available computational resources and application domain. As a result, it is preferred over other types of performance measures. We shall begin by presenting the two main approaches to approximating oracles, *proxy models* and *index-based methods*.

We will study the inference approximation problem from the static, non-streaming, batch workload perspective. All the works that we will examine assume that all the data is available at query time. What is more, we also assume that the data is static and does not change between queries in the same workload. As a result, we will not present works associated with the streaming case, e.g. the InQuest system [14]. The aim of this chapter is to introduce the reader to different approaches to approximating oracle inference, so we also will not delve into possible approaches combining these two methods, e.g., index-driven proxy model training or using empirical approximate methods as a middleware for methods with statistical guarantees.

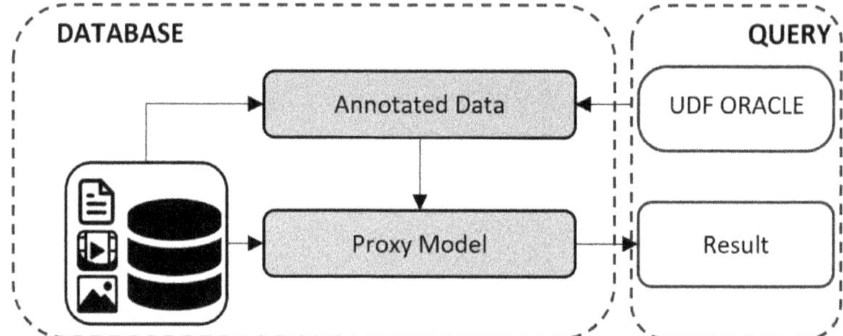

Fig. 12.1 Representation of an Inference Query approximation through a proxy model. The oracle is used to train the proxy model which is then used as a replacement in the query plan

12.2.1 Proxy Models

In order to avoid expensive oracle calls during the execution of analytical queries, multiple systems focus on constructing cheap, so-called *proxy* models that offer orders of magnitude faster inference times at the cost of having lower accuracy than the main oracle model. Since they are computationally cheaper, predicates based on these models can be executed on the whole dataset and provide proxy scores that can be used instead of executing the oracle model [1–3, 15]. A typical workflow for training and deploying a proxy model is the following [3]: First, the oracle model is executed on a subset of the data such that the combined cost of target model execution, proxy model training and inference is within the available computational budget limits. The resulting dataset along with its labels generated by the execution of the target model, is then used as a training set for the proxy model. Once the training is complete, the proxy model is typically executed on the whole dataset, providing proxy scores that can be utilized by specific methods that, for example, provide guarantees on the error produced by the proxy model.

Concretely, the workflow for training and deploying a proxy model is as follows. After the user issues a query with a desired oracle, a sample of m data points is drawn from the database and passed to the oracle. The oracle inference results are then used as labels in order to train the proxy model. Once the model is trained, the next step is to decide the *proxy threshold*, the point after which proxy inference results are regarded as positive, meaning that they satisfy the oracle-based filter. As we will see, this threshold is important for determining the statistical guarantees of the query result or for acting as filters themselves, reducing the overall number of oracle calls during query processing by discarding data not satisfying the proxy threshold. Injecting the proxy model as a filter into the query plan of an inference query aims to reduce the oracle calls down to $m + r_p * n$ where $r_p \in [0, 1]$ is the ratio of data points that pass the filter and n is the number of points in the database. Of course, the injected proxy model can replace the oracle, resulting in just m oracle calls, as seen in Fig. 12.1.

Relation to Active Learning.
The deployment of a proxy model can initially be compared with the field of Active Learning, where a model is trained with specific data points that maximize training efficiency. However, the goals of the two approaches are different. Proxy models do not aim to maximize inference accuracy; instead, a proxy model is deployed in order to maximize performance with respect to query time and computational resources, while maintaining result quality.

12.2.2 Index-Based Methods

Indexing is a key aspect of data management systems and data processing engines in general and has been the subject of extensive study and experimentation by the database systems community [16]. The increasing availability of vector databases in the past few years [17] has made the accumulation, storage and processing of embedding vectors for unstructured data increasingly feasible. In the case of queries involving expensive predicates such as those associated with oracle inference that is of interest in this chapter, the goal of indexing is to organise such data embeddings so that similar data points that are grouped together in the index structure also exhibit similarity in their oracle inference results. To better understand the intuition behind this concept, we shall first define the *Lipschitz continuity* of a function.

Definition 12.1 A function $f(x)$ exhibits *Lipschitz continuity* if for each $x, y \in R^n$ there exists a constant L such that:

$$\|f(x) - f(y)\| \leq L\|x - y\| \tag{12.1}$$

By considering that oracle models can be assumed to be Lipschitz continuous functions, we can utilize Definition 12.1 to derive a powerful result: the distance similarity between two data points dictates their oracle inference results, and points that are closer in the metric space will have identical inference results. In practice, oracles are best viewed as *locally Lipschitz continuous*, which means that the continuity property holds for some neighborhood around each point. When combined with an appropriate index structure that partitions data into groups so that similar data points belong to the same group, this means that we only need *one point per group* to be sent to the oracle in order to derive inference results for the whole dataset. Oracle results for data points in the same group can be easily inferred via a suitable propagation function, e.g. directly sharing the inference result of their representative point annotated by the oracle. This indexing and propagation process is presented in Fig. 12.2.

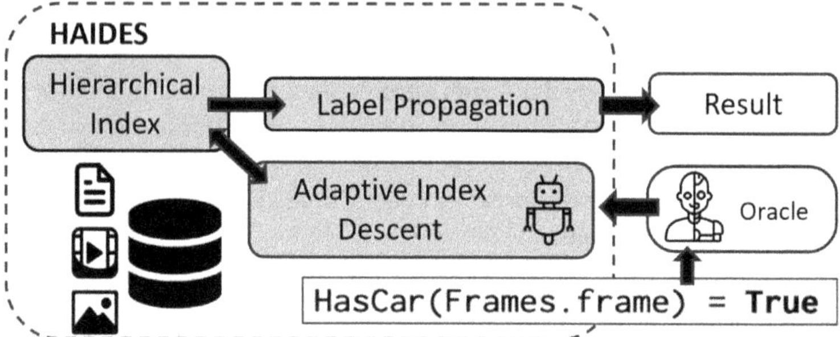

Fig. 12.2 Representation of oracle approximation through index-based processing. The representative points selected through the indexing process are annotated with the oracle, and the final result is derived via inference propagation

Concretely, the problem of interest for index-based methods and systems is *how to define such a partition* and the associated representative points to be processed by the oracle. Since there is no prior knowledge about the oracle or its statistical properties, this partition is oracle-agnostic approach which means that indexes can be constructed before the actual query workload is submitted. Once an index has been constructed, it can be reused across different queries and oracles, since each different oracle is a separate function that still exhibits Lipschitz continuity.

12.2.3 Robust Adaptive Oracle Approximation

So far, we have discussed the two main approaches for approximating oracles. In general, both proxies and indexes require an initial budget of m oracle calls in order to be constructed, either for training the proxy model or for annotating the representative points of an index. A special case for both types of methods is whether the method takes advantage of *intermediate oracle results*, meaning that the approximation is modified *during query processing* by using feedback from the oracle as it is being executed. This leads to *adaptive approximations*, that make better use of the available computational resources and available oracle calls budget.

The main overall use of proxies and indexes, whether adaptive or not, leads to strong empirical accuracy for a fraction of the oracle calls required for full inference. However, it is often the case that users also require guarantees on the approximation result, so that the approximation error is bounded given the limitation on the available computational resources allocated to the query. The desired outcome is often cast as an Approximate Query Processing (AQP) template, where the approximation result is within some error bound with a given probability. The main challenge for guaranteeing the approximation result of an oracle is that the result is often measured by different metrics depending on the problem. For example, given a proxy model, we may wish to obtain results that are above a certain Precision or Recall threshold, which requires a tailor-made method for providing guarantees on each of the two measures [15]. Statistical guarantees may also be imposed on indexes. As we will see, these guarantees specify the number of points in each group that should be annotated by the oracle in order to reach a specific accuracy threshold with high probability.

12.3   State of the Art

With the problem domain and the specific issues that we wish to examine established, we now present representative approaches to handle *Approximate Inference Queries*. A representative taxonomy of the research landscape regarding the problem of approximate inference queries is shown in Fig. 12.3. We further present properties that are of interest, such as the specific target query types and capability of supporting multiple modalities in Table 12.1. We will now proceed with briefly describing proxy models and index-based approaches as well as how empirical methods differ from methods that come with statistical guarantees on the query results.

12.3.1   Empirical Proxy Models

NoScope [1]. NoScope was one of the first systems to target the inference query problem from a data management perspective. It is a video processing system, designed to take advantage of the unique characteristics of the video modality, namely the autocorrelative properties of contiguous frames. The NoScope system injects three

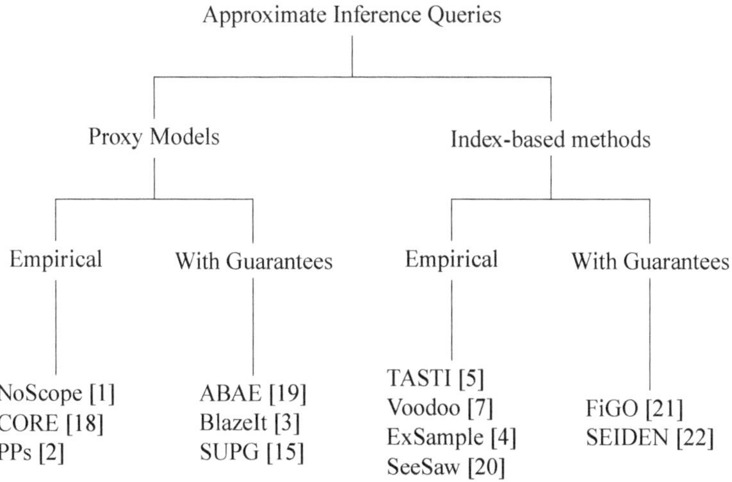

Fig. 12.3 Taxonomy of Approximate methods for Inference Queries

Table 12.1 System properties of examined works regarding support for different query types, modalities and adaptive processing

Solution	Query type support			Modality support		Adaptive processing	
	Selection	Aggregate	Limit	Single	Multiple	Yes	No
NoScope [1]	✓	✓	✓	✓			✓
CORE [18]	✓				✓		✓
PPs [2]	✓				✓		✓
ABAE [19]		✓			✓		✓
Blazelt [3]		✓	✓	✓		✓	
SUPG [15]	✓				✓		✓
TASTI [5]	✓	✓	✓		✓		✓
Voodoo [7]		✓			✓	✓	
ExSample [4]		✓			✓	✓	
SeeSaw [20]	✓			✓		✓	
FiGO [21]	✓			✓	✓	✓	
SEIDEN [22]	✓	✓	✓	✓			✓

types of specialized filters into the query plan of a video processing query with an object detector as the oracle. The first filter is a rule-based time-skip filter that samples from the video at pre-determined frame intervals. Secondly, a trainable difference detector on the contents of two frames is deployed as a proxy model, filtering out frames that do not contain video context changes. Finally, a proxy model is trained on the oracle-annotated set of frames, and filtering thresholds are chosen by a cost-based optimization based on the desired accuracy set by the user. The aim of the NoScope system is to reduce the total oracle cost down to $f_d C_p + f_d f_p C_O$ where $f_d, f_p \in [0, 1]$ are the fractions of the data points that pass through the difference detector and proxy model filters, respectively.

Probabilistic Predicates [2]. Probabilistic predicates (PPs) is another proxy injection system, designed with the objective to train and inject proxy model filters into arbitrary query plans that contain oracles of any type and modality. PPs are trained per oracle UDF (not per query) and feature different possible proxy model architectures, namely SVMs [23], kernel density estimators [24], and neural networks. These different models have both different training and deployment costs and they can lead to variable data reduction rates when deployed as filters in the query plan. The main challenge then is to select the best architecture for each proxy model, taking into account the individual training and testing costs on the oracle-annotated set, as well as the data reduction rates for each architecture that result in a proxy filter with a specified filtering accuracy. This means that the architecture selected is the one resulting in the highest ratio of correctly non-filtered out data from the proxy model.

CORE [18]. CORE is a query optimizer designed to take advantage of the correlation between different UDF oracles in a query plan. Correlation between oracles means that subsequent proxy models can be trained on the data produced by applying the first proxy/oracle filter in the query plan, leading to higher reduction rates while maintaining accuracy. The CORE optimizer uses a branch-and-bound [25] method to determine the best order for the (possibly) correlated oracle UDFs in the query plan, given cost and proxy accuracy bounds set by the user. It is important to note that, although the core optimizer requests a desired accuracy from the user, this accuracy is only validated on a hold-out set annotated by each oracle in the query plan. As a result, *the desired accuracy in CORE does not correspond to a guaranteed accuracy on the approximation result.*

12.3.2 Proxy Models with Statistical Guarantees

So far, we have examined systems that provide strong empirical results when using proxy models during inference query processing. However, none of the systems above provides a guarantee on the result of the proxy models, neither when used as filters nor when used as direct replacements of the oracle. We will next examine the systems that are able to provide guarantees on the approximation of the oracle filter by the proxy model.

BlazeIt [3]. BlazeIt is a video processing system that is able to provide statistical guarantees on results of aggregation queries over object detectors, e.g. when the query needs to compute the average number of cars in a frame. The main contribution of BlazeIt is treating the proxy model not as a filter, as previous systems that we have seen so far, but as a *control variate* [26], formulating the relationship between the oracle and the proxy model as follows. For an oracle m and a proxy model p, we can construct an auxiliary variable for m as $\hat{m} = m + c(p - E[p])$. Provided that the proxy model results are correlated with the oracle, $\hat{m}$ has less variance than the oracle itself. The BlazeIt optimizer takes advantage of this property by sampling from the oracle with an Empirical Bernstein Stopping criterion [27], which adaptively limits the number of samples needed to provide a desired error guarantee when the variance of the sampled variable is small.

SUPG [15]. Selection queries are in general harder to handle than aggregates, due to the different metrics that may be associated with the query, such as queries that need to satisfy a minimum recall versus queries that need to satisfy a minimum precision. The SUPG system defines routines for both precision and recall metrics that are able to provide statistical guarantees regarding the possible deviation from the true metric when sampling during a selection query with an oracle. In practice, the SUPG optimizer aims to set an appropriate threshold for the proxy model filter so that the empirical precision or recall for the data that satisfy the proxy filter is within the desired range.

ABAE [19]. Finally, we take a closer look at aggregate queries over all types of unstructured data. ABAE is an optimizer that calculates confidence intervals for aggregate queries where the aggregate function is based on oracle invocations. ABAE's optimizer uses proxy models' outputs to stratify the data. The optimizer then calculates the proportional oracle sampling allocation for each data group, in order to provide a confidence interval for the estimate of the aggregate quantity that respects the user's error and desired confidence. To compute the intervals, the optimizer uses the bootstrap method [28], resampling from the original sample generated by the main ABAE method. Although bootstrapping requires extensive resampling and can be expensive in the original AQP setting, expensive oracles make it a much cheaper method, as the oracle will only have to process and produce results for the original data sample.

12.3.3 Empirical Index-Based Methods

Voodoo [7]. The problem of appropriately selecting the amount of samples to use in an inference query is of high importance in LIMIT queries, where the query must select a specified amount of data that correspond to positive oracle results. Assuming that the optimizer had perfect knowledge about the results of processing each data point with the oracle, the query would only need to call the oracle the number of times set in the LIMIT clause. However, in the baseline case, the optimizer does not have access to the oracle's results or some prior statistic related to the oracle itself so it conducts random sampling and processing with the oracle until the LIMIT quantity is fulfilled.

The Voodoo optimizer presents a two-stage approach to quickly identify groups of data that contain points that satisfy the oracle. In the first stage, the data is clustered according to some oracle-agnostic objective, like vectorized pixel values in images. The main contribution of the Voodoo optimizer comes in the second stage, where the optimizer adaptively samples from groups of data in the index constructed in the first stage. It does so by using a Multi-Armed Bandit strategy. The Multi-Armed Bandits (MAB) problem is a classic problem that addresses the exploration versus exploitation dilemma in sequential decision making [29]. Given K actions (or groups of data in Voodoo's case) with different unknown oracle reward distributions v_K, Voodoo uses a policy to identify the data group with the best mean reward. A popular policy for similar problems is the Upper Confidence Bound Policy [30] (UCB). In practice, groups of data that have low mean oracle rewards but are not highly explored will still occasionally be selected by the algorithm due to their bounds contributing to a higher UCB bound. Voodoo additionally makes use of the distances between the groups of data in the index, using a version of the UCB bound that incorporates distances between the actions to be considered [31].

ExSample [4]. In a similar vein to Voodoo, ExSample is also an adaptive query optimizer designed to sample parts of a video in order to discover a number of distinct events set by the LIMIT clause. Events are detected through a corresponding oracle which processes groups of frames and confirms whether a unique event has occurred. ExSample takes advantage of the temporal dimension of a video to split a video into chunks without the need for an indexing pre-processing step. It then uses Thomson Sampling [32] to estimate which chunks of the video are more likely to provide new detections, in order to proportionally sample more from these chunks with the given oracle during the rest of the query processing.

TASTI [5]. TASTI is an index-based method for approximating full oracle inference results that proposes constructing indexes based on the semantic similarity of data point embeddings [5]. The indexing approach in TASTI consists of two stages. First, data points are projected into an embedding space either through available pre-trained embeddings from a pre-trained network or through specialized embedding models. The specialized embedding models produce embeddings tailored to a boolean objective function supplied by the user at query time. For example, for an object detector oracle, the user can provide an objective function that takes two video frames as input, and is positive when there are the same number of objects in each frame that additionally are close together in the frame. At the indexing stage, the embeddings produced in the first stage are clustered using the Furthest Point First (FPF) algorithm [33]. The oracle that needs to be approximated during query execution is evaluated on the cluster representatives only, and inference results are propagated from the representatives' labels to the rest of each cluster's data points.

SeeSaw [20]. Finally, SeeSaw is a system that operates on large vector stores that store and index cross-modal embeddings [34]. In this case, the oracle is the actual user, who issues textual queries of the contents of images that are of interest to them. One such example is to find a number of images containing wheelchairs. The main challenges here are associated with the quality of the embeddings themselves, as a task-agnostic pre-training approach can produce embeddings that cannot form semantic groups when indexed, i.e. images with wheelchairs may be mixed with images containing simple chairs. The second challenge is that the ad-hoc textual query may produce a vector that is not similar to the group of objects of interest, completely missing the required objects to successfully answer the query. The SeeSaw system aims to improve queries on pre-trained embedding indexes by introducing a learned query aligner component to the query process. Concretely, SeeSaw modifies the query vector with a few-shot method based on logistic regression. This method takes advantage of the few examples that the user (oracle) has annotated so far, and uses a label propagation process to produce labels for the whole dataset of indexed vectors. This index-based proxy result is then used to learn the logistic regression model that modifies the user-issued query vector on each new query.

12.3.4 Index-Based Methods with Statistical Guarantees

For the final section of this brief literature review, we shall examine systems that are index-based, but are also able to provide statistical guarantees on the oracle approximation result.

FIGO [21]. FIGO is a query optimizer that, given a series of oracle models, selects which part of a video to process with a specific oracle. As these oracles are different with respect to their accuracy and processing times, the FIGO optimizer must select which oracle to use in a specific part of the video, so as not to violate the user's given accuracy constraints. Concretely, the family of models used as the oracles is the collection of EfficientDNet models [8]. The accuracy is set with respect to processing the video with the most accurate, expensive oracle model available. FIGO makes two key contributions. The first is a profiling procedure that selects the cheapest oracle to process a set of contiguous video frames, given an accuracy constraint. This profiling process produces query plans that dictate which oracle from the available oracle models will process a specific part of the video, while adhering to accuracy and time constraints. The authors in FIGO prove that this process will respect the accuracy constraints with high probability. At the second stage, the optimizer applies the profiling process by starting with the whole video as a single chunk. It then repeats the profiling process if the profiling process in two chunks resulting from splitting the initial chunk result in cheaper models needed to fulfill the accuracy constraint.

SEIDEN [22]. As we have seen, video can be cheaper to index than general unstructured data vectors, due to the temporal continuity property of the frames. Moreover, the video format naturally defines representative points in the form of anchor-frames, frames that provide reference points for video compression. Each other frame in the video is encoded as the difference to the previous and/or next frame. Making these observations, the SEIDEN system constructs video indexes with minimal indexing overhead, in order to provide AQP systems with fast but efficient index-generated proxy scores. During query processing, SEIDEN first selects all anchor frames as representative points for the index. Given an oracle and a sampling budget, SEIDEN annotates a subset of these anchor frames. It then splits the video in chunks along the temporal dimension, with each annotated anchor frame acting as the representative point for the chunk. As query processing continues, SEIDEN uses a MAB sampling method in order to sample more from video chunks that contain frames that satisfy the oracle. Once the oracle sampling budget is exhausted, SEIDEN generated proxy scores for the whole frame dataset, by propagating scores from sampled frames to their temporally close frames. These proxy scores are then used as input to an appropriate AQP system such as the BlazeIt system that we have examined.

12.4   Open Problems and Future Directions

The challenge of efficient processing of Inference Queries is a relatively new research domain within the data management community. Although efficient training of AI models as well as efficient approximate processing of large volumes of data have both been extensively studied in their respective scientific communities, their intersection that constitutes the Inference Query paradigm is still a fruitful area of research. In this section, we will outline the most important open problems regarding Inference Queries as well as provide possible promising research directions.

12.4.1   Streaming Inference Queries

In this chapter, we have examined systems and methods that assume full (batch) access to the dataset, as well as the fact that neither the dataset nor the query objective changes during query processing. Thus, it is interesting to consider changes in the inference process, when arriving data cause changes in the underlying data distribution [35]. The first issue, regarding changes in data distribution, is that the approximation methods that are being deployed during the query will stop having high correlation with the oracle. Specifically, proxy models will need to be retrained, and the optimizer should choose when to retrain a proxy model, in order to inject the new model into the query plan. For indexes, new index structures that can adapt to changing distributions of the streaming data should be considered, otherwise the groups generated by the index structure may become useless in the case of new data being inserted into the index. Schemes for potential re-indexing are also of interest, although these should be carefully designed so as not to introduce high computational overhead.

12.4.2   Operators Involving Multiple Oracles

So far in this chapter, we considered oracles as single boolean operators. However, many queries require other oracle-based operators such as, for example, running a join on the results of two oracles. A seemingly simple solution to this problem would be to consider the operator as a single oracle, and simply sample from the results of this operator. However, approximating these composite oracles can prove to be a difficult task. Firstly, each oracle may have a different corresponding dataset. Even worse, these datasets may not even belong to the same domain. As a result, it is difficult to construct a single proxy model, or to index the underlying data vectors since their embedding spaces are different.

A simpler solution would be to consider each oracle separately, constructing proxy models or indexes separately. Even this seemingly simple solution, however, requires carefully considering how the proxy scores that are generated by the approximation methods will be used. The AQP interfaces that we have seen so far will have to be extended in order to support multi-oracle operators.

12.4.3 Non-uniform Oracle Cost

Our problem formulation involves the assumption that the oracles that we consider all have uniform processing costs, i.e., all data points have the same cost when being evaluated by the oracle. While this is a plausible assumption, e.g., when considering frames in video and object detectors as the oracle, other domains and tasks have inherently different oracle costs depending on the data points used as the input. For example, Large Language Models that process text prompts fluctuate in terms of latency depending on the prompt. Likewise, human annotators may also take different times to annotate an image. The machine learning community has made attempts to develop cost-aware methods for training models with few data points [36]. It is therefore important to develop methods for approximating inference queries under variable oracle cost. One promising research direction here is to adapt cost-aware active learning methods to our performance-aware database setting. Another fruitful direction is to create adaptive sampling methods for indexes based on cost-aware bandit policies [37].

12.4.4 New Data and Application Domains

As a final, promising future research direction, we point out the opportunities that arise when considering different application domains. As seen in this chapter, a lot of active research efforts have gone into applications considering the video domain, both for index-based and proxy model methods. However, the Inference Query paradigm is not specific to video, and other applications and data domains can greatly benefit from progress made thus far.

As a representative example, we consider the case of the time-series domain. Time-series data features huge volumes and unique applications like anomaly detection, which incorporates the use of expensive oracles based on deep learning models [38–40]. As a result, there is a growing need for efficient methods in order to be able to effectively and efficiently query huge amounts of time-series data, and the use of these deep learning oracle models presents a great opportunity to further extend the methods and systems associated with Inference Queries.

12.5 Conclusions

In this chapter, we examined the promising research domain of Inference Queries, which constitute queries over large amounts of unstructured data that integrate a black-box oracle in the query processing workload. Oracles can range from deep neural networks, such as object detectors and LLMs, to human annotators or black-box UDF procedures. In this context, oracles are viewed as expensive operators that cannot be handled by a traditional query optimizer due to the lack of statistics about the oracle at query time. Therefore, research efforts have focused on approximating these expensive oracle operators within the query plan.

We provided insight on the two main categories of oracle approximations, namely proxy models and index-based inference methods that can provide both strong empirical results or results with statistical guarantees, following the AQP paradigm. In the last part of this chapter, we shared promising research directions based on our analysis of the most prominent works in the area of inference queries, which include possible extensions of the inference query paradigm to new application domains, data modalities and query paradigms.

References

1. Kang D, Emmons J, Abuzaid F, Bailis P, Zaharia M (2017) Noscope: optimizing deep cnn-based queries over video streams at scale. Proc VLDB Endow 10(11):1586–1597
2. Lu Y, Chowdhery A, Kandula S, Chaudhuri S (2018) Accelerating machine learning inference with probabilistic predicates. In: Gautam Das PB Christopher Jermaine (ed) Proceedings of the 2018 international conference on management of data, SIGMOD conference 2018, Houston, TX, USA, 10–15 June 2018. ACM, Houston, TX, USA, pp 1493–1508
3. Kang D, Bailis P, Zaharia M (2019) Blazeit: optimizing declarative aggregation and limit queries for neural network-based video analytics. Proc VLDB Endow 13(4):533–546
4. Moll O, Bastani F, Madden S, Stonebraker M, Gadepally V, Kraska T (2022) Exsample: efficient searches on video repositories through adaptive sampling. 2022 IEEE 38th international conference on data engineering (ICDE). IEEE, Kuala Lumpur, Malaysia, pp 2956–2968
5. Kang D, Guibas J, Bailis PD, Hashimoto T, Zaharia M (2022) Tasti: semantic indexes for machine learning-based queries over unstructured data. In: Proceedings of the 2022 international conference on management of data, association for computing machinery New York, Philadelphia, PA, USA, pp 1934–1947
6. Anderson MR, Cafarella MJ (2016) Input selection for fast feature engineering. 32nd IEEE international conference on data engineering, ICDE 2016, Helsinki, Finland, 16–20 May 2016. IEEE Computer Society, Helsinki, Finland, pp 577–588
7. He W, Anderson MR, Strome M, Cafarella MJ (2020) A method for optimizing opaque filter queries. In: Maier D, Pottinger R, Doan A, Tan W, Alawini A, Ngo HQ (eds) Proceedings of the 2020 international conference on management of data, SIGMOD conference 2020, online conference [Portland, OR, USA], 14–19 June 2020. ACM, Portland, OR, USA, pp 1257–1272
8. Tan M, Pang R, Le QV (2020) Efficientdet: scalable and efficient object detection. In: 2020 IEEE/CVF conference on computer vision and pattern recognition, CVPR 2020, Seattle, WA, USA, 13–19 June 2020. Computer Vision Foundation/IEEE, pp 10778–10787
9. Wan Z, Wang X, Liu C, Alam S, Zheng Y, Liu J, Qu Z, Yan S, Zhu Y, Zhang Q, Chowdhury M, Zhang M (2024) Efficient large language models: a survey. Trans Mach Learn Res

10. Foufoulas Y, Simitsis A (2023) Efficient execution of user-defined functions in SQL queries. Proc VLDB Endow 16(12):3874–3877
11. Foufoulas Y, Simitsis A (2023) User-defined functions in modern data engines. 39th IEEE international conference on data engineering, ICDE 2023, Anaheim, CA, USA, 3–7 Apr 2023. IEEE, Anaheim, California, The United States, pp 3593–3598
12. Touvron H, Lavril T, Izacard G, Martinet X, Lachaux M, Lacroix T, Rozière B, Goyal N, Hambro E, Azhar F, Rodriguez A, Joulin A, Grave E, Lample G (2023) Llama: open and efficient foundation language models. arxiv:abs/2302.13971:1–27
13. Garofalakis MN, Gibbon PB (2001) Approximate query processing: taming the terabytes. In: Apers PMG, Atzeni P, Ceri S, Paraboschi S, Ramamohanarao K, Snodgrass RT (eds) Proceedings of the 27th international conference on very large data bases. Morgan Kaufmann Publishers Inc, San Francisco, CA, USA, p 725
14. Russo M, Hashimoto T, Kang D, Sun Y, Zaharia M (2023) Accelerating aggregation queries on unstructured streams of data. Proc VLDB Endow 16(11):2897–2910
15. Kang D, Gan E, Bailis P, Hashimoto T, Zaharia M (2020) Approximate selection with guarantees using proxies. Proc VLDB Endow 13(11):1990–2003
16. Ullman JD (1982) Principles of database systems, 2nd edn. Computer Science Press
17. Pan JJ, Wang J, Li G (2024) Survey of vector database management systems. VLDB J 33(5):1591–1615
18. Yang Z, Wang Z, Huang Y, Lu Y, Li C, Wang XS (2022) Optimizing machine learning inference queries with correlative proxy models. Proc VLDB Endow 15(10):2032–2044
19. Kang D, Guibas J, Bailis P, Hashimoto T, Sun Y, Zaharia M (2021) Accelerating approximate aggregation queries with expensive predicates. Proc VLDB Endow 14(11):2341–2354
20. Moll OR, Favela M, Madden S, Gadepally V, Cafarella MJ (2023) Seesaw: interactive ad-hoc search over image databases. Proc ACM Manag Data 1(4):260:1–260:26
21. Cao J, Ramyad Hadidi KS, Arulraj J, Kim H (2022) Figo: fine-grained query optimization in video analytics. In: SIGMOD '22: international conference on management of data, Philadelphia, PA, USA, 12–17 June 2022. ACM, Philadelphia, PA, USA, pp 559–572
22. Bang J, Kakkar GT, Chunduri P, Mitra S, Arulraj J (2023) SEIDEN: revisiting query processing in video database systems. Proc VLDB Endow 16(9):2289–2301
23. Suthaharan S, Suthaharan S (2016) Machine learning models and algorithms for big data classification thinking with examples for effective learning. Supp Vector Mach 1:207–235
24. Davis RA, Lii KS, Politis DN (2011) Remarks on some nonparametric estimates of a density function. Sel Works Murray Rosenblatt 1:95–100
25. Kohler WH, Steiglitz K (1974) Characterization and theoretical comparison of branch-and-bound algorithms for permutation problems. J ACM 21(1):140–156
26. Hammersley JM, Handscomb DC (1964) General principles of the monte carlo method. In: Carlo M (ed) Chapman, Hall. Springer, Methods, pp 50–75
27. Mnih V, Szepesvári C, Audibert J (2008) Empirical bernstein stopping. In: Cohen WW, McCallum A, Roweis ST (eds) Machine learning. In: Proceedings of the twenty-fifth international conference (ICML 2008), Helsinki, Finland, 5–9 June 2008. ACM, Helsinki, Finland, pp 672–679
28. Efron B, Tibshirani RJ (1994) An introduction to the bootstrap. Chapman and Hall/CRC
29. Slivkins A (2019) Introduction to multi-armed bandits. Found Trends Mach Learn 12(1–2):1–286
30. Auer P, Cesa-Bianchi N, Fischer P (2002) Finite-time analysis of the multiarmed bandit problem. Mach Learn 47(2):235–256
31. Kleinberg R, Slivkins A, Upfal E (2019) Bandits and experts in metric spaces. J ACM 66(4):30:1–30:77
32. Agrawal S, Goyal N (2012) Analysis of thompson sampling for the multi-armed bandit problem. In: Mannor S, Srebro N, Williamson RC (eds) COLT 2012—the 25th annual conference on learning theory, 25–27 June 2012, Edinburgh, Scotland, JMLR.org, Edinburgh, Scotland, pp 39.1–39.26

33. Gonzalez TF (1985) Clustering to minimize the maximum intercluster distance. Theoret Comput Sci 38:293–306
34. Radford A, Kim JW, Hallacy C, Ramesh A, Goh G, Agarwal S, Sastry G, Askell A, Mishkin P, Clark J, Krueger G, Sutskever I (2021) Learning transferable visual models from natural language supervision. In: Proceedings of the 38th international conference on machine learning, ICML 2021, 18–24 July 2021, Virtual Event, PMLR, pp 8748–8763
35. Gama J, Zliobaite I, Bifet A, Pechenizkiy M, Bouchachia A (2014) A survey on concept drift adaptation. ACM Comput Surv 46(4):44:1–44:37. https://doi.org/10.1145/2523813
36. Krishnamurthy A, Agarwal A, Huang TK, Daumé H III, Langford J (2019) Active learning for cost-sensitive classification. J Mach Learn Res 20(65):1–50
37. Sinha D, Sankararaman KA, Kazerouni A, Avadhanula V (2021) Multi-armed bandits with cost subsidy. In: Banerjee A, Fukumizu K (eds) International conference on artificial intelligence and statistics. PMLR, Virtual Event, pp 3016–3024
38. Darban ZZ, Webb GI, Pan S, Aggarwal C, Salehi M (2025) Deep learning for time series anomaly detection: a survey. ACM Comput Surv 57(1):15:1–15:42
39. Liu Q, Boniol P, Palpanas T, Paparrizos J (2024) Time-series anomaly detection: overview and new trends. Proc VLDB Endow 17(12):4229–4232
40. Schmidl S, Wenig P, Papenbrock T (2022) Anomaly detection in time series: a comprehensive evaluation. Proceedings of the VLDB endowment 15(9):1779–1797

Chapter 13
Analysis of Unconstrained Trajectories, the Case of AIS

Song Wu, Kristian Torp, Mahmoud Sakr, and Esteban Zimányi

Abstract In recent years, the widespread deployment of sensors and communication technologies has enabled the collection of large-scale trajectory data in open, unconstrained environments. One prominent example is the Automatic Identification System (AIS), which transmits real-time navigational data such as location, speed, and heading between moving entities like ships and external observers, including terrestrial stations and satellites. The accumulation of such data opens new avenues for analyzing mobility patterns and supports a variety of applications-ranging from environmental impact assessments to activity detection and transport modeling. This chapter uses AIS as a case study to explore the potential of trajectory data analytics. We focus on two illustrative applications: detecting domain-specific activities (e.g., fishing behavior) and estimating CO_2 emissions. In addition, we address a major challenge that arises across many free-range tracking systems—the presence of large spatial or temporal gaps in the collected trajectories. We outline common causes of such missing data and review methods for addressing them, including trajectory imputation and the integration of complementary data sources such as coastal cameras. We conclude with a discussion of open research problems and promising directions for future work in free-range trajectories.

Keywords Free-range trajectory · AIS · Mobility · Ship · Fishing · CO_2 emissions

S. Wu (✉) · M. Sakr · E. Zimányi
Department of Computer & Decision Engineering (CoDE), Université libre de Bruxelles, Brussels, Belgium
e-mail: song.wu@ulb.be

M. Sakr
e-mail: mahmoud.sakr@ulb.be

E. Zimányi
e-mail: esteban.zimanyi@ulb.be

K. Torp
Department of Computer Science, Aalborg University, Aalborg, Denmark
e-mail: torp@cs.aau.dk

© The Author(s) 2026
G. Dejaegere et al. (eds.), *Data Engineering for Data Science*,
https://doi.org/10.1007/978-3-032-18765-9_13

13.1 Background

Nowadays, the proliferation of various positioning devices has led to huge amounts of trajectory data being collected, and particularly, there is a growing interest in analyzing free-range trajectories in the research community. Free-range trajectories are not constrained by infrastructure such as road networks and typically exhibit more flexibility and diversity. In particular, this chapter presents an analysis of free-range trajectories, exemplified through ship movements recorded by the widely used Automatic Identification System (AIS). In the modern world, 90% of global trade is fulfilled by maritime shipping, and ships play a pivotal role in various human activities on the sea.[1] Due to adverse factors such as harsh weather and considerable difficulty of sea rescues, ensuring navigation and crew life safety has always been a top priority in the maritime industry. Accordingly, a necessary and fundamental task is to continuously track the position of ships. In the past two decades, this task has been greatly facilitated with the advent of the Automatic Identification System (AIS).

Since 2004, the International Maritime Organization (IMO) has mandated that AIS transponders be fitted on board all ships above 300 gross tonnage engaged in international voyages, cargo ships above 500 gross tonnage not engaged on international voyages, and all passenger ships regardless of size.[2] Although AIS was initially conceived as a technical means for collision avoidance, it has also been increasingly employed for various other applications over time, given the ship-tracking nature of AIS data [1]. Basically, AIS allows ships to broadcast their identification, location, and other navigation data to other nearby ships, terrestrial stations, and even satellites. On one hand, these data help the crew on board build situational awareness about the nearby ship traffic and thus can support their decision-making. On the other hand, this data also enables coastal and/or port authorities to better monitor the passing ship traffic and enhance their response capabilities in case of emergencies or accidents.

The past two decades have witnessed an extensive deployment of AIS worldwide, which has led to huge amounts of AIS data being collected every day. For example, MarineTraffic—the leading ship-tracking service provider—receives more than 1 billion AIS signals per day and tracks more than 400 thousand ships daily.[3] As a result, AIS has become an indispensable component in the landscape of modern maritime industry. In nowadays era of big data, this massive AIS data contains an immense potential in value creation for numerous crucial applications.

This chapter aims to provide an overview of AIS data particulars, present two major use cases of AIS data, and discuss a significant challenge in utilizing AIS data. The rest of this chapter in organized as follows: Sect. 13.2 gives a more detailed introduction of AIS data; Sect. 13.3 reviews applications that benefit from AIS data and presents the latest methods in two representative applications—fishing activity

---

[1] https://www.emsa.europa.eu/eumaritimeprofile.html.

[2] https://www.imo.org/en/OurWork/safety/navigation/ais.aspx.

[3] https://www.kpler.com/product/maritime.

Table 13.1 Common public AIS data sources

Data source	Spatial coverage	Temporal coverage	Data format	Access method
Danish Maritime Authority	Danish waters	2006–present	csv files	Downloadable
Marine Cadastre	U.S. coastal waters	2009–present	csv files	Downloadable
Norwegian Coastal Administration	Norwegian waters	Unspecified	Json	Restful API
Fintraffic	Finnish waters	Unspecified	Json	restful API
Australian Maritime Safety Authority	Australia's Search and Rescue Region	2012–present	Shapefile	Downloadable

detection and CO_2 emissions estimation; Sect. 13.4 then turns to discuss one of the significant challenges that remains in utilizing AIS data, namely large spatial and/or temporal gaps in AIS signals, and review methods to mitigate this challenge. Next, Sect. 13.5 discusses open problems and promising future research directions. Finally, Sect. 13.6 concludes this chapter.

13.2 Preliminaries of AIS Data

Generally, AIS data contains three categories of information: (1) static information about the ship itself, such as a ship identifier, the ship type, and the ship dimensions; (2) dynamic information about the ship position and movement; and (3) voyage-related information such as destination and estimated time of arrival.[4] AIS data can be obtained either by purchasing from commercial data providers or by accessing public AIS data sources. Table 13.1 shows a list of common public AIS data sources.

Taking, for example, the AIS data published by the Danish Maritime Authority, 26 columns exist in the CSV files.[5] Among these, the most used are described below.
Static Information:

- **MMSI** is short for Maritime Mobile Service Identity, which is usually used as the ship's identity. It is a 9-digit number, where the first 3 digits represent the country code. Example values of MMSI are 209968000, 538010234, etc.
- **IMO** refers to the International Maritime Organization number, which is composed of 7 digits. This unique number remains the same during the entire life of a ship

---

[4] https://unstats.un.org/wiki/display/AIS/Overview+of+AIS+dataset.

[5] https://web.ais.dk/aisdata/.

and can also be used to identify a ship. Example values are 9624550, 7719832, etc.

- **Ship type** takes values like "Fishing", "Pilot", "Passenger", "Cargo", etc.[6]
- **Width** of the ship in meters. Example values are 10, 16, etc.
- **Length** of the ship in meters. Example values are 104, 36, etc.
- **Size A, Size B, Size C, Size D**. These four columns represent the onboard location of the AIS device, and they refer to distances from the device to the bow, to the stern, to the starboard side, and to the port side, respectively.
- **Type of mobile** is the type of target from which the AIS message is generated. Example values are "Class A", "Base Station", "AtoN", "Class B", "SAR Airborne", "Search and Rescue Transponder", etc.

Dynamic Information:

- **Timestamp** indicates the time at which an AIS message is received by the AIS base station [2]. An example of values in this column is "18/09/2023 10:03:24".
- **Latitude** is the reported latitude of the ship, which ranges from $-90°$ to $90°$. Example values are 54.650243, 57.099767, etc.
- **Longitude** is the reported longitude of the ship, which ranges from $-180°$ to $180°$. Example values are 14.152783, 9.134452, etc.
- **Navigational status**. Example values are "Under way using engine", "Engaged in fishing", "Restricted maneuverability", "Moored", etc.
- **Rate of turn** ranges from 0 to 720 degrees per minute, e.g., -37.5, 21.6, etc.
- **Speed over ground** ranges from 0 to 102 knots with a 0.1-knot resolution. One knot means one nautical mile (approximately 1.852 km) per hour.
- **Course over ground** is the movement direction of the ship relative to due north. It comes in a 0.1° resolution.
- **Heading** is the direction pointed by the bow of the ship relative to due north. It ranges from 0° to 359°. Example values are 76, 268, etc.

Voyage-Related Information:

- **Draught** of the ship ranges, which ranges from 0.1 to 25.5 m.
- **Estimated time of arrival** at the destination if available.
- **Destination** manually inserted by the captain.

The reporting frequency of AIS depends on the AIS transponder type, the ship speed over ground, and whether the ship is making a turn, etc.[7] Tables 13.2 and 13.3 show the details in various cases.

A variety of tools are available that can handle AIS data, among which some popular ones are introduced below.

- MobilityDB [3] is a mainstream moving object database system that provides rich functionality for managing and analyzing spatiotemporal trajectory data. Due to its

---

[6] https://api.vtexplorer.com/docs/ref-aistypes.html.

[7] https://www.milltechmarine.com/faq.htm.

Table 13.2 Reporting frequency of Class A AIS transponder

Ship moving status	Dual channel receiver	Single channel receiver
Ship anchored or moored	3 min	6 min
Speed between 0–14 knots	10 s	20 s
Speed between 0–14 knots and making a turn	3.3 s	6.6 s
Speed between 14–23 knots	6 s	12 s
Speed between 14–23 knots and making a turn	2 s	4 s
Speed above 23 knots	2 s	4 s
Ship static information	6 min	12 min

Table 13.3 Reporting frequency of Class B AIS transponder

Ship moving status	Dual channel receiver	Single channel receiver (min)
Speed below 2 knots	3 min	6
Speed above 2 knots	30 s	1
Ship static information	6 min	12

compact geospatial data storage, it supports the processing of large-scale trajectory data. It is open-source and easy to use through its SQL interface.

- MEOS (Mobility Engine, Open Source) [4] is a C library that provides a set of APIs for manipulating temporal and spatiotemporal data. Also, it is the core component of MobilityDB and has bindings available such as PyMEOS.[8]

- MovingPandas [5] is an open-source Python library for movement data exploration and analysis.[9] It focuses on spatio-temporal data exploration with rich functions for data manipulation and analysis.

- QGIS is an open-source software for spatial data visualization and processing.[10] Its functionality can be easily extended and enhanced by various plugins. For example, using a plugin like Move [6], QGIS supports integration with MobilityDB and can automatically display and/or animate query results from MobilityDB.

- Tableau Public allows fast exploration and visualization of AIS data with low-code.[11] Also, dashboards can be easily created in Tableau Public for more comprehensive analysis of AIS data. Figure 13.1 shows two trajectory examples that are visualized in Tableau Public.

- Folium is a popular Python library for creating interactive map visualizations.[12] It allows both binding data to a map and showing various markers on the map.

---

[8] https://github.com/MobilityDB/PyMEOS.

[9] https://github.com/movingpandas/movingpandas.

[10] https://www.qgis.org/.

[11] https://www.tableau.com/products/public.

[12] https://python-visualization.github.io/folium/latest/.

13.3 Applications of AIS Data

AIS data records detailed movement of ships on the sea and serves as an indispensable data source for understanding ship activities. Many applications have benefited from AIS data, including but not limited to: collision avoidance [7–9], maritime transport network construction [10–12], maritime anomaly detection [13–15], fishing activity analysis [16–18], data fusion for traffic surveillance [19, 20], ship emissions estimation [21, 22], and so on.

In the following subsections, we focus on two of the most important applications of AIS data—fishing activity detection and CO_2 emissions estimation—and review the latest methods in the literature.

13.3.1 Fishing Activity Detection

Fishery is a traditional and one of the most important human activities on the sea. It is estimated that 10 to 12% of the world's population relies on fishing for a living.[13] However, in recent years, the sustainable utilization of fishery resources has been more and more challenged by the ever-increasing practices of illegal, unreported, and unregulated (IUU) fishing. IUU fishing not only poses a serious threat to sustainable fishing but also damages the marine environment. Therefore, some authorities, such as the EU, have set up regulations to strengthen control over the fishery activities and combat IUU fishing.[14] To enforce these regulations, a first step is to know when and where a ship may have conducted fishing activities [18].

Definition: AIS Trajectory. An AIS trajectory T is defined as a temporally-ordered sequence of AIS points $\{a_1, a_2, ..., a_n\}$. Each AIS point a_i $(1 \leq i \leq n)$ is a quadruple (t, p, s, c), where t is the timestamp, p is the ship location at t, s is the ship speed at t, and c is the ship's course over ground at t.

Problem Definition: Fishing Activity Detection. Given an AIS trajectory T, the goal of fishing activity detection is to identify which portions of T correspond to fishing activities.

The main challenge of this problem is that there exists a variety of fishing gear types, such as trawls, longlines, and seines.[15] As a result, these different gear types lead to different movement patterns in fishing boat trajectories. Figure 13.2 shows a taxonomy of the methods for fishing activity detection using AIS data, where two categories of methods exist:

---

[13] https://www.imo.org/en/MediaCentre/SecretaryGeneral/Pages/FAOCOFI.aspx.

[14] https://oceans-and-fisheries.ec.europa.eu/fisheries/rules/illegal-fishing_en.

[15] https://fish-commercial-names.ec.europa.eu/fish-names/fishing-gears_en.

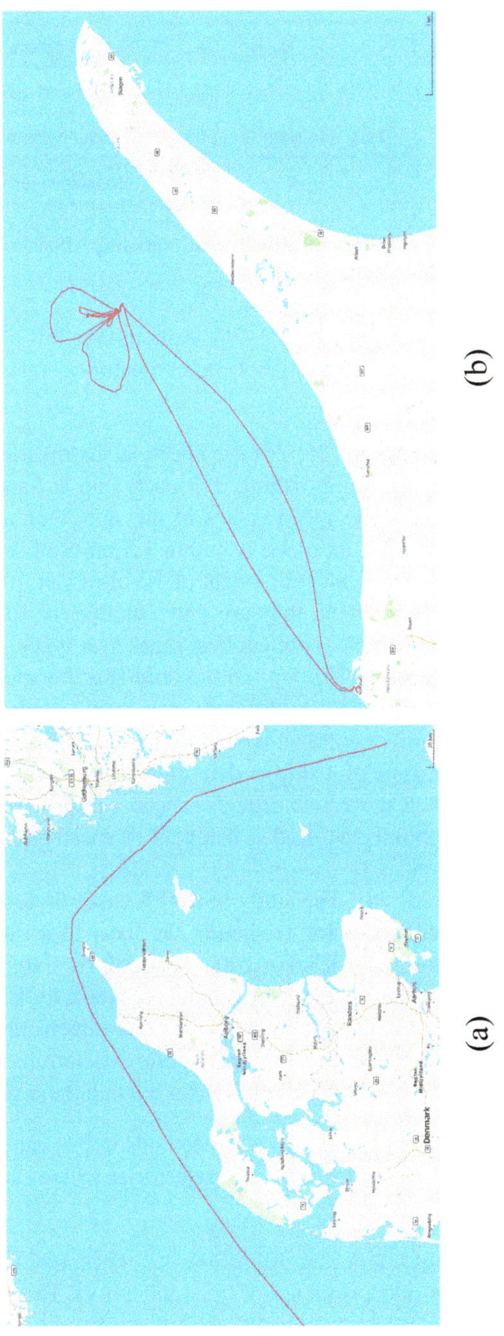

(a) (b)

Fig. 13.1 Two examples of visualizing AIS ship trajectories: **a** trajectory of a cargo ship that moves from the North Sea to the Baltic Sea; **b** trajectory of a fishing boat that starts and ends at the port of Hirtshals in Denmark

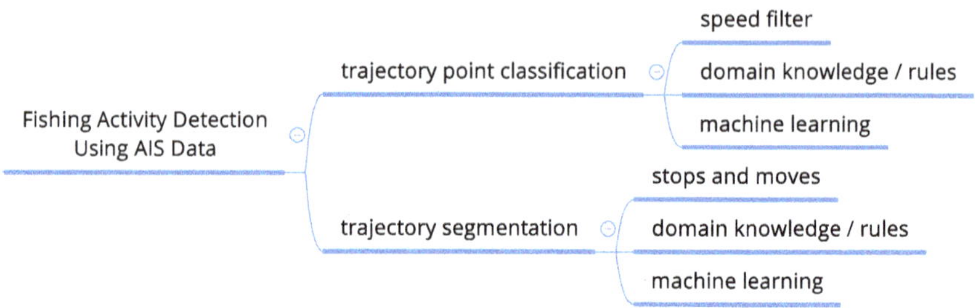

Fig. 13.2 Taxonomy of methods for fishing activity detection using AIS data

- **Trajectory point classification** [16, 23–26]. A method in this category performs point-level classification and tries to label each trajectory point as fishing or non-fishing.

  - **Speed filters**. The studies in [25, 26] use speed as the most important feature to distinguish fishing from non-fishing. For each ship at hand, they first fit a bi-modal distribution for the speed values of the ship, then a trajectory point will be labeled as fishing if its speed is within 1.5 times of standard deviation from the mean value of the first component of the bi-modal. Although methods like this are easy to implement, they are only suitable for some specific gear types like trawlers, and are not applicable to other gear types.
  - **Domain knowledge/rules** [16]. Unlike the ship-specific speed thresholds in [25, 26], the study in [16] defines different gear-level speed intervals for four gear types: small bottom otter trawl, large bottom otter trawl, pelagic pair trawl, and rapido. The gear type of a ship is determined by matching AIS data with the European Fleet Register which provides specific information for ships.[16] In this way, a trajectory point is labeled as fishing if its speed falls within the speed interval of its gear type.
  - **Machine learning** [23, 24]. The study in [24] focuses on longliner ships and trains an autoencoder model for predicting the label of a trajectory window. When making predictions, the label of a trajectory point is assigned as the label of the trajectory window that centers at the point. In contrast, the study in [23] trains a trajectory window classification model with a one-dimensional convolutional neural network, and the label of a trajectory point is determined using majority voting between all the trajectory windows that contain the trajectory point.

- **Trajectory segmentation** [17, 18, 27–30]. Methods in this category aim to split the whole trajectory into a sequence of sub-trajectories where each sub-trajectory can be labeled as fishing or non-fishing.

  - **Stops and moves**. Studies in [29, 30] follow the conceptual model in [31] and try to split a trajectory into a sequence of stops and moves. The obtained stops are

---

[16] https://webgate.ec.europa.eu/fleet-europa/search_en.

then considered to be places where a ship performs fishing activities. The study in [30] uses direction change as the main criterion for trajectory partitioning, whereas the study in [29] combines both speed and direction change for splitting trajectories. However, these methods may fail when fishing activities at sea exhibit complex movement patterns and cannot be simply treated as stops or moves [18]. For example, the complete fishing activity in Fig. 13.3 contains both high-speed points and low-speed points.

- **Domain knowledge/rules** [17]. The study in [17] focuses on squid in the South-west Atlantic and detects fishing activities using rules designed by domain experts. For example, fishing is expected to occur between dusk and dawn, because the squid fishing technique works by attracting them with powerful incandescent lights at night. However, it is worth noting that such rules are usually very specific and cannot be generalized to other types of fish.
- **Machine learning** [18, 27]. The study in [18] aims at detecting complete fishing activities. It first proposes a technique based on visualization to help design features distinguishing fishing movement from non-fishing movement. The designed features are then used to train a random forest model that classifies a trajectory window as fishing or not. Furthermore, a merging technique based on run-length encoding is proposed in [18] to account for occasional classification mistakes, thus making the algorithm more tolerant and robust. The study in [27] determines the trajectory partitioning positions as follows: an error signal indicating the deviation from the expected location is first computed for each trajectory point, then the error signals are used to train a classifier that judges whether a trajectory window contains a partitioning position or not. Finally, whether a trajectory point is a partitioning position is determined by majority voting between all trajectory windows that contain the point. The limitation of this approach is that the returned segments do not have associated labels indicating whether fishing is performed during the trajectory segment.

Figure 13.4 shows the returned segmentation results for the example trajectory in Fig. 13.3 by four segmentation algorithms WBS-RLE [18], CB-SMoT [31], Warped K-Means [32], and SWS [28]. In Fig. 13.4, each color represents a different trajectory segment. It can be observed that the results by WBS-RLE are the closest and most realistic w.r.t. the ground truth.

Despite the great efforts already done in the community, several challenges remain to be solved for detecting fishing activities using AIS data. First, high-quality labeled data are crucial for developing and validating novel solutions, but such data are usually hard to obtain and require close collaboration with domain experts. Second, many types of fishing gear exist in fishing practices, and ship movement patterns during fishing differ from one gear type to another. Therefore, general solutions are desired that can be applied to as many gear types as possible.

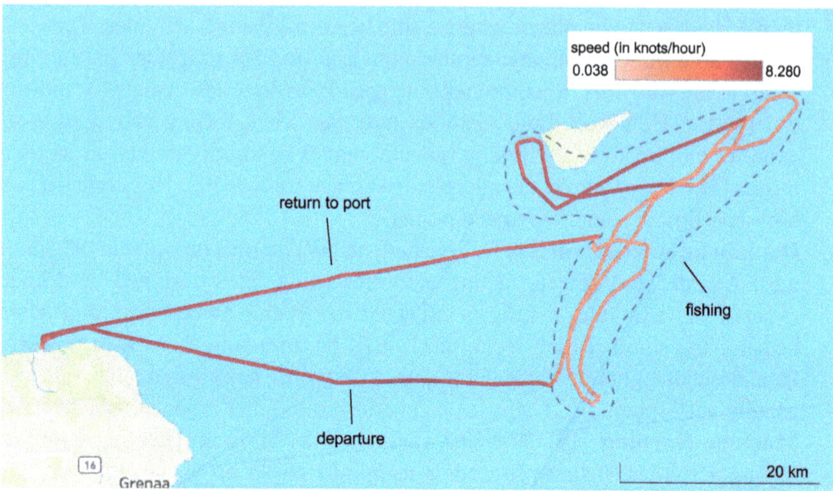

Fig. 13.3 A fishing trip in three stages: heading to the fishing ground, fishing, and returning to the port

13.3.2 Ship CO_2 Emissions Estimation

In the past years, climate change and environmental protection have drawn more and more attention around the world, and these issues pose threats to the sustainable development of human society in the coming decades. Among others, one particular challenge is the so-called greenhouse effect, which is largely caused by the excessive CO_2 emissions from human activities such as transportation and industry.[17] As the dominant means of transport for global trade, maritime shipping gives rise to a significant amount of CO_2 emissions every year. For example, the study in [33] estimates that shipping CO_2 emissions increased from 962 million tons in 2012 to 1,056 million tons in 2018. In light of this, many studies on shipping CO_2 emissions have emerged recently (e.g. [34–36]), and AIS data comes as an invaluable data source for this application since it is easily available and contains detailed historical movement of ships.

Shipping CO_2 emissions are directly determined by the fuel type and the amount of fuel consumed [33]. Fuel consumption of a ship is usually recorded in the so-called noon report data [37, 38], which is created by the crew every 24 h at noon and contains the daily fuel consumption of main and auxiliary engines. Although it can be used as ground truth for fuel consumption, the drawback of the noon report data is that it has low frequency and is subject to human errors [39].

To estimate the fuel consumption of ships, three categories of models exist in the literature [37, 39]:

---

[17] https://tinyurl.com/83hr2fsn.

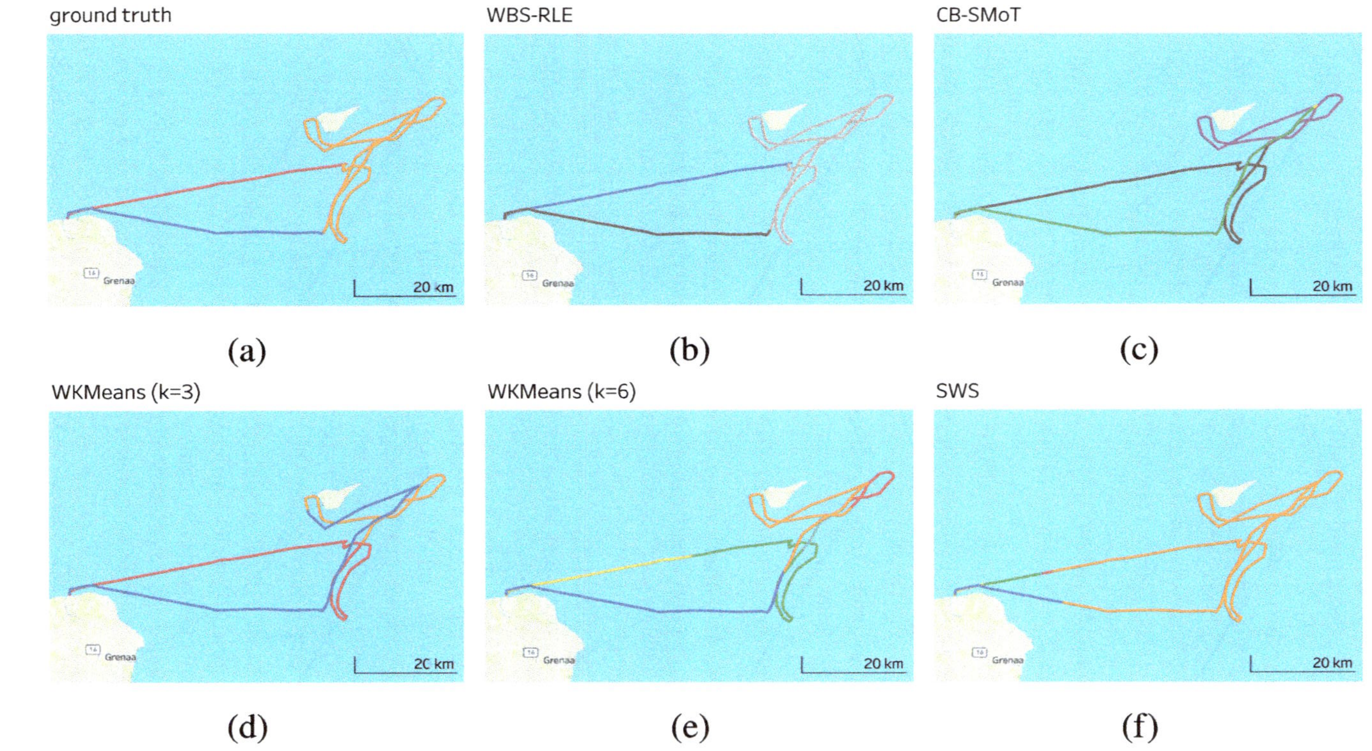

Fig. 13.4 Segmentation results of four algorithms

- **White-box models (WBM).** Using physics laws and hydrodynamics, the idea of WBMs is to estimate various resistances that a ship encounters during sailing. These resistances will be combined with the estimated propulsion efficiency to compute a ship's real-time power demand [34, 40]. Since speed is generally considered as the most important factor for fuel consumption, many studies (e.g. [41, 42]) simplify the impact of resistance as a fixed coefficient. As a result, the fuel consumption rate would be proportional to the cube of a ship's speed. The advantage of WBMs is that they are suitable for the early stages of ship design as all parameters are known a priori [39]. However, the accuracy of WBMs depends largely on the various assumptions [39, 43].
- **Black-box models (BBM).** BBMs are data-driven and predict fuel consumption using statistical approaches and machine learning models [38, 44]. For example, the study in [38] uses a Gaussian process to predict fuel consumption using seven variables. The advantage of BBMs is that they usually achieve a higher accuracy than WBMs, but they require large amounts of data and ground truth fuel consumption data are difficult to obtain at large scale [39].
- **Grey-box models (GBM).** Basically, GBMs combine the advantages of WBMs and BBMs, and they can be constructed in two ways [39]. The first way is to use a BBM to fine-tune the parameters in a WBM [37], whereas the second way is to integrate a WBM's prior knowledge into a BBM [45]. However, developing GBMs is a non-trivial task, and GBMs are not used as frequently as BBMs [43].

Nevertheless, two limitations exist in most existing models. First, they are usually validated on a few specific ships, and their suitability for other ships is unclear. Second, these models need access to the real fuel consumption data, which is usually proprietary and difficult to obtain at large scale. More importantly, as far as we are concerned, none of the review studies (e.g. [21, 39, 43]) on ship fuel consumption models are data-driven, and only a few studies exist that conduct experimental comparisons of different shipping CO_2 emissions models [22, 46, 47]. However, both studies in [46, 47] are restricted to a small area around the Strait of Gibraltar. The study in [46] assumes a ship's speed to be constant in the study area, and the study in [47] only considers one Ro-Pax ship (roll-on/roll-off passenger ship). Therefore, more detailed comparisons in a larger geographic region and on more ships are needed such as the one performed in [22].

Figure 13.5 shows the general framework proposed in [22] for comparing different shipping CO_2 emissions models. This framework includes three analysis modules:

1. The grid-based analysis module focuses on the spatiotemporal distribution of CO_2 emissions estimated by different models, and such analysis can be useful for decision-makers to develop regulation policies.
2. The trajectory-based analysis module targets representative individual trajectories, which are of interest for relevant practitioners such as ship owners.
3. The MRV-based validation module aims to check if the CO_2 emissions estimates comply with the MRV dataset. MRV refers to the EU's Monitoring, Reporting, and Verification system [48] which requires ships above 5,000 gross tonnage to report their CO_2 emissions data for their transport activities within the EU waters.

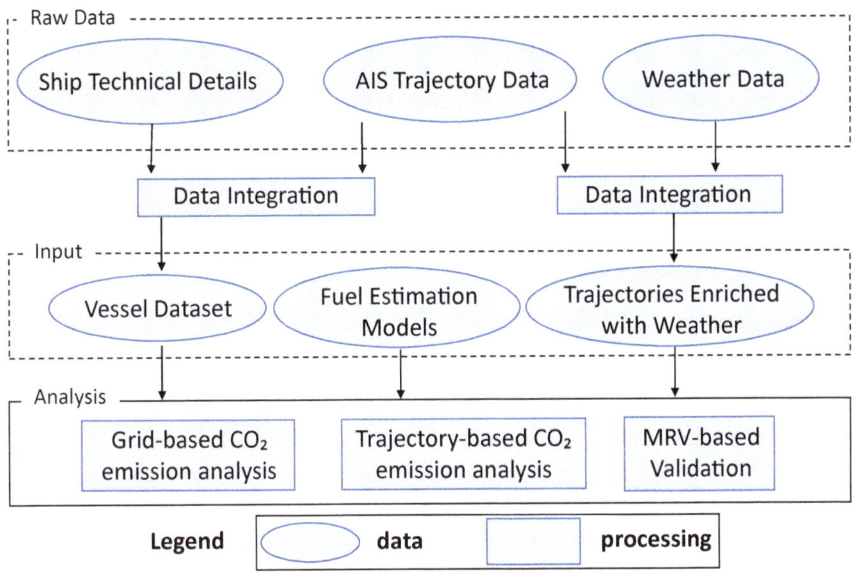

Fig. 13.5 The framework in [22] for comparing shipping CO_2 emissions models

A verified annual report for each ship is then released on the MRV's website for public access.[18] As a result, the MRV dataset is used in [22] as the ground-truth data for ship-level CO_2 emissions.

By applying the framework to six CO_2 emissions models using one-month AIS data from May 2022 around Denmark, a handful of results are presented in [22]. For example, Fig. 13.6 shows the spatial distribution of ranking for each model at a resolution of 5×5 km, where each model exhibits its own particular patterns. For complete results and analysis, we refer interested readers to [22].

13.4 Challenges of AIS Data

Despite the great benefits of AIS data for analyzing ship activities, challenges still exist in the utilization of AIS data. Among others, a significant challenge is the presence of large gaps in AIS data. These gaps manifest themselves either as the large spatial distance [49] or as the long time interval [50] between two consecutive AIS points in a ship trajectory. For example, Fig. 13.7 shows a trajectory that contains a 12-h gap, and this gap has a spatial distance of 13 km. As a result, the whereabouts of this ship during this gap are unknown from AIS data. Large gaps like this can be caused by several reasons: (1) AIS signals from ships are not captured by any receiver

---

[18] https://mrv.emsa.europa.eu/#public/emission-report.

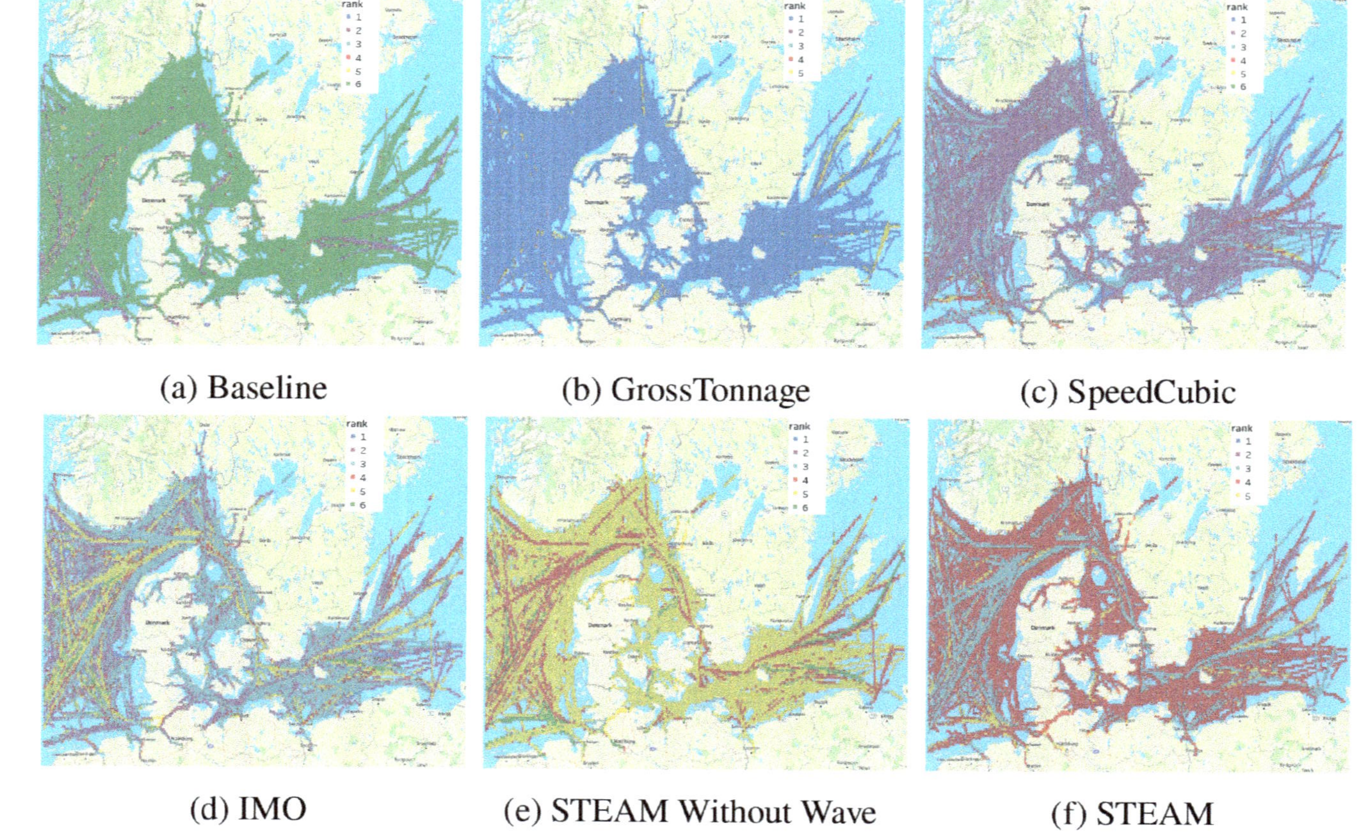

Fig. 13.6 Spatial ranking distribution of six models based on their cell-level CO_2 emissions, where '1' means that a model gives the highest emissions at a cell, and '6' means that a model gives the lowest emissions at a cell (taken from [22])

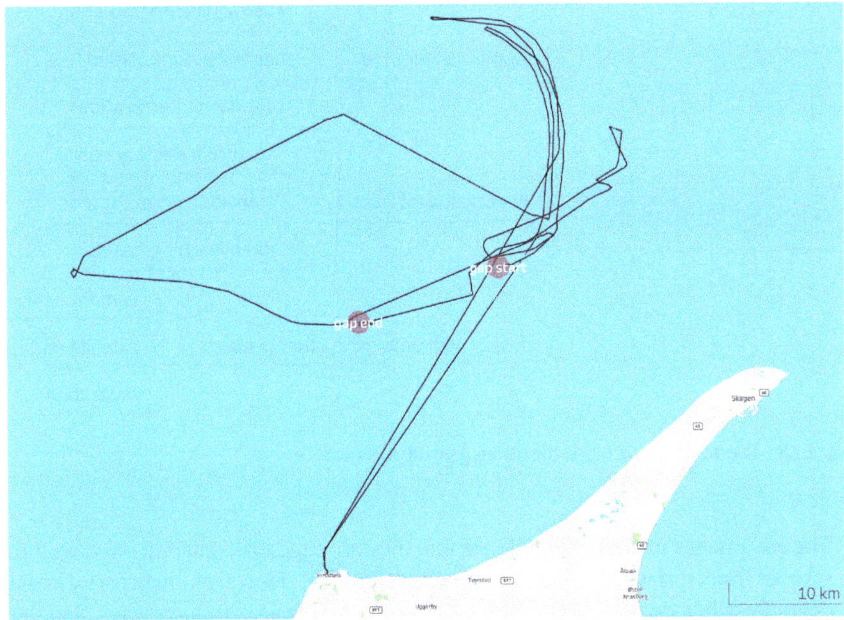

Fig. 13.7 A real trajectory near Denmark. The two red circles depict where a 12-h gap happens

due to poor weather conditions, high ship density in an area, or limited coverage of AIS receivers [49]; (2) a ship may intentionally switch off its AIS signals in order to hide its location from other vessels. In such a case, the ship is likely to conduct some illegal or secret activities. Therefore, it is important to detect AIS gaps and reconstruct trajectories to improve and ensure the quality of AIS data before any further processing.

Definition: AIS Gap. Given an AIS trajectory T, an AIS gap is defined as any two consecutive AIS points gap^{start} and gap^{end} in T satisfying:

$$gap^{start}.t - gap^{end}.t \geq \theta_t$$

where θ_t is a time threshold.

Problem Definition: Trajectory Imputation. Given an AIS gap, the goal of trajectory imputation is to restore the missing ship movement during this gap by adding points between gap^{start} and gap^{end}.

In the literature, multiple methods have been proposed for AIS trajectory imputation, and Fig. 13.8 shows a taxonomy of these methods.

- **Numerical methods**. These methods mainly use mathematical formulas or kinematic information to fill AIS gaps, such as the cubic spline interpolation in [51], the Lagrange interpolation in [52], and the improved kinematic interpolation in [53].

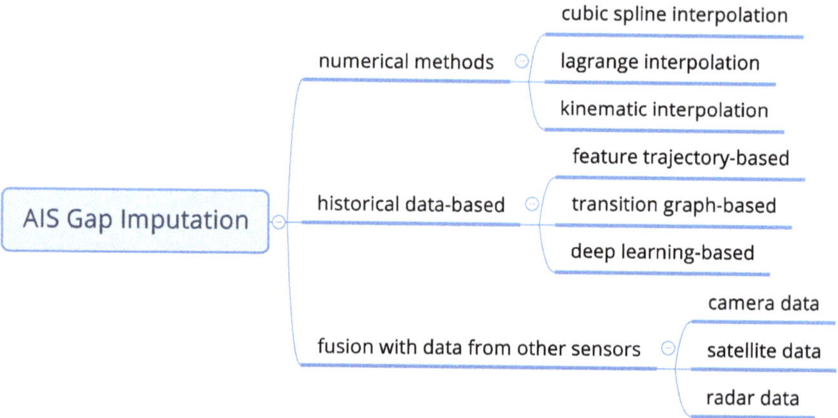

Fig. 13.8 Taxonomy of methods for filling gaps in AIS data

The advantages of such methods are that they are time-efficient and work well for smooth trajectories in non-restrictive scenarios [49]. However, their performance degrades when dealing with complex ship movement or very large gaps, and they may return unrealistic trajectories that cross land.

- **Historical data-based methods**. The idea behind this category of methods is to use massive historical AIS data to interpolate AIS gaps, and historical AIS data can be used in different ways.

  - **Feature trajectory-based methods**. The study in [54] interpolates an AIS gap as follows. First, a set of potential trajectories that pass near the gap is searched. Second, one potential trajectory is selected as the feature trajectory that has the smallest DTW (Dynamic Time Warp) distance to the other potential trajectories. Finally, the AIS gap is filled by translating and scaling the trajectory points in this feature trajectory. It is worth noting that this method fails when no historical trajectories are passing near the AIS gap.
  - **Transition graph-based methods**. These methods fill AIS gaps by computing, over a transition graph, the shortest path that connects the gap start point and the gap end point. The transition graph is built using historical AIS data and is thus assumed to reflect the ship movement patterns in an area of interest. In the study [49], nodes in the transition graph are AIS points, and edges connect AIS points that satisfy the specified neighborhood criterion and direction criterion. In contrast, nodes in the graph built in [55] are spatial cells of size 1×1 km; and neighboring cells are connected using edges whose weight combines the transition probability and the spatial distance between cells. After the transition graph is built, both studies [49, 55] apply the A* algorithm to find the shortest path connecting the gap start and the gap end points.
  - **Deep learning-based methods**. Given the big success of the deep learning technology, some methods have also emerged that apply deep learning for filling

AIS gaps [50, 56]. The study in [56] employs the convolutional network U-Net [57] for ship trajectory reconstruction, whereas the study in [50] proposes a physics-guided probabilistic diffusion model for long-term ship trajectory imputation.

- **Fusion with data from other sensors**. The assumption under historical data-based methods is that ships will follow common traffic patterns. Therefore, these methods fall short when a ship exhibits movement characteristics unseen in the past. In such cases, it becomes necessary to fill AIS gaps through data fusion with other types of sensors.

  - **Camera data**. Surveillance cameras provide a visual means of monitoring ship traffic. For the purpose of filling AIS gaps, a critical step is to compute the real-world coordinates of the detected ship targets in videos captured by cameras. To this end, several approaches exist in the literature.

    · **Transformation matrix-based**. These methods convert a pixel coordinate to the corresponding longitude/latitude coordinate by using a transformation matrix [58–60]. This matrix is usually established by manually preparing a set of reference point pairs, where each pair contains a pixel coordinate in the image space and its corresponding longitude/latitude coordinate in the real world. This process is commonly referred to as camera calibration.

    · **Pinhole model-based**. Another popular approach for coordinate conversion is the pinhole imaging model [20]. This model works by performing trigonometric calculations to obtain the relative horizontal and vertical angle of a ship w.r.t. a camera. However, this approach requires detailed knowledge about the camera parameters, such as camera height, fields of view of cameras, and camera heading etc.

    · **Extended pinhole model-based**. One aspect ignored by the transformation matrix-based methods and pinhole model-based methods is that, in fact, there is no one-to-one correspondence between discrete pixel coordinates and continuous longitude/latitude coordinates. The study in [61] takes this aspect into account and extends the pinhole model by mapping each pixel coordinate into a spatial polygon rather than one longitude/latitude coordinate. Furthermore, the study in [61] targets a multi-camera scenario and aims to reduce the uncertainty of ship location estimation through the intersection of polygons from multiple cameras. Extensive experiments in [61] show that using multiple cameras can improve the accuracy of ship location estimation by one order of magnitude compared with using one camera.

  - **Satellite data**. The study in [55] proposes a pipeline that fuses AIS data with satellite image data captured by the Sentinel-1 [62] and Sentinel-2 missions [63]. First, this pipeline detects ships appearing in satellite images. Second, the geo-referenced coordinates of the detected ships are computed. Finally, fusion is performed to correlate AIS data with ships detected in satellite images. The advantage of fusion with satellite data is that a global coverage can be achieved using satellites. However, satellite data have a significantly higher temporal sparsity than AIS data, e.g., the revisit time of Sentinel satellites is 2–3 days in high coverage areas [55].

- **Radar data**. Radars are also broadly used for surveillance purposes [64, 65], and the data from a radar can provide accurate bearing and distance of the objects in its vicinity [59]. The main drawback of radar-based ship tracking is that it is prone to be interfered with [66, 67] by the clutter of irrelevant objects such as buildings, bridges, etc. Also, the deployment of radars is usually costly [66] and may be prohibited in some areas due to the electromagnetic radiation [59].

13.5 Open Problems and Future Directions

Despite significant progress in AIS-based analysis, several open problems remain. One important challenge is **AIS spoofing**, which occurs when AIS data does not comply with actual ship activities. For example, a ship may falsify its identity or location to hide its whereabouts from other ships [68, 69]. Another example is that although the 'navigational status' attribute in AIS data is very important for supervised machine learning applications, this attribute is not always as reliable as expected [70]. Therefore, how to effectively solve this data veracity issue remains an open problem and needs more efforts. Another unresolved issue relates to **ships sharing MMSI numbers**. MMSI numbers are generally used as the ship identity to extract its trajectory from all AIS signals. However, an open problem arises when multiple ships share the same MMSI number [71]. This phenomenon gives rise to trajectories that appear anomalous because movements from different ships are mixed together. Based on the OPTICS clustering algorithm [72], the study in [71] proposes a visual analysis method for separating these mixed trajectories. Nevertheless, the main limitation of this visual method is that considerable human intervention is needed for the density parameter selection in OPTICS. Therefore, automatic separation of mixed trajectories remains a challenging task.

Looking forward, several promising directions for future research can be identified. One such direction is **mobility pattern mining involving multiple ships**. A ship's behavior at sea is likely to be influenced by other ships in its vicinity. An interesting direction that has not been sufficiently studied is how to effectively and efficiently explore and analyze mobility patterns involving multiple ships. Currently, only a few relevant studies exist in the literature, such as the work in [73], which analyses ship escort and convoy operations in ice conditions.

Another key direction is **large-scale multi-source data fusion**. A noticeable trend in the literature is that multi-source data are being used in more and more studies. For effective analysis, AIS data alone is sometimes not enough, and it becomes necessary to incorporate other contextual data. For example, the study in [74] reveals insights into three events in 2022 (e.g., the Nord Stream explosions[19]) by combining AIS data, satellite data, bathymetry data, weather conditions, and more.

Finally, there is growing interest in **explainable deep-learning models**. Recently, deep learning methods have been increasingly used in various tasks such as ship trajectory imputation [50]. Although these methods yield remarkable results, they are often used as black boxes and cannot be easily interpreted by end users [75]. For use cases such as autonomous surface vehicles, the explainability of deep learning models is essential for stakeholders to trust the reasoning results of such models. Therefore, more research efforts are needed to enhance the explainability and transparency of these models.

13.6   Conclusion

This chapter provides, through AIS data, an example of analysis of free-range trajectories. First, we give an introduction to AIS data, which is widely used for ship tracking in the maritime domain. Afterwards, methods are reviewed for two important applications of AIS data, i.e., fishing activity detection and shipping CO_2 estimation. Later on, one of the main challenges in utilizing AIS data—large spatial and/or temporal gaps—is discussed, and approaches are presented that can help mitigate this challenge. Finally, some open problems related to AIS data are discussed, and we point out several research directions that need more attention and efforts in the future. It is envisioned that AIS data will continue to play a critical role in various maritime applications.

References

1. Yang D, Wu L, Wang S, Jia H, Li KX (2019) How big data enriches maritime research-a critical review of automatic identification system (AIS) data applications. Transp Rev 39(6):755–773
2. Patroumpas K, Alevizos E, Artikis A, Vodas M, Pelekis N, Theodoridis Y (2017) Online event recognition from moving vessel trajectories. GeoInformatica 21:389–427
3. Zimányi E, Sakr M, Lesuisse A (2020) MobilityDB: a mobility database based on PostgreSQL and PostGIS. ACM Trans Database Syst 45(4):1–42
4. Zimányi E, Duarte MM, Diví V (2024) MEOS: an open source library for mobility data management. EDBT. EDBT, Italy, pp 810–813
5. Graser A (2019) Movingpandas: efficient structures for movement data in python. GIForum 1:54–68
6. Schoemans M, Sakr MA, Zimányi E (2022) MOVE: interactive visual exploration of moving objects. In: EDBT/ICDT Workshops, EDBT/ICDT Workshops, UK
7. Nguyen M, Zhang S, Wang X (2018) A novel method for risk assessment and simulation of collision avoidance for vessels based on AIS. Algorithms 11(12):204

---

[19] https://tinyurl.com/2nnjbt7w.

8. Zhang M, Montewka J, Manderbacka T, Kujala P, Hirdaris S (2021) A big data analytics method for the evaluation of ship-ship collision risk reflecting hydrometeorological conditions. Reliab Eng Syst Saf 213:107674

9. Zhang W, Goerlandt F, Montewka J, Kujala P (2015) A method for detecting possible near miss ship collisions from AIS data. Ocean Eng 107:60–69

10. Bläser N, Magnussen BB, Fuentes G, Lu H, Reinhardt L (2024) MATNEC: AIS data-driven environment-adaptive maritime traffic network construction for realistic route generation. Transp Res Part C Emerg Technol 169:104853

11. Eljabu L, Etemad M, Matwin S (2022) Spatial clustering method of historical AIS data for maritime traffic routes extraction. In: 2022 IEEE international conference on big data. IEEE, Japan, pp 893–902

12. Yan Z, Xiao Y, Cheng L, He R, Ruan X, Zhou X, Li M, Bin R (2020) Exploring AIS data for intelligent maritime routes extraction. Appl Ocean Res 101:102271

13. Kontopoulos I, Chatzikokolakis K, Zissis D, Tserpes K, Spiliopoulos G (2020) Real-time maritime anomaly detection: detecting intentional AIS switch-off. Intl J Big Data Intell 7(2):85–96

14. May Petry L, Soares A, Bogorny V, Brandoli B, Matwin S (2020) Challenges in vessel behavior and anomaly detection: From classical machine learning to deep learning. In: Advances in artificial intelligence: 33rd Canadian conferences on artificial intelligence. Springer, Canada, pp 401–407

15. Mazzarella F, Vespe M, Alessandrini A, Tarchi D, Aulicino G, Vollero A (2017) A novel anomaly detection approach to identify intentional AIS on-off switching. Expert Syst Appl 78:110–123

16. Brandoli B, Raffaetà A, Simeoni M, Adibi P, Bappee FK, Pranovi F, Rovinelli G, Russo E, Silvestri C, Soares A et al (2022) From multiple aspect trajectories to predictive analysis: a case study on fishing vessels in the Northern Adriatic sea. GeoInformatica 26(4):551–579

17. Pons Recasens G, Bilalli B, Abelló Gamazo A, Blanco Sánchez S (2022) Learning fishing information from AIS data. In: Proceedings of the 2nd ACM SIGSPATIAL international workshop on animal movement ecology and human mobility. ACM, USA, pp 9–18

18. Wu S, Zimányi E, Sakr M, Torp K (2022) Semantic segmentation of AIS trajectories for detecting complete fishing activities. In: 2022 23rd IEEE international conference on mobile data management (MDM). IEEE, Cyprus, pp 419–424

19. Lu Y, Ma H, Smart E, Vuksanovic B, Chiverton J, Prabhu SR, Glaister M, Dunston E, Hancock C (2021) Fusion of camera-based vessel detection and AIS for maritime surveillance. In: 2021 26th international conference on automation and computing. IEEE, UK, pp 1–6

20. Qu J, Liu RW, Guo Y, Lu Y, Su J, Li P (2023) Improving maritime traffic surveillance in inland waterways using the robust fusion of AIS and visual data. Ocean Eng 275:114198

21. Wang K, Wang J, Huang L, Yuan Y, Wu G, Xing H, Wang Z, Wang Z, Jiang X (2022) A comprehensive review on the prediction of ship energy consumption and pollution gas emissions. Ocean Eng 266:112826

22. Wu S, Torp K, Sakr M, Zimányi E (2023) Evaluation of vessel CO_2 emissions methods using AIS trajectories. In: Proceedings of the 18th international symposium on spatial and temporal data. ACM, Canada, pp 65–74

23. Arasteh S, Tayebi MA, Zohrevand Z, Glässer U, Shahir AY, Saeedi P, Wehn H (2020) Fishing vessels activity detection from longitudinal AIS data. In: Proceedings of the 28th international conference on advances in geographic information systems. ACM, USA, pp 347–356

24. Jiang X, Silver DL, Hu B, de Souza EN, Matwin S (2016) Fishing activity detection from AIS data using autoencoders. In: Advances in Artificial Intelligence: 29th Canadian Conferences on Artificial Intelligence, Springer, Canada, pp 33–39

25. Natale F, Gibin M, Alessandrini A, Vespe M, Paulrud A (2015) Mapping fishing effort through AIS data. PLoS ONE 10(6):e0130746

26. Vespe M, Gibin M, Alessandrini A, Natale F, Mazzarella F, Osio GC (2016) Mapping EU fishing activities using ship tracking data. J Maps 12(sup1):520–525

27. Etemad M, Etemad Z, Soares A, Bogorny V, Matwin S, Torgo L (2020) Wise sliding window segmentation: A classification-aided approach for trajectory segmentation. In: Advances in artificial intelligence: 33rd Canadian conferences on artificial intelligence. Springer, Canada, pp 208–219
28. Etemad M, Soares A, Etemad E, Rose J, Torgo L, Matwin S (2021) SWS: an unsupervised trajectory segmentation algorithm based on change detection with interpolation kernels. GeoInformatica 25:269–289
29. Mazzarella F, Vespe M, Damalas D, Osio G (2014) Discovering vessel activities at sea using AIS data: mapping of fishing footprints. In: 17th international conference on information fusion. IEEE, Spain, pp 1–7
30. Rocha JAM, Times VC, Oliveira G, Alvares LO, Bogorny V (2010) DB-SMoT: a direction-based spatio-temporal clustering method. 2010 5th IEEE International Conference on Intelligent Systems. IEEE, UK, pp 114–119
31. Palma AT, Bogorny V, Kuijpers B, Alvares LO (2008) A clustering-based approach for discovering interesting places in trajectories. In: Proceedings of the 2008 ACM symposium on applied computing. ACM, Brazil, pp 863–868
32. Leiva LA, Vidal E (2013) Warped K-Means: an algorithm to cluster sequentially-distributed data. Inf Sci 237:196–210
33. Organization IM (2020) Fourth IMO greenhouse gas study
34. Hensel T, Ugé C, Jahn C (2020) Green Shipp Using AIS Data Assess Glob Emiss 28(1):39–47
35. Schwarzkopf DA, Petrik R, Matthias V, Quante M, Majamäki E, Jalkanen JP (2021) A ship emission modeling system with scenario capabilities. Atmos Environ X 12:100132
36. Wang Y, Watanabe D, Hirata E, Toriumi S (2021) Real-time management of vessel carbon dioxide emissions based on automatic identification system database using deep learning. J Mar Sci Eng 9(8):871
37. Yang L, Chen G, Rytter NGM, Zhao J, Yang D (2019) A genetic algorithm-based grey-box model for ship fuel consumption prediction towards sustainable shipping. Ann Oper Res 1–27
38. Yuan J, Nian V (2018) Ship energy consumption prediction with Gaussian process metamodel. Energy Procedia 152:655–660
39. Yan R, Wang S, Psaraftis HN (2021) Data analytics for fuel consumption management in maritime transportation: status and perspectives. Transp Res Part E Logist Transp Rev 155:102489
40. Kim SH, Roh MI, Oh MJ, Park SW, Kim II (2020) Estimation of ship operational efficiency from AIS data using big data technology. Intl J Nav Arch Ocean Eng 12:440–454
41. Han W, Yang W, Gao S (2016) Real-time identification and tracking of emission from vessels based on automatic identification system data. In: 2016 13th international conference on service systems and service management. IEEE, China, pp 1–6
42. Jalkanen JP, Brink A, Kalli J, Pettersson H, Kukkonen J, Stipa T (2009) A modelling system for the exhaust emissions of marine traffic and its application in the baltic sea area. Atmos Chem Phys 9(23):9209–9223
43. Fan A, Yang J, Yang L, Wu D, Vladimir N (2022) A review of ship fuel consumption models. Ocean Eng 264:112405
44. Wang S, Ji B, Zhao J, Liu W, Xu T (2018) Predicting ship fuel consumption based on LASSO regression. Transp Res Part D: Transp Environ 65:817–824
45. Coraddu A, Oneto L, Baldi F, Anguita D (2017) Vessels fuel consumption forecast and trim optimisation: a data analytics perspective. Ocean Eng 130:351–370
46. Moreno-Gutiérrez J, Calderay F, Saborido N, Boile M, Valero RR, Durán-Grados V (2015) Methodologies for estimating shipping emissions and energy consumption: a comparative analysis of current methods. Energy 86:603–616
47. Moreno-Gutiérrez J, Pájaro-Velázquez E, Amado-Sánchez Y, Rodríguez-Moreno R, Calderay-Cayetano F, Durán-Grados V (2019) Comparative analysis between different methods for calculating on-board ship's emissions and energy consumption based on operational data. Sci Total Environ 650:575–584
48. Fenhann JV (2017) CO_2 emissions from international shipping. Technical University of Denmark, Technical report

49. Magnussen BB, Bläser N, Lu H (2023) DAISTIN: a data-driven AIS trajectory interpolation method. In: Proceedings of the 18th international symposium on spatial and temporal data. ACM, Canada, pp 75–84
50. Zhang Z, Fan Z, Lv Z, Song X, Shibasaki R (2024) Long-term vessel trajectory imputation with physics-guided diffusion probabilistic model. In: Proceedings of the 30th ACM SIGKDD conference on knowledge discovery and data mining. ACM, Spain, pp 4398–4407
51. Zhang D, Li J, Wu Q, Liu X, Chu X, He W (2017) Enhance the AIS data availability by screening and interpolation. In: 2017 4th international conference on transportation information and safety. IEEE, Canada, pp 981–986
52. Kontopoulos I, Varlamis I, Tserpes K (2021) A distributed framework for extracting maritime traffic patterns. Intl J Geograph Inf Sci 35(4):767–792
53. Guo S, Mou J, Chen L, Chen P (2021) Improved kinematic interpolation for AIS trajectory reconstruction. Ocean Eng 234:109256
54. Liang M, Su J, Liu RW, Lam JSL (2024) AISClean: AIS data-driven vessel trajectory reconstruction under uncertain conditions. Ocean Eng 306:117987
55. Troupiotis-Kapeliaris A, Zissis D, Bereta K, Vodas M, Spiliopoulos G, Karantaidis G (2023) The big picture: an improved method for mapping shipping activities. Remote Sens 15(21):5080
56. Li S, Liang M, Wu X, Liu Z, Liu RW (2020) AIS-based vessel trajectory reconstruction with U-Net convolutional networks. In: 2020 IEEE 5th international conference on cloud computing and big data analytics. IEEE, China, pp 157–161
57. Ronneberger O, Fischer P, Brox T (2015) U-Net: Convolutional networks for biomedical image segmentation. In: Medical image computing and computer-assisted intervention-MICCAI 2015: 18th international conference. Springer, Germany, pp 234–241
58. Carrillo-Perez B, Barnes S, Stephan M (2022) Ship segmentation and georeferencing from static oblique view images. Sensors 22(7):2713
59. Guo Y, Liu RW, Qu J, Lu Y, Zhu F, Lv Y (2023) Asynchronous trajectory matching-based multimodal maritime data fusion for vessel traffic surveillance in inland waterways. IEEE Trans Intell Transp Syst 24(11):12779–12792
60. Huang Z, Hu Q, Mei Q, Yang C, Wu Z (2021) Identity recognition on waterways: A novel ship information tracking method based on multimodal data. J Navig 74(6):1336–1352
61. Wu S, Troupiotis-Kapeliaris A, Zissis D, Torp K, Zimányi E, Sakr M (2024) Uncertainty-aware ship location estimation using multiple cameras in coastal areas. In: 2024 25th IEEE international conference on mobile data management. IEEE, Belgium, pp 109–118
62. Geudtner D, Torres R, Snoeij P, Davidson M, Rommen B (2014) Sentinel-1 system capabilities and applications. In: 2014 IEEE geoscience and remote sensing symposium. IEEE, Canada, pp 1457–1460
63. Gascon F, Cadau E, Colin O, Hoersch B, Isola C, Fernández BL, Martimort P (2014) Copernicus Sentinel-2 mission: products, algorithms and cal/val. Earth Observing Systems XIX. SPIE, USA, pp 455–463
64. Cope S, Tougher B, Zetterlind V, Gilfillan L, Aldana A (2023) Building a practical multi-sensor platform for monitoring vessel activity near marine protected areas: case studies from urban and remote locations. Remote Sens 15(13):3216
65. Wu Y, Chu X, Deng L, Lei J, He W, Królczyk G, Li Z (2022) A new multi-sensor fusion approach for integrated ship motion perception in inland waterways. Measurement 200:111630
66. Loomans MJ, de With PH, Wijnhoven RG (2013) Robust automatic ship tracking in harbours using active cameras. In: 2013 IEEE international conference on image processing. IEEE, Australia, pp 4117–4121
67. Zwemer MH, Wijnhoven RG, de With PH (2018) Ship detection in harbour surveillance based on large-scale data and CNNs. VISIGRAPP (5: VISAPP). SciTePress, Portugal, pp 153–160
68. Kessler GC, Zorri DM (2024) AIS spoofing: A tutorial for researchers. In: 2024 IEEE 49th conference on local computer networks. IEEE, France, pp 1–7
69. Kontopoulos I, Spiliopoulos G, Zissis D, Chatzikokolakis K, Artikis A (2018) Countering real-time stream poisoning: an architecture for detecting vessel spoofing in streams of AIS data. In: 2018 IEEE 16th international conference on dependable, autonomic and secure computing, 16th

international conference on pervasive intelligence and computing, 4th international conference on Big data intelligence and computing and cyber science and technology congress. IEEE, Greece, pp 981–986

70. Andrienko N, Andrienko G (2021) Visual analytics of vessel movement. Springer International Publishing, Cham, pp 149–170

71. Lei J, Chu X, He W (2021) Trajectory data restoring: a way of visual analysis of vessel identity base on OPTICS. J Web Eng 20(2):413–430

72. Ankerst M, Breunig MM, Kriegel HP, Sander J (1999) OPTICS: ordering points to identify the clustering structure. ACM SIGMOD Rec 28(2):49–60

73. Goerlandt F, Montewka J, Zhang W, Kujala P (2017) An analysis of ship escort and convoy operations in ice conditions. Saf Sci 95:198–209

74. Soldi G, Gaglione D, Raponi S, Forti N, d'Afflisio E, Kowalski P, Millefiori LM, Zissis D, Braca P, Willett P, Maguer A, Carniel S, Sembenini G, Warner C (2023) Monitoring of critical undersea infrastructures: the nord stream and other recent case studies. IEEE Aerosp Electron Syst Mag 38(10):4–24

75. Troupiotis-Kapeliaris A, Kastrisios C, Zissis D (2025) Vessel trajectory data mining: a review. IEEE Access 13:4827–4856

Chapter 14
Network-Constrained Trajectory Data for Traffic Analytics

Rodrigo Sasse David⊙, Kristian Torp⊙, Mahmoud Sakr⊙, and Esteban Zimányi⊙

Abstract Network-constrained trajectory analysis is the study of movement constrained by an underlying network, such as a road system. Such analysis has become increasingly important in addressing contemporary urban challenges. With the growing availability of Global Navigation Satellite System (GNSS) data, it is now possible to analyze mobility patterns at scale and with high precision. This chapter explores how GNSS data can be leveraged to support urban planning, with a particular focus on two key applications: road traffic management and sustainable mobility. By examining vehicle trajectories, we highlight methods for quantifying congestion, monitoring traffic dynamics, and promoting environmentally friendly practices such as eco-driving and eco-routing.

Keywords Network-constrained trajectory · GNSS · Eco-driving · Eco-routing · Traffic monitoring

14.1 Introduction

Over the last three decades, the world has witnessed rapid economic growth, which, among other things, has led to a large increase in motorization and urbanization. As a consequence, the heavy traffic caused by the increased number of vehicles has

R. S. David (✉) · K. Torp
Department of Computer Science, Aalborg University, Aalborg, Denmark
e-mail: rsd@cs.aau.dk

K. Torp
e-mail: torp@cs.aau.dk

M. Sakr · E. Zimányi
Department of Computer and Decision Engineering, Université Libre de Bruxelles, Bruxelles, Belgium
e-mail: mahmoud.sakr@ulb.ac.be

E. Zimányi
e-mail: esteban.zimanyi@ulb.ac.be

G. Dejaegere et al. (eds.), *Data Engineering for Data Science*,
https://doi.org/10.1007/978-3-032-18765-9_14

become a major issue across the globe—heavy traffic not only increases the time people spend commuting, but it also degrades the air quality, poses a significant safety risk, and may even increase the general feeling of inequity in society [1].

The transportation sector is a significant contributor to greenhouse gas (GHG) emissions, accounting for approximately 25% of total emissions in developed countries. This substantial impact stems from various modes of transport, including freight, public transit, and personal vehicles. However, passenger vehicles are particularly noteworthy, as they alone are responsible for nearly 45% of emissions within this sector [2]. The reliance on fossil fuels for powering these vehicles is a major contributor to this high level of emissions. Additionally, the increasing number of vehicles on the road, along with urban sprawl and inefficient traffic management, exacerbates the situation. As cities grow and populations increase, the demand for personal transportation continues to rise, leading to higher emissions [3].

In order to better understand and improve road traffic, municipalities and transit authorities are interested in collecting information about the different traffic participants. With this information, they can provide various services that enable smoother, safer, and environmentally friendly transportation [4]. Although quite valuable, traffic monitoring is among the most challenging problems in transportation [5].

Initially, traffic was monitored by humans manually counting the number of vehicles traveling through a road or intersection of interest [6]. Recently, new technology has allowed junctions to be monitored with loop detectors, cameras, or wireless devices [7]. Such devices automate the task, but with major drawbacks, such as low coverage and high cost [8]. Since such devices have to be installed at specific locations, another issue is identifying where and when monitoring should take place [9].

More recently, network-constrained trajectory data has presented itself as a resourceful alternative for traffic monitoring and analytics [10]. First, it's relatively cheaper to obtain and maintain compared to many conventional traffic monitoring technologies [11]. Second, it offers a much broader and more comprehensive coverage of the network [12]. Third, the sheer volume and granularity of network-constrained trajectory data can enable more advanced analytics, such as predictive traffic modeling and route optimization [13].

In this chapter, we synthesize and present how network-constrained trajectory data can be used for traffic analytics. The particularity of such data is that, in contrast to free-range trajectories presented in Chap. 13, the positions represented by this data are constrained by an underlying road infrastructure. This infrastructure can be represented by a network and exploited to enhance the analysis. It is worth noting that the following review does not seek to provide an exhaustive overview of the literature, but highlights some of the relevant work in order to illustrate applications of network-constrained trajectory data in different areas.

The remainder of the chapter is organized as follows. Section 14.2 presents a background of network-constrained trajectory data. Section 14.3 introduces key definitions. Section 14.4 presents the state of the art. Section 14.5 discusses open problems and future directions. And Sect. 14.6 concludes the chapter.

14.2 Background

In the early 1980s, the Global Positioning System (GPS) was made available for civilian use, and in the late 1990s, GPS satellites began transmitting two additional signals to be used to improve aircraft safety. Since then, demand for satellite-based navigation has grown steadily across many sectors and the technology has become a critical part of everyday life, especially as it has been integrated into a wide array of devices, from handheld navigation units to smartphones.

Private vehicle owners were among the earliest adopters, and by the early 2000s, in-car navigation systems became a common feature in many consumer vehicles. This trend only accelerated with the rise of smartphones, which combined Global Navigation Satellite System (GNSS) with mapping applications like Google Maps and Apple Maps, giving drivers turn-by-turn directions and real-time traffic updates. Nowadays, privately owned vehicles represent the dominant consumers of GNSS technology, with billions of people around the world relying on it for navigation. Beyond transportation, GNSS is now used in everything from fitness tracking and agriculture to logistics, disaster response, and even scientific research, making it one of the most widely adopted and influential technologies of the 21st century.

The quick development of wireless communication and data acquisition technologies, combined with the evolution of technologies that enable storing and processing large data volumes, has contributed to the significant growth of applications that deal with trajectory data. Trajectory data records the object's location in space at a certain instant of time. According to Zheng [14], there are four categories of trajectory data: mobility of people, mobility of vehicles, mobility of animals, and mobility of natural phenomena.

Dealing with large-scale spatial data is a research topic called Spatial Big Data [15], in which the issues related to Big Data applications are handled to enable the development of geographical information systems. Since the volume of trajectory data is usually very large, it is necessary to deploy an infrastructure that can analyze this massive data properly, solve complex queries, extract relevant insights, and support the decision-making process.

Over the last two decades, a lot of effort has been put into developing commercial solutions to address the former challenges. An example of such a system is MobilityDB [16]. MobilityDB is an open-source spatiotemporal database extension for PostgreSQL and PostGIS. It provides advanced functionality for managing and analyzing temporal and spatiotemporal data. Designed for applications that deal with dynamic geospatial data, such as moving objects or time-evolving geographies, MobilityDB extends PostgreSQL's capabilities with additional data types, operators, and functions.

The field of trajectory data is extensive and encompasses many aspects. So many that there is no review capable of addressing them all at the same time. We highlight some recent surveys that focus on different aspects of trajectory data. The survey of Parent et al. [17] presents an analysis of mobility data management, listing and discussing the main techniques for building, enriching, mining, and extracting knowledge from trajectory data. The survey of Kong et al. [18] presents trajectory applications and data from travel behavior, travel patterns, trajectory data service description in terms of transport management, and other aspects. The survey of Bian et al. [13] presents a set of trajectory clustering techniques, classifying them into three categories: unsupervised, supervised, and semi-supervised. The survey of Feng and Zhu [19] presents some trajectory mining applications, such as pathfinding, location prediction, and mobile object behavior analysis. The survey of Alsahfi et al. [20] reviews existing studies on storing, managing, and analyzing trajectory data, and proposes a framework to provide the requirements for building trajectory data warehouses. Finally, Chap. 13 of this book proposes a study of free-range trajectories focused on the analysis of AIS data.

Large-scale trajectory data are increasingly being collected in urban areas, thanks to the widespread adoption of technologies such as GNSS and mobile devices. Those data offer a rich and detailed representation of the movement of people, vehicles, and goods within the city, making them a valuable tool for various applications, particularly in the context of real-time smart cities and timely decision-making. Smart cities aim to enhance the quality of life for their inhabitants by leveraging technology to improve urban infrastructure, services, and governance. Large-scale trajectory data play a crucial role in enabling real-time insights and informed decisions, thereby contributing to the optimization of urban environments.

In this review, we focus on two key applications of network-constrained trajectory data in urban planning: road traffic management and green transport. These areas are critical to the efficient functioning of cities and are closely tied to sustainability and livability.

14.3 Definitions

In this section, we present and discuss fundamental terminologies in the field of trajectory data analytics.

14.3.1 GNSS Data

GNSS is a navigation system that provides geolocation and time information to a GNSS receiver anywhere on or near the Earth where there is an unobstructed line of sight to four or more satellites. As of 2024, four global systems are operational: the United States Global Positioning System (GPS), Russia's Global Navigation Satellite System (GLONASS), China's BeiDou Navigation Satellite System (BDS), and the European Union's Galileo.

A GNSS receiver determines the position of the object it is tracking by translating satellite signals into geographic coordinates. It can also offer some details about the movement of the object being tracked. In this context, a GNSS point p

Table 14.1 GNSS data and its attributes

vid	lat	lon	altitude	heading	speed	accuracy	timestamp
1	55.654923	11.299352	12.9	271	33	2.1	2017-05-27 13:25:34
1	55.65493	11.299348	13.0	284	37	1.8	2017-05-27 13:25:35
1	55.654957	11.299348	12.9	295	41	1.9	2017-05-27 13:25:36
1	55.65493	11.29935	12.8	296	39	2.1	2017-05-27 13:25:37
1	55.654925	11.299358	13.0	293	35	1.9	2017-05-27 13:25:38

can be represented as a tuple (vid, lat, lon, $altitude$, $heading$, $speed$, $precision$, $timestamp$), in which vid is the identifier of the vehicle being tracked, lat is the latitude coordinate, lon is the longitude coordinate, $altitude$ is the elevation above sea level, $heading$ is the direction of movement in degrees, $speed$ is the speed, $accuracy$ is the precision of the horizontal measurement, and $timestamp$ is the timestamp of when the point was sampled.

GNSS data, known for its rich detail and relatively low cost, has been widely used in mobility-based applications. These include monitoring wildlife migration patterns [21], tracking human movement [22], coordinating fleet management [23], and even studying the movement of natural events like hurricanes [24]. In commercial sectors, GNSS data has driven innovations across various industries, including transportation, agriculture, environmental science, and public safety [25].

Table 14.1 presents a sample dataset of GNSS readings collected from a vehicle, illustrating key parameters such as altitude (measured in meters), heading (in degrees), speed (in kilometers per hour), and accuracy (in meters).[1]

14.3.2 OBD Data

The on-board diagnostics (OBD) is a system built into vehicles that monitors and reports on the performance of various engine components and other critical systems. Initially introduced to improve emissions control, OBD now provides a comprehensive way for technicians and vehicle owners to diagnose issues.

Modern OBD systems, known as OBD-II, have evolved to include a broader range of diagnostics and can process information such as engine revolutions, vehicle speed and acceleration, fuel usage, and exhaust emissions. In 1996, the United States made OBD-II mandatory for all passenger cars and petro-powered light trucks. In 2001, the European Union made OBD-II mandatory for all petrol vehicles sold in the union.

---

[1] https://www.gps.gov/systems/gps/performance/accuracy/.

In 2006, all vehicles manufactured in Australia and New Zealand were required to be OBD-II compliant.

In addition to the traditional role of vehicle diagnostics, modern OBD-II systems have become an integral part of vehicle performance analysis, particularly regarding fuel consumption estimation and optimization. Studies focused on fuel efficiency often integrate OBD-II data with GNSS data to create more accurate and context-specific models of fuel consumption [26].

14.3.3 Road Network

A road network is an advanced mapping system that uses digital data to represent and manage the road infrastructure in a given area. This network includes detailed information about roads, intersections, traffic signals, speed limits, and other relevant features of the transportation system.

A road network can be seen as a graph $G = (V, E)$, in which V is a set of vertices v representing the intersections and terminal points of the road segments, E is a set of edges e representing road segments, and each vertex has a unique ID ranging from 1 to $|V|$. Figure 14.1 illustrates an example of the road network.

In the context of road networks, road segments are essential for modern transportation and mobility systems. Each segment is typically annotated with key information, including its length, category, and speed limit. Additionally, road segments can be assigned weights, such as average speed, travel time, or fuel consumption, which help enhance routing algorithms. These annotations enable algorithms like Dijkstra or A* to calculate optimal routes based on various factors, such as distance, time, or fuel efficiency.

Building and maintaining up-to-date road networks is an exhausting and costly task. This is one of the reasons why most digital maps available, such as Azure Maps[2] or Google Maps,[3] require a fee or subscription for their premium services. Among the free and open-source options, OpenStreetMap is the most popular and widely used in the literature [27].

14.3.4 Trajectory and Segment Data

A trajectory is a sequence of geographic coordinates that captures the movement of an object over time. Given the widespread use of GNSS technology, most trajectories in the literature are derived from GNSS data, reflecting the position of the object at various points in time. In this context, a trajectory T, often referred to as a raw trajectory, is essentially a time-ordered sequence of GNSS points $\{p_1, p_2, ..., p_n\}$,

---

[2] https://azure.microsoft.com/en-us/pricing/details/azure-maps/.

[3] https://mapsplatform.google.com/pricing/.

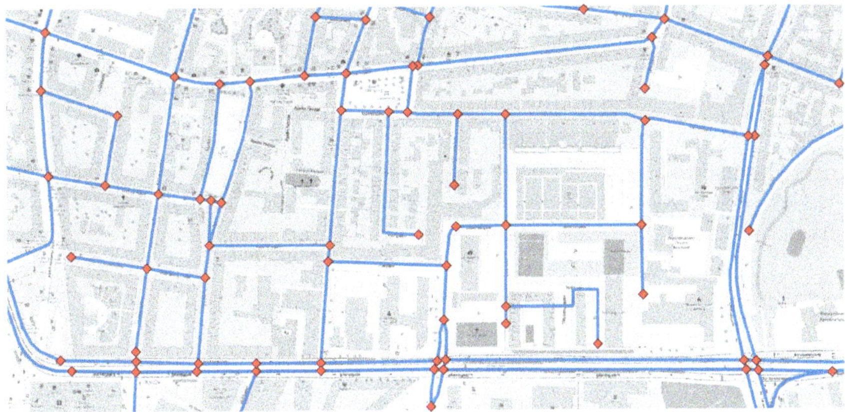

Fig. 14.1 An example of a road network. In the figure, the red squares are intersections and terminal points; the blue lines are road segments

where each point p represents a specific location at a given moment and is expressed as a tuple of geographic coordinates, latitude and longitude, and a timestamp. The length of a trajectory T refers to the total number of points it contains, while the sampling rate indicates how frequently these points are recorded, typically measured in samples per unit of time (such as seconds).

Trajectories provide detailed records of the movement of objects, but in many applications, it may be necessary to further refine the data to represent actual routes taken, especially when the trajectories span over complex road networks. This is when a process called map-matching comes into play [28]. During map-matching, each GNSS point p in the trajectory is aligned with the nearest road segment in a digital map, effectively translating the raw trajectory T into a network-constrained trajectory, denoted as T_{net}. The resulting trajectory T_{net} is now a sequence of connected road segments, in which each segment maps to an edge of the road network G. Also, when a trajectory is transformed, other important information can be annotated to each segment, e.g., the travel time on the segment, the average speed, and the average fuel consumption. This way, a network-constrained trajectory can be seen as an ordered sequence of segments, and each segment can be seen as $\{vid, tid, seg\_no, edge\_id, travel\_time, avg\_speed, consumption\}$, in which vid is the identifier of the vehicle that generated the trajectory, tid is the identifier of the trajectory, seg_no is the sequential number of the segment, $edge\_id$ is the identifier of the edge, $travel\_time$ is how long the vehicle took to travel through the segment, avg_speed is the average speed of the vehicle as it traveled through the segment, and $consumption$ is how much fuel (or energy) it is used to travel through the segment. Table 14.2 shows an example of a network-constrained trajectory. In the table, $travel\_time$ is in seconds, avg_speed is in kilometers per hour, and $consumption$ is in kilowatt-hour.

Table 14.2 An example of a network-constrained trajectory

vid	tid	seg_no	edge_id	travel_time	avg_speed	consumption
1	1	1	307810	15.2	34.1	0.052
1	1	2	445043	58.1	5.4	−0.026
1	1	3	307811	34.7	10.5	0.032
1	1	4	307812	14.5	32.7	0.035
1	1	5	605369	15.0	31.1	0.029

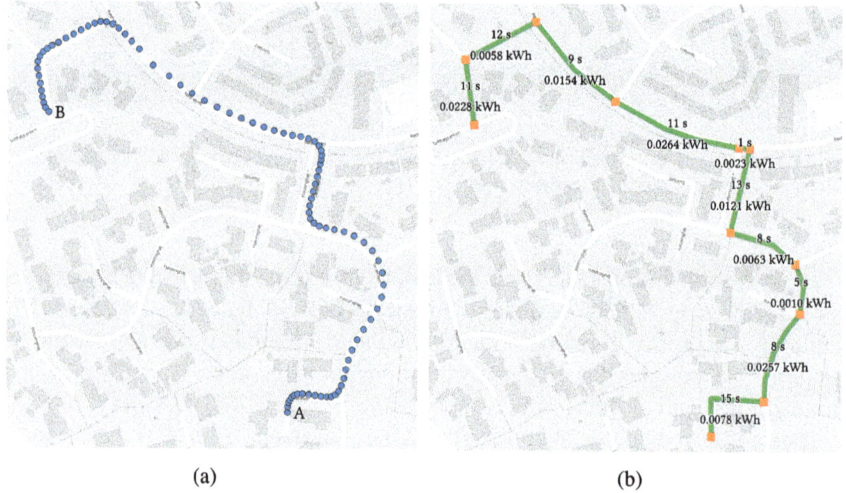

Fig. 14.2 An example of a raw trajectory (**a**) and a network-constrained trajectory (**b**). In **a** the vehicle travels from A to B. In **b** each green line is a segment annotated with the travel time and energy consumption of the vehicle; the orange squares are the intersections of the road network

This transformation from a raw trajectory to a network-constrained trajectory enhances the utility of the data for applications such as navigation, traffic analysis, and transportation planning, where understanding the actual path taken is critical. Figure 14.2 shows an example of a raw trajectory and a network-constrained trajectory. In the figure, the accuracy (or imprecision) of the GNSS device is clearly illustrated.

14.4 State of the Art

In this section, we provide an overview of the current state of the art regarding the use of trajectory data for traffic analytics. The review is centered around two key areas: road traffic management and green transport.

For road traffic applications, we specifically address traffic monitoring as well as the analysis of traffic congestion and flow. In the context of green transport, we explore applications related to eco-driving and eco-routing.

14.4.1 Applications in Road Traffic

Since trajectories capture the movement of vehicles on the road, numerous visualization systems have been developed to observe and monitor both past and real-time traffic trends and movement patterns based on trajectory data. These systems allow users to interactively explore traffic conditions in specific areas or along particular roads, and take control of traffic management if needed.

Real-time traffic monitoring enables analysts to track the usage of specific roads or regions, helping to detect traffic jams and notify commuters through digital traffic signs or GNSS applications. By proactively managing traffic flow, these systems contribute to reducing congestion, minimizing delays, and enhancing overall road safety. Furthermore, they provide valuable insights that can inform long-term infrastructure planning and the optimization of public transportation systems.

In this subsection, we review key studies in the areas of traffic monitoring, as well as traffic jam and flow analysis, all of which utilize trajectory data.

14.4.1.1 Traffic Monitoring

Wang et al. [29] presented an interactive system for visual analysis of urban traffic congestion based on trajectory data. The system measures the average speed on each road segment and concatenates spatially and temporally related events in so-called traffic jam propagation graphs. The graphs form a high-level description of a traffic jam and its propagation in time and space.

Guo et al. [1] presented an interactive visual analytics system named TripVista, for exploring and analyzing complex traffic trajectory data. The authors developed visualization methods to investigate and analyze traffic patterns and abnormal behaviors. By simultaneously inspecting the time-varying trajectory data from spatial, temporal, and multi-dimensional views, the system provides powerful tools for data comprehension and exploration.

David et al. [30] tackled the challenging problem of monitoring roundabouts and signalized intersections by presenting a framework for the automated identification and monitoring of all junctions in a road network. The framework utilizes detailed trajectory data or high-level segment-based data to compute the travel time and energy consumption for all turn directions. These metrics are then aggregated per junction to enable a fair comparison between different types of junctions. The aggregated metric is used to provide an overview of all junctions and to pinpoint those performing poorly. Finally, the authors used the framework to analyze 1,394 junctions and quantify the different benefits of roundabouts and intersections.

Borresen et al. [31] created an interactive system to analyze intersections using trajectory data. With the system, users select an intersection of interest in a map and select a point before and after the intersection to mark where to start and end measuring the travel time. Then, the system issues a strict path query [32] with the location of these points to find all trajectories traveling through them. Finally, the system builds a travel time matrix showing the average travel time and number of traversals for the intersection. The system also allows users to filter by time of day, weekdays, and months.

Wang et al. [33] demonstrated an intelligent traffic analytics system called T4. The system allows for intelligent analytics over real-time and historical trajectories from vehicles. At the front end, the system allows for the visualization of the current traffic flow and result trajectories of different types of queries, e.g., only for a certain period of the day or certain days of the week, as well as the histograms of traffic flow and traffic lights. At the back end, the system is able to support multiple types of common queries over trajectories. The authors demonstrated the system working in four scenarios: trajectory search, trajectory visualization, traffic prediction, and traffic light control.

Pang et al. [34] proposed an efficient pattern mining approach for spatiotemporal traffic data from taxis in order to detect outliers in Beijing, China. The authors also proposed two statistical models, which encompass the generic features of anomalous patterns. In their case studies, their model is able to detect regions with an emerging number of taxis that can be validated by known major traffic events.

Wang et al. [35] proposed a three-layer framework for the recognition and visualization of multiscale traffic patterns. The first layer computes the middle-tier synopses at fine spatial and temporal scales, which are indexed and stored in a geodatabase. The second layer uses synopses to efficiently extract multiscale traffic patterns. The third layer supports real-time interactive visual analytics for intuitive explorations by end users. The authors conducted an experiment in Shenzhen, China, using trajectories from taxis. The proposed visualization interfaces show various traffic patterns in different forms intuitively, which facilitates comparison, perception, and understanding of citywide road traffic.

14.4.1.2 Traffic Jam and Flow Analysis

Kan et al. [8] proposed an approach for detecting traffic congestion from GNSS trajectories from taxis at the turn level. Based on analyzing features of GNSS trajectories and identifying valid trajectory segments, the proposed approach detects congested trajectory segments of three different intensities—regional, road, and lane levels. Then, it identifies congestion events in each turning direction through a clustering approach. Finally, the authors analyzed congestion intensity, time of the day when congestion occurred, and queue length in each turning direction at a road intersection in Wuhan, China. The results showed that the proposed approach can identify valid trajectory segments accurately 95.4% of the time.

Wang et al. [36] used sparse trajectory data together with social media data to estimate traffic congestion of various arterial roads in Chigaco, United States. For the trajectory data, the authors used 2.4 million GNSS probe readings, and for the social media data, the authors used official traffic authority accounts on Twitter (now X) and traffic-related tweets from regular users. In the framework they proposed, they first map the GNSS and social media data to a road segment. Then, they construct two congestion matrices. Finally, the matrices are coupled and factorized. The authors also used the same pipeline with historical data to perform anomaly detection in the road network.

D'Andrea and Marcelloni [37] proposed a system for detecting traffic congestion and incidents from real-time GNSS data collected from GNSS trackers or drivers' smartphones. First, the system processes the GNSS points and matches them onto the road network. Then, it assigns to each road segment of the map a traffic state based on the speed of the vehicles. Finally, it sends to the users traffic alerts based on a spatiotemporal analysis of the classified segments. Each traffic alert contains the affected area, a traffic state, and the estimated velocity of vehicles in the area. The authors performed an analysis using a combination of simulated GNSS data and real GNSS data from the city of Pisa, Italy. The results showed a detection rate of 91.6% and an average detection time lower than 7 min.

Sun et al. [38] used trajectory data to predict traffic congestion in a road network. The authors used trajectory data to cope with the low coverage of the network offered by fixed detection devices, e.g., cameras or speed sensors. In their work, they first match the GNSS points to road segments. Then, they calculate the average driving speed of the network and convert the results into time sequences and spatiotemporal images. Then, they use convolutional and recurrent neural network models to predict the average driving speed of the network. Finally, the traffic congestion levels are classified according to the traffic congestion level standard. The results show that the convolutional network performed better than the recurrent network.

Chen et al. [39] proposed a model to uncover urban traffic congestion propagation patterns. Their model aims to discover how congested road segments spread traffic congestion to adjacent segments and how these new congested roads continue to propagate the congestion. The authors modeled the congestion propagation phenomenon with a spatiotemporal congestion subgraph. The authors used sparse taxi GNSS trajectory data. They proposed two strategies to fill missing congestion edges from both temporal and spatial viewpoints.

14.4.2 Applications in Green Transport

The use of trajectory data goes beyond merely monitoring traffic or analyzing flow—it also provides valuable insights into driver behavior, which is essential for promoting more sustainable transportation. By studying how drivers accelerate, brake, and navigate different routes, we can identify patterns that lead to fuel inefficiency and increased emissions.

Understanding these behaviors enables the development of targeted interventions, such as eco-driving and eco-routing. Eco-driving involves providing drivers with simple recommendations–such as accelerating gently or maintaining a consistent speed–to reduce fuel consumption and lower emissions. Eco-routing, on the other hand, focuses on optimizing travel routes to minimize fuel use, emissions, and overall environmental impact, while still factoring in considerations like travel time and convenience.

In this subsection, we review key studies in the areas of eco-driving and eco-routing, all of which utilize trajectory data.

14.4.2.1 Eco-Driving

Jakobsen et al. [40] compared five different eco-driving advice—accelerate moderately, maintain a steady speed, drive at or below the speed limit, anticipate traffic flow and avoid frequent starts and stops, and eliminate idling. The authors used real-world GNSS trajectories of four different vehicles over a period of eight months. The authors compared how good each driver was at following the advice and ranked them—the driver that managed to follow the advice more often showed better fuel consumption. The results also showed that, although no single advice is capable of reducing consumption too much, it is the many small improvements that lead to a good fuel economy.

Xu et al. [41] used GNSS trajectory data from buses collected by two local urban transit agencies in Atlanta, United States. The authors proposed an algorithm to balance top speed and acceleration rates using the modal emissions modeling framework. They also presented results from three scenarios—implementation of eco-driving cycles with the existing fleet, assessment of the purchase of new vehicles to replace the fleet, and a combination of implementation of eco-driving with new fleet purchase. The results showed that eco-driving can reduce fuel consumption by 5% in local transit service and 7% in express bus service.

Fleming et al. [42] proposed a real-time predictive eco-driving assistant. The system architecture has three layers—perception, decision, and action. The perception layer encompasses sensors like GNSS and OBD. The decision layers implement fuel consumption and driver behavior models. And the action layer offers a visual interface for the user. The assistant recommends the best speed for the driver to minimize fuel consumption considering the road geometry. The results showed a decrease of 6% in fuel consumption and 6.5% in travel time when using the assistance system versus unassisted eco-driving.

David et al. [43] investigated how the energy consumption of electric vehicles is affected when the vehicle is at a steady speed. Driving at a steady speed is part of the so-called eco-driving, a collection of simple advice aimed at changing driving behavior in order to reduce both energy consumption and emissions. The authors used a large collection of GNSS trajectories to quantify, on a segment level, how much energy can be saved by driving at a steady speed. The authors also investigated how different road segments and seasonality can affect energy consumption. Results showed that drivers can save up to 42% of energy, while increasing the travel time by just 10%.

14.4.2.2   Eco-Routing

Guo et al. [44] proposed a general framework named EcoMark 2.0, for eco-weight assignment to enable eco-routing. An eco-weight is how much fuel (or electricity) a vehicle consumes to travel through a road segment. The framework combines six instantaneous and five aggregated fuel consumption models to estimate vehicular environmental impact and uses travel information derived from GNSS trajectories for speed and acceleration, as well as actual fuel consumption data from vehicles. The authors compared how well each model performs on different circumstances. The results indicated which models are appropriate for assigning eco weights. They also showed that high-frequency GNSS data can be reliably used to estimate fuel consumption at the segment level.

David et al. [45] used trajectory data to investigate how roundabouts affect the energy consumption of EVs. The authors analyzed over 1.000 roundabouts all over Denmark and how different characteristics, such as radius, number of arms, and category, as well as factors such as seasonality and speed affect energy consumption. The results showed that there can be a difference of 25–33% in consumption inside roundabouts depending on their category. They also showed that the consumption highly correlates with the speed the vehicles are at right before they enter the roundabouts.

Zhu et al. [46] proposed a framework to identify fuel-saving opportunities through green routing. Their framework uses an enhanced pre-trip consumption rate estimation model to calculate fuel consumption over potential routes that do not have driving data, enabling estimation over routes that were not actually driven. It also uses a large-scale, real-world travel data set to quantify expected fuel savings from green routing. The framework then computes several routes from an origin to a destination and evaluates both fuel consumption and travel time between them. The results showed that 31% of actual routes show an opportunity for fuel savings and about two-thirds of the green routing savings come from routes estimated to save time as well.

Andersen et al. [47] proposed an eco-routing system named EcoTour. The authors used GNSS and OBD-II data to estimate the fuel consumption of vehicles and annotate road segments with their eco-weights. This allows the system to estimate not only the fastest and the shortest routes but also the route that uses less fuel, referred to as the eco-route. The system also displays the estimated travel time, distance, and fuel consumption of each route. The authors tested their system with the road network of Denmark.

Ding et al. [48] proposed a framework named GreenPlanner, for modeling personalized fuel consumption and recommending fuel-efficient routing for drivers. The framework is divided into two modules. The first module creates a fuel consumption model, taking into account the drivers' reactive and proactive behaviors. The second module uses real-time traffic information to estimate the fuel consumption among different routes for each driver, and suggest fuel-efficient routes. The authors used a dataset of taxi GNSS trajectories from Beijing, China. The results of the case study showed that drivers can save up to 20% on fuel on average if they follow the suggested routes.

Zeng et al. [49] proposed a routing algorithm to find a path that consumes the minimum amount of fuel while ensuring that the travel time satisfies a travel time budget and an on-time arrival probability. First, the authors created a model that estimates the fuel consumption for every road segment based on vehicle dynamics. They tested the algorithm in the road network of Toyota, Japan, using a dataset of GNSS trajectories from 150 probe vehicles. The results showed that there can be significant savings in fuel consumption, but as demand increases, the savings start to decrease, indicating that an eco-routing strategy can have a greater impact on fuel saving when traffic congestion is mild.

14.5 Challenges and Future Directions

In the last section, we presented and discussed various studies that effectively leveraged trajectory data for traffic monitoring, congestion detection, and traffic flow analysis. By leveraging trajectory data, researchers and practitioners have been able to develop more efficient traffic management strategies, detect and mitigate congestion hotspots, and improve overall road network performance.

Additionally, we presented and discussed studies that applied trajectory data to green transportation initiatives, with a particular emphasis on eco-driving and eco-routing. These approaches aim to optimize driving behavior and route selection to minimize fuel consumption and reduce emissions, contributing to a more sustainable and environmentally friendly transportation system.

From the discussions presented in Sect. 14.4, it is evident that trajectory data has emerged as a cost-effective yet highly reliable source of positional information. Its versatility and accessibility make it a valuable asset across a wide range of applications, ranging from urban traffic management to sustainable mobility solutions. As technology advances and data collection methods continue to improve, the role of trajectory data in shaping intelligent and eco-friendly transportation systems is expected to expand even further.

Despite the widespread use of trajectory data across various applications, several challenges still need to be overcome, and new directions remain to be assessed. In this section, we present and discuss some challenges and future directions of the field.

14.5.1 Challenges

Although many challenges exist, ranging from data cleaning to performance benchmarking, we choose to highlight three key challenges that are particularly relevant to the use of trajectory data for traffic analytics.

The first challenge is the lack of openly accessible real-time applications [14]. Most real-time applications are privately owned, catering to specific commercial interests. These systems often do not share data openly or integrate with public infrastructure. This lack of accessibility limits the potential benefits of trajectory data, as public agencies, researchers, and policymakers struggle to access the information needed for effective traffic management, urban planning, and sustainability initiatives. By enabling greater transparency and interoperability, openly accessible real-time applications can significantly enhance mobility solutions, improve resource allocation, and contribute to smarter, more sustainable transportation systems.

The second challenge is privacy [18]. Since trajectory data often involves tracking individuals or vehicles, it is essential to implement strong anonymization and security measures to protect user information. Without robust safeguards, there is a risk of data misuse, unauthorized access, and public distrust, which could hinder the adoption of such applications. In regions like the European Union, the use of tracking data is strictly regulated under laws such as the General Data Protection Regulation (GDPR), ensuring that privacy and data protection remain a top priority. As real-time applications continue to evolve, maintaining compliance with privacy regulations and building user trust will be crucial for their widespread acceptance and success.

Finally, the third challenge is data integration [10]. Although we focus on GNSS data in this chapter, other sources of position data exist, such as cameras, loop detectors, and wireless technologies like Bluetooth and Radio Frequency Identification (RFID). Integrating data from these diverse sources can enhance the accuracy, reliability, and comprehensiveness of traffic monitoring and analysis. Additionally, metadata plays a crucial role in real-world applications, providing valuable contextual information such as speed limits, transit timetables, and public service operating hours. While such metadata is commonly available in commercial applications, it remains scarce in publicly available datasets typically used by the research community. Addressing this gap could significantly improve the usability and effectiveness of open transportation data for academic and public sector applications.

14.5.2 Future Directions

Similar to the previous discussion, there are many directions yet to be explored or further investigated. We select three promising areas that are already attracting attention and show significant potential for the near future.

The first future direction is improving Vehicle-to-Vehicle (V2V) communications [50]. In V2V systems, vehicles exchange real-time information about their positions, speeds, and trajectories, allowing them to communicate with each other to enhance safety and coordination on the road. By sharing trajectory data, vehicles can anticipate each other's movements, reducing the risk of collisions, improving traffic flow, and enabling more efficient decision-making. In self-driving vehicles, trajectory data is vital for navigating and interacting with the surrounding environment. It allows autonomous vehicles to predict the movements of other vehicles, pedestrians, and obstacles, making real-time adjustments to their path. By integrating trajectory data with sensors, mapping systems, and machine learning algorithms, self-driving vehicles can operate safely and efficiently in dynamic traffic conditions.

The second future direction is improving safety [51]. Trajectory data can help prevent accidents by identifying dangerous driving behaviors, such as speeding or erratic movements, in real-time. This enables early intervention and improved traffic flow, reducing the risk of collisions. Furthermore, by integrating trajectory data with advanced safety technologies, such as collision avoidance systems and smart traffic signals, it is possible to create more responsive and proactive safety measures that protect both drivers and pedestrians. These advancements could significantly contribute to reducing traffic-related fatalities and injuries in the near future.

Finally, the third future direction is enriching trajectories [19]. While a trajectory, as a sequence of geographical locations and timestamps, is sufficient for many applications in traffic analytics, this representation fails to contextualize human dynamics, as it does not include any semantics regarding the movement. New trajectory representations are being proposed and explored, notably in the works of Spaccapietra et al. [52] and Su et al. [53]. The former proposes representing trajectories with three dimensions: space, time, and semantics, thereby enhancing the understanding of movement patterns. The latter suggests segmenting traditional trajectories and assigning a short, human-readable summary to each segment, offering a more intuitive interpretation of behavior. These enriched representations could unlock deeper insights into human mobility, improving the accuracy of traffic predictions, behavioral analysis, and the design of personalized transportation systems.

14.6 Conclusion

This chapter highlights the critical role of trajectory data in traffic analytics, emphasizing its use in traffic monitoring, traffic jam detection, traffic flow analysis, and green transport initiatives such as eco-driving and eco-routing. Derived mainly from GNSS position data, trajectory data has become a valuable resource for traffic-related applications, offering insights that enhance the efficiency and safety of urban mobility systems.

First, we discuss applications in road traffic management. We present several studies that utilize trajectory data to power advanced visualization systems that track road usage, identify congestion hotspots, and optimize traffic flow. These systems enable predictive traffic modeling and inform infrastructure planning, ultimately contributing to smoother traffic operations and improved safety. By leveraging trajectory data, municipalities can address issues like traffic congestion and inefficient traffic flow more effectively.

Second, we examine applications in green transport initiatives, particularly eco-driving and eco-routing. We present several studies that use trajectory data to analyze driving behaviors and identify patterns that lead to reduced fuel consumption and lower emissions. Strategies such as maintaining steady speeds, optimizing travel routes, and designing energy-efficient road infrastructure have shown great potential for energy savings and environmental benefits.

Despite the significant progress and advantages of trajectory data, several challenges remain. In this chapter, we identify and discuss three key issues: the lack of openly accessible real-time applications, privacy concerns, and data integration complexities. Addressing these challenges is essential to unlocking the full potential of trajectory data for both research and real-world applications. Additionally, we highlight three promising future directions: vehicle-to-vehicle communication, improved safety measures, and enriched trajectory representations. These advancements have the potential to further enhance traffic management, reduce accidents, and provide deeper insights into human mobility patterns.

Through an extensive review of the current state of the art, this chapter provides a comprehensive understanding of how trajectory data is shaping modern transportation systems. As technology evolves, continued research and collaboration will be crucial in overcoming existing challenges and exploring new opportunities to make transportation smarter, more sustainable, and safer for all.

References

1. Guo H, Wang Z, Yu B, Zhao H, Yuan X (2011) Tripvista: triple perspective visual trajectory analytics and its application on microscopic traffic data at a road intersection. In: Battista GD, Fekete JD, Qu H (eds) 2011 IEEE pacific visualization symposium. IEEE, China, pp 163 170
2. Van Fan Y, Perry S, Klemeš JJ, Lee CT (2018) A review on air emissions assessment: transportation. J Clean Prod 194:673–684
3. Schiller PL, Kenworthy J (2017) An introduction to sustainable transportation: policy, planning and implementation. Routledge
4. Bernas M, Płaczek B, Korski W, Loska P, Smyła J, Szymała P (2018) A survey and comparison of low-cost sensing technologies for road traffic monitoring. Sensors 18(10):3243
5. Zheng B, Lin CW, Shiraishi S, Zhu Q (2019) Design and analysis of delay-tolerant intelligent intersection management. ACM Trans Cyber-Phys Syst 4(1):1–27
6. Dinh H, Tang H (2017) Development of a tracking-based system for automated traffic data collection for roundabouts. J Mod Transp 25:12–23
7. Jain NK, Saini RK (2019) A review on traffic monitoring system techniques. Soft Comput Theor Appl Proc SoCTA 742:569–577
8. Kan Z, Tang L, Kwan MP, Ren C, Liu D, Li Q (2019) Traffic congestion analysis at the turn level using taxis' gps trajectory data. Comput Environ Urban Syst 74:229–243
9. Zheng J, Liu HX (2017) Estimating traffic volumes for signalized intersections using connected vehicle data. Transp Res Part C Emerg Technol 79:347–362

10. Wang S, Bao Z, Culpepper JS, Cong G (2021) A survey on trajectory data management, analytics, and learning. ACM Comput Surv (CSUR) 54(2):1–36
11. Byon Y, Shalaby A, Abdulhai B (2006) Travel time collection and traffic monitoring via gps technologies. In: 2006 IEEE intelligent transportation systems conference. IEEE, Canada, pp 677–682
12. Wang J, Wang C, Song X, Raghavan V (2017) Automatic intersection and traffic rule detection by mining motor-vehicle gps trajectories. Comput Environ Urban Syst 64:19–29
13. Bian J, Tian D, Tang Y, Tao D (2019) Trajectory data classification: a review. ACM Trans Intell Syst Technol (TIST) 10(4):1–34
14. Zheng Y (2015) Trajectory data mining: an overview. ACM Trans Intell Syst Technol (TIST) 6(3):1–41
15. Shekhar S, Gunturi V, Evans MR, Yang K (2012) Spatial big-data challenges intersecting mobility and cloud computing. In: Proceedings of the eleventh ACM international workshop on data engineering for wireless and mobile access, association for computing Machinery, USA, pp 1–6
16. Zimányi E, Sakr M, Lesuisse A (2020) Mobilitydb: a mobility database based on postgresql and postgis. ACM Trans Database Syst (TODS) 45(4):1–42
17. Parent C, Spaccapietra S, Renso C, Andrienko G, Andrienko N, Bogorny V, Damiani ML, Gkoulalas-Divanis A, Macedo J, Pelekis N et al (2013) Semantic trajectories modeling and analysis. ACM Comput Surv (CSUR) 45(4):1–32
18. Kong X, Li M, Ma K, Tian K, Wang M, Ning Z, Xia F (2018) Big trajectory data: a survey of applications and services. IEEE Access 6:58295–58306
19. Feng Z, Zhu Y (2016) A survey on trajectory data mining: techniques and applications. IEEE Access 4:2056–2067
20. Alsahfi T, Almotairi M, Elmasri R (2020) A survey on trajectory data warehouse. Spat Inf Res 28(1):53–66
21. Calenge C, Dray S, Royer-Carenzi M (2009) The concept of animals' trajectories from a data analysis perspective. Eco Inform 4(1):34–41
22. Rudenko A, Palmieri L, Herman M, Kitani KM, Gavrila DM, Arras KO (2020) Human motion trajectory prediction: a survey. Int J Robot Res 39(8):895–935
23. Fan J, Fu C, Stewart K, Zhang L (2019) Using big gps trajectory data analytics for vehicle miles traveled estimation. Transp Res Part C Emerg Technol 103:298–307
24. Ejigu YG, Teferle FN, Klos A, Bogusz J, Hunegnaw A (2021) Monitoring and prediction of hurricane tracks using gps tropospheric products. GPS Solut 25(2):76
25. Lin M, Hsu WJ (2014) Mining gps data for mobility patterns: a survey. Pervasive Mob Comput 12:1–16
26. Malekian R, Moloisane NR, Nair L, Maharaj BT, Chude-Okonkwo UA (2016) Design and implementation of a wireless obd ii fleet management system. IEEE Sens J 17(4):1154–1164
27. Minghini M, Frassinelli F (2019) Openstreetmap history for intrinsic quality assessment: is osm up-to-date? Open Geospatial Data Softw Stand 4(1):1–17
28. Newson P, Krumm J (2009) Hidden markov map matching through noise and sparseness. In: Proceedings of the 17th ACM SIGSPATIAL international conference on advances in geographic information systems. Association for Computing Machinery, USA, pp 336–343
29. Wang Z, Lu M, Yuan X, Zhang J, Van De Wetering H (2013) Visual traffic jam analysis based on trajectory data. IEEE Trans Visual Comput Graph 19(12):2159–2168
30. David RS, Torp K, Sakr M, Zimányi E (2024) A framework for automated junction monitoring. In: Proceedings of the 32nd ACM international conference on advances in geographic information systems. Association for Computing Machinery, USA, pp 304–313
31. Borresen JL, Andersen O, Jensen CS, Torp K (2017) Interactive intersection analysis using trajectory data. In: Proceedings of the 25th ACM SIGSPATIAL international conference on advances in geographic information systems. Association for Computing Machinery, USA, pp 1–4
32. Krogh B, Pelekis N, Theodoridis Y, Torp K (2014) Path-based queries on trajectory data. In: Proceedings of the 22nd ACM SIGSPATIAL international conference on advances in geographic information systems. Association for Computing Machinery, USA, pp 341–350

33. Wang S, Shen Y, Bao Z, Qin X (2019) Intelligent traffic analytics: from monitoring to controlling. In: Proceedings of the twelfth ACM international conference on web search and data mining. Association for Computing Machinery, USA, pp 778–781
34. Pang LX, Chawla S, Liu W, Zheng Y (2013) On detection of emerging anomalous traffic patterns using gps data. Data Knowl Eng 87:357–373
35. Wang Q, Lu M, Li Q (2020) Interactive, multiscale urban-traffic pattern exploration leveraging massive gps trajectories. Sensors 20(4):1084
36. Wang S, Zhang X, Cao J, He L, Stenneth L, Yu PS, Li Z, Huang Z (2017) Computing urban traffic congestions by incorporating sparse gps probe data and social media data. ACM Transa Inf Syst (TOIS) 35(4):1–30
37. D'Andrea E, Marcelloni F (2017) Detection of traffic congestion and incidents from gps trace analysis. Expert Syst Appl 73:43–56
38. Sun S, Chen J, Sun J (2019) Traffic congestion prediction based on gps trajectory data. Int J Distrib Sens Netw 15(5):1550147719847440
39. Chen Z, Yang Y, Huang L, Wang E, Li D (2018) Discovering urban traffic congestion propagation patterns with taxi trajectory data. IEEE Access 6:69481–69491
40. Jakobsen K, Mouritsen SC, Torp K (2013) Evaluating eco-driving advice using gps/canbus data. In: Proceedings of the 21st ACM SIGSPATIAL international conference on advances in geographic information systems. Association for Computing Machinery, USA, pp 44–53
41. Xu Y, Li H, Liu H, Rodgers MO, Guensler RL (2017) Eco-driving for transit: an effective strategy to conserve fuel and emissions. Appl Energy 194:784–797
42. Fleming J, Yan X, Allison C, Stanton N, Lot R (2021) Real-time predictive eco-driving assistance considering road geometry and long-range radar measurements. IET Intel Transp Syst 15(4):573–583
43. David RS, Zimányi E, Torp K, Sakr M (2022) Speed and energy consumption for electrical vehicles. In: Proceedings of the 15th ACM SIGSPATIAL international workshop on computational transportation science. Association for Computing Machinery, USA, pp 1–10
44. Guo C, Yang B, Andersen O, Jensen CS, Torp K (2015) Ecomark 2.0: empowering eco-routing with vehicular environmental models and actual vehicle fuel consumption data. GeoInformatica 19:567–599
45. Sasse David R, Zimányi E, Torp K, Sakr M (2023) Roundabouts and the energy consumption of electrical vehicles. In: Proceedings of the 16th ACM SIGSPATIAL international workshop on computational transportation science. Association for Computing Machinery, Germany, pp 9–18
46. Zhu L, Holden J, Wood E, Gender J (2017) Green routing fuel saving opportunity assessment: a case study using large-scale real-world travel data. In: 2017 IEEE intelligent vehicles symposium (IV). IEEE, USA, pp 1242–1248
47. Andersen O, Jensen CS, Torp K, Yang B (2013) Ecotour: Reducing the environmental footprint of vehicles using eco-routes. 2013 IEEE 14th International Conference on Mobile Data Management. IEEE, Italy, pp 338–340
48. Ding Y, Chen C, Zhang S, Guo B, Yu Z, Wang Y (2017) Greenplanner: planning personalized fuel-efficient driving routes using multi-sourced urban data. In: 2017 IEEE international conference on pervasive computing and communications (PerCom). IEEE, USA, pp 207–216
49. Zeng W, Miwa T, Morikawa T (2020) Eco-routing problem considering fuel consumption and probabilistic travel time budget. Transp Res Part D: Transp Environ 78:102219
50. Zhang Z, Li X, Liu D, Luo T, Zhang Y (2020) Trajectory data driven v2v/v2i mode switching and bandwidth allocation for vehicle networks. IEEE Wirel Commun Lett 9(6):795–798
51. Xie K, Ozbay K, Yang H, Li C (2019) Mining automatically extracted vehicle trajectory data for proactive safety analytics. Transp Res Part C Emerg Technol 106:61–72
52. Spaccapietra S, Parent C, Damiani ML, de Macedo JA, Porto F, Vangenot C (2008) A conceptual view on trajectories. Data Knowl Eng 65(1):126–146
53. Su H, Zheng K, Zeng K, Huang J, Sadiq S, Yuan NJ, Zhou X (2015) Making sense of trajectory data: a partition-and-summarization approach. In: 2015 IEEE 31st international conference on data engineering. IEEE, Korea, pp 963–974

Conclusion

The chapters in this volume offer a comprehensive view of the current landscape of data engineering for data science, a field that has grown rapidly in both academic research and industrial practice. As a collective work, the book brings together perspectives from disciplines and methodologies. This diversity reflects the very essence of data science: an interdisciplinary endeavour where innovation is possible only through collaboration.

In these concluding remarks, we do not simply restate the individual contributions but rather seek to weave them into a coherent picture of the information management lifecycle. We revisit the four parts of the book, governance, storage and processing, preparation, and analysis, to draw out the common threads and tensions that run through them. We also look ahead, identifying the major directions in which the field is likely to evolve and the challenges that researchers, practitioners, and policy makers will need to address.

Governance: From Control to Enabling Use

The first part of the book addresses issues of data governance. At first glance, governance might seem to be a matter of control: defining access policies, ensuring compliance with regulations, or enforcing security. Yet the chapters here demonstrate that governance is more fundamentally about **making data usable**.

The integration of textual and structured sources shows how governance mechanisms allow heterogeneous data to be meaningfully combined. Data fusion illustrates the need to reconcile conflicts and build trust across sources. Privacy-aware relational synthesis demonstrates that without careful governance, sensitive data cannot be shared, replicated, or reused, thereby preventing scientific progress.

Governance, then, is not an external layer imposed after the fact. It is the **basis on which all subsequent work depends**. Without metadata and quality control, storage

© The Editor(s) (if applicable) and The Author(s) 2026
G. Dejaegere et al. (eds.), *Data Engineering for Data Science*,
https://doi.org/10.1007/978-3-032-18765-9

infrastructures cannot function effectively. Without privacy guarantees, preparation and analysis cannot proceed responsibly. Governance is therefore best seen not as a restriction, but as the enabler of everything that follows.

Storage and Processing: Efficiency, Scale, and Sustainability

The second part of the book focuses on the problems of storage and processing under the prism of handling modern data volumes and velocity. Data engineering has always been about managing scale, but in the past decade the **dimensions of scale** have multiplied: not just more records, but more features, more frequency, more temporality.

Feature selection illustrates how **dimensionality reduction remains a fundamental challenge**, not only for efficacy and efficiency but also for interpretability. Time series management systems exemplify the complexity of storing and querying data that arrives continuously and at high frequency. MLOps systems highlight that storage and processing are not only about passive data management, but about supporting iterative experimentation, reproducibility, and lifecycle tracking in machine learning.

At the lower lovel of the compute stack, there is **the critical interplay of software and hardware co-design**. The need for massive parallelism in data science applications renders the use of specialized accelerators, such as the GPUs, as an important component of the modern systems architecture. However, how to integrate such accelerators into the design is far from trivial as the last chapter of the second part illustrates.

Preparation: Trustworthy Data for Responsible Learning

The third part examines data preparation, an often underestimated but indispensable step in the pipeline. The chapters of this part demonstrate how preparation is not only about cleaning or transforming data, but also about ensuring trust, privacy, and transparency.

The exploration of blockchain-based federated learning shows how preparation intersects with **decentralization and security**, allowing sensitive data to be harnessed without central collection. Example-based explainability highlights the need for interpretability not only in final models but also during preparation, where training data must be scrutinized and understood. The survey of table search in data lakes reminds us that access to the right data is itself a preparation challenge: before one can clean, transform, or explain data, one must first find it.

Together, these chapters point to a central message: preparation is the **bridge between raw data and meaningful analysis**. Without robust preparation, even the most advanced analytical techniques risk producing biased, opaque, or unusable results.

Analysis: From Methods to Applications

The fourth part focuses on analysis, showcasing both **methodological advances and real-world applications**. The diversity of topics, ranging from adversarial learning in fraud detection, to approximate methods for inference queries, and to trajectory analysis of ships and vehicles, illustrates the breadth of data science today.

Despite this diversity, a unifying theme emerges: **analysis pushes the boundaries of what earlier stages make possible**. Fraud detection systems require secure and reliable governance to resist adversarial attacks. Inference queries over unstructured data depend on efficient storage and processing infrastructures. Trajectory analysis illustrates how preparation techniques, such as imputation or data augmentation, are prerequisites for reliable modeling.

The analysis chapters also remind us of the stakes. When analyzing traffic or emissions, the implications extend to urban sustainability and climate change. When detecting fraud, the consequences are financial and societal. **Analysis is thus where the value of data engineering and data science most visibly manifests, but it is also where their limitations are most consequential.**

Common Threads Across the Lifecycle

Looking across all four parts, several **recurring challenges** stand out:

First, **scalability** remains at the forefront. Whether in integration, time series, federated learning, or trajectory modeling, solutions must cope with data that is larger, faster, and more complex than ever before.

Second, **trust and transparency** are essential. Data fusion requires trust across sources; preparation requires transparent processes; analysis requires explainable models. Without trust and transparency, scalability alone is insufficient.

Third, **privacy and security** are universal. They appear in governance (data synthesis), in preparation (federated learning), and in analysis (adversarial fraud detection). The adversarial dimension reminds us that data systems are not only technical artifacts but also targets of attack.

Finally, **integration and interdisciplinarity** are key. The variety of data types (e.g., structured, unstructured, temporal, spatial) demands approaches that can cross boundaries. Similarly, the variety of perspectives in computer science, statistics, operations research, and domain expertise, demands collaboration.

Looking Ahead: The Future of Data Engineering for Data Science

What, then, lies ahead for the field? Several trajectories can already be discerned.

The trend toward **integration and adaptability** will intensify. Data systems will need to combine heterogeneous sources more seamlessly, adapt dynamically to new formats and requirements, and interoperate across platforms and domains.

At the same time, the demand for **responsible data science** will continue to grow. Regulatory frameworks, ethical expectations, and societal scrutiny are becoming stricter. Privacy-preserving and fair techniques, explainable AI, and accountable infrastructures will no longer be optional but mandatory.

Another development is the **automation of data engineering**. MLOps represents only the beginning. Smart and automated software-hardware co-design is a necessity. We are moving toward AI-assisted systems that can configure, monitor, and optimize themselves, reducing human effort in routine tasks and enabling researchers and practitioners to focus on higher-level decision-making.

Finally, **sustainability** will increasingly shape research agendas. Efficiency, resource optimization, and ecological impact will become central criteria for evaluating new systems, placing environmental responsibility alongside scalability and performance as core design goals.

Closing Note

This book has sought to map the contours of data engineering for data science as it stands today. It has surveyed a wide spectrum of challenges and solutions, from foundational governance to cutting-edge analytics. More importantly, it has illustrated how these elements interconnect, forming a lifecycle in which every stage plays an essential role.

We hope that the reader takes away not only an understanding of current techniques and issues, but also a sense of the open questions that remain. Progress in this field will depend on collective effort between institutions, disciplines, and sectors. In this spirit, this book is not a conclusion but an invitation: to build on the work presented here, to push the boundaries further, and to contribute to data ecosystems that are scalable, transparent, sustainable, and above all, beneficial to society.

GPSR Compliance
The European Union's (EU) General Product Safety Regulation (GPSR) is a set
of rules that requires consumer products to be safe and our obligations to
ensure this.

If you have any concerns about our products, you can contact us on

ProductSafety@springernature.com

In case Publisher is established outside the EU, the EU authorized
representative is:

Springer Nature Customer Service Center GmbH
Europaplatz 3
69115 Heidelberg, Germany

www.ingramcontent.com/pod-product-compliance
Ingram Content Group UK Ltd.
Pitfield, Milton Keynes, MK11 3LW, UK
UKHW021545090726
473054UK00007B/250